Gypsum & Gypsum Products

Their Science & Technology

Gypsum & Gypsum Products

Their Science & Technology

Dr. Manjit Singh

Former Scientist 'G'
Head Environmental Science & Technology
and Clay Products Division
Central Building Research Institute, Roorkee

CRC Press is an imprint of the
Taylor & Francis Group, an **informa** business

First published 2023
by CRC Press
4 Park Square, Milton Park, Abingdon, Oxon, OX14 4RN

and by CRC Press
6000 Broken Sound Parkway NW, Suite 300, Boca Raton, FL 33487-2742

CRC Press is an imprint of Informa UK Limited

British Library Cataloguing-in-Publication Data
A catalogue record for this book is available from the British Library

Library of Congress Cataloging-in-Publication Data
A catalog record has been requested

ISBN: 9781032384269 (hbk)
ISBN: 9781032384276 (pbk)
ISBN: 9781003345008 (ebk)

DOI: 10.1201/9781003345008

Typeset in Arial, Helvetica, MinionPro, Symbol, Times New Roman
by Manakin Press, Delhi

Brief Contents

Detailed Contents

Preface

Development of useful building materials from different types of wastes and the associated research activities are essential in the present context. It's also desirable to take developed technologies from laboratory and research phase to full commercial implementation.

CSIR - Central Building Research Institute (CBRI), a Constituent Laboratory of Council of Scientific and Industrial Research, at Roorkee, was established to carry out research activities on building science with particular emphasis on building materials and products.

Scientific & technical efforts were made in CSIR - CBRI to develop cost effective and efficient building materials and components from wastes such as fly ash, fluorogypsum, phosphogypsum, red mud, clay, etc. to cater to the needs of industry. Several materials & technologies developed by CSIR - CBRI have been adopted by building industries.

Amongst various materials studied at CSIR - CBRI, gypsum based building materials have received attention for producing gypsum plaster and the building products. A large quantity of natural and by-product gypsum (wastes from phosphoric acid and hydrofluoric acid industries) are available in India. This variety of gypsum is contaminated with certain harmful impurities of phosphates and fluorides which have to be removed/reduced before using gypsum in an effective manner. Different technological processes developed for by-product gypsum have been commercialized.

There has been a long standing demand from the entrepreneurs to bring out a manual on the products related to gypsum containing all the available information on the subject for the benefit of the people working in this area. Effectively, the basic aim of the book is to fulfil the long awaited requirement of people working in the area of gypsum and related aspects to have a complete scientific & technical knowledge on the subject hitherto not available in a concise form.

This book will be first of its kind and will help manufacturers, entrepreneurs, planners, decorators, builders, architects, students working in the area of gypsum worldwide. The book contains eleven chapters encompassing different aspects of gypsum, its use in industry, in forming building products, in making a particular type of cement, etc. Stipulations of the provisions of various National and International standards on gypsum and its applications have been compiled in this book in one of the chapters for the benefit of the users of this material.

The book in its present form can be expected to be of immense use to various stakeholders in the industry and institutions.

Dr. Manjit Singh

Acknowledgements

To our chagrin, our father passed away before he could pen acknowledgements for this book – his dream project. Effectively, we feel compelled to thank all the people who were instrumental in the development of this book on his behalf. We are aware that our father was most grateful to the Department of Science & Technology (DST), New Delhi for supporting the development of this book financially, under their USERS Scheme.

We are sure he would want this book to be dedicated to our late mother and his eternal companion Daljit Kaur (1949 to 2010). We are certain that without our mother's presence, our father wouldn't have found the inspiration or even the devotion to write this book. She was a charming, gifted, considerate, and sociable individual who took care of the whole family with effusive affection and dedication and was a source of all our strengths. She has been and will always be standing next to him celebrating his victories in our eyes. Our father was especially affectionate of his son-in-laws Mayank and Rajiv who he relied and depended upon throughout the development process of the book.

We know that he would have liked to further his appreciation to Prof. S.K. Bhattacharya, Director of CSIR – Central Building Research Institute - CBRI (Roorkee) for his expert inputs and essential suggestions along with access to facilities that proved crucial to this book. Dr M.O. Garg, Director of I.I.P. (Dehradun) & Former Director of CBRI (Roorkee) also deserves a vote of thanks and appreciation for showing great interest in the creation of this book. Our father was also profusely thankful of the efforts of Sewa Ram, Former Personal Secretary (PS), CBRI and stated clearly that he cannot praise his contributions enough.

The compilation of the manuscript for this book would have been nigh impossible without the support and efforts of Dr L. P. Singh, Sr. Scientist, Jaswinder Singh and Naresh Kumar, Sr. Tech. Officers, CBRI. D. Deshmukh, Advisor (Projects) and his colleagues V. K. Agrawal (Manager) and Vijay Pal (Addl. Chief Engineer), Rashtriya Chemicals & Fertilizers Ltd. (RCF), Mumbai made critical contributions towards this book as well by providing data on 'Rapid Wall Gypsum Panels'. Thanks are also due to scores of scientists and researchers working at CSIR, CBRI, and other institutions inside and outside the country, whose work has been attributed in the book.

Finally, this book would have taken much longer than it did to be available in print without the all-round cooperation and support of the publishing division at Manakin Press Pvt. Ltd. A big kudos to them for all their devoted and dedicated endeavours towards making our father's dreams a reality!

Param Agarwal
Mandeep Kaur

1

About Gypsum

INTRODUCTION

Gypsum and lime were used as mortar in olden times. Gypsum and anhydrite are widely distributed in the earth crust. Only in volcanic regions, the gypsum and anhydrite are completely absent. Large quantities of gypsum and anhydrite are available as a by-product of industrial chemical processes when flue gases are desulphurised or calcium salts are reacted with sulphuric acid. In this case, calcium sulphate is obtained as a moist fine powder.

1.1 The History of Gypsum and Gypsum Plaster

The earliest hard evidence can be found in the town of Catal Huyuk in Asia Minor, where a gypsum plaster was used as a base for decorative board/tile. The findings have been dated around 9000 BC. The name originates from the Greek word "gypsos" which has given us the modern-day terms 'Gips' in German and 'gypsum' in English. In English 'gypsum' is used for both raw mineral and 'plaster' as the building agent or material.

Gypsum is found naturally as a sedimentary rock of common occurrence, precipitated from evaporating sea water over million of years. In Germany, gypsum deposits are mainly found in the geological formations of Muschelkalk, Keuper and Zechstein, which were formed some 285 million years ago. The chemical name of gypsum plaster is calcium sulphate hemihydrate. Calcining causes the chemically combined water in the gypsum rock gradually released, with various forms (phases) of gypsum being produced depending on the calcining temperature. If water is added to the calcined (and fine-ground) gypsum, the paste hydrates (hardens) to calcium sulphate dihydrate and forms solid gypsum again with different engineering properties. This cycle was discovered 10,000 to 20,000 years ago, it supposed that gypsum rocks were used to construct places for fires, the heat of which dehydrated gypsum, made it brittle and caused it to disintegrate. When wetted by the rain, it turned into a plaster which hardened again into solid hardened mass[1].

The ancient Egyptians, whose culture dates back to the third millennium BC, were also familiar with gypsum as a binding agent. Tests on mortars from the pyramid of Chefren (c.2000 BC) show that they are made from a mixture of gypsum and lime; mortar mixtures of gypsum and lime (used for filling the cavities for vertical joints) were used also for certain work in building the Sphenix. The heavy limestone blocks, weighing 2.5 tonnes, on average from which the pyramids are constructed were generally laid without bonding agents.

1.1.1 Minoans to the Romans

From the Egyptians, the knowledge of production travelled to Crete. In the palace of Knossos, outside walls were made of gypsum blocks. The Greeks received the knowledge of manufacture and use of gypsum from the Egyptians and Babylonians. The Greek philosopher and naturalist Theoptus (372–287 BC), a follower of Aristotle, devoted a chapter on gypsum in his treatise on rocks and minerals thus giving us detailed description of manufacture and use of gypsum products in Greece around 400 BC.

The knowledge of gypsum application was later passed on by the Greeks to the Romans, Pliny the Elder (23–79 AD). The use of gypsum in making the human face and subsequent duplication by filling the mould with wax, was made as long as ago as 350 BC. Romans only used gypsum in interior works, not in outside construction due to its low water resistance.

1.1.2 Gypsum in the Middle Ages

The knowledge of gypsum later reached to Central and Northern Europe. The Merovingians in particular, who inhabited the kingdom of Franks, were skilled in the manufacture and working of gypsum plaster. From the end of the 6th century onwards, they used gypsum plaster for making sarcophagi. Casting with plaster slurry enabled reproduction of decoration and various motifs, which are found on several sarcophagi in various catacombs in Paris[2].

After Romans withdrew from Central Europe, the knowledge of the manufacture and use of gypsum plaster fell into oblivion – as happened with Romans in north of Alps. During the Romanesque period, addition of straw fibres or horse hair to the gypsum plaster became fashionable for use in mortars for use in interior walls[1].

When cities suffered fires (*e.g.*, Basel in 1417 or London in 1212, 1666 and 1794), the fire retardant properties of gypsum plaster were recognised[3]. After greatest London fire in 1666, King Louis XIV instructed his architects and advisors to search for the most effective material. From 18th August, 1667, it became obligatory, by royal decree, to cover the wooden framework of the houses "both inside and outside" with nailed laths and plaster to make them fire resistant.

In the early Middle Ages, gypsum was used in Germany in the Herz Mountains, the entire Thuringian Basin and occasionally in the Thuringian Forest in the

Form of flooring plaster, as mortar and on the Lambardic pattern, for relief decoration of walls, architectural elements, graves and monuments[3]. Masonry and floor screeds constructed with the gypsum mortars and screeds made at that time are notable for their excellent durability without loosing their strength, etc. In the imperial palace of Tilleda, in Sangerthausen district in the state of Saxony - Anhalt, the remains of gypsum floor screeds have been preserved. The strength of floor screeds has been found to be of standard concrete.[4]

The midlevel technology of gypsum mortars and floor screeds showed extremely low water demand due to their low finenesses. Their water/binder ratios was less than 0.4. The mortars had low air void contents but relatively high bulk densities around 2 g/cc.

1.1.3 The Gypsum Plaster in Baroque and Rococo Period

Little is known about the origins and early development of stucco marble.[5] In the 17th century, there were craft centers in Southern Germany and Northern Italy. The technique of stucco marble did not spread into other parts of Europe-Britain, France, Austria and Switzerland until the 18th century. It reached its heyday in the Baroque and Rococo period. The earliest example of stucco marble work in Southern Germany, dating from 1590–1615, was found in the Residenz in Munich.

Stucco marble is made from gypsum plaster, glue size and pigments[6] or stucco marble only alabaster or moulding plaster is used, which is ultra pure white and possesses greater hardness than normal plaster of Paris. Bone glue helps in hardening and retarding the setting of plaster both. The plaster was coloured by light and fast pigments, earth pigments and also some mineral pigments. The three components were mixed to give a firm paste. A separate coloured paste is made for each shade.

Slices of this gypsum paste about 1 cm thick are cut, placed on the base and pressed on the base and then pressed down with special trowel. Coloured gypsum plaster mixed with very dilute size is used as the bedding mortar. This layer of stucco marble left until it can be easily worked with a scraper. The gloss of typical marble is attained after the stucco has completely hardened by means of 6 to 8 polishing operations with water and various abrasives. Finally to protect the surface against moisture a beeswax polish is applied. This technique of making stucco marble is still adhered today for the reconstruction of Opera House (Semper-Oper) in Dresden, originated in 19th century. In 17th and 18th century, lard tree oils were in practice to be used as surface agent for protecting surfaces of stucco marble[5].

1.1.4 Gypsum in Modern Times

From the mid 19th century onwards, a distinction made between hemihydrate and anhydrite plasters and there was an awareness of the relationship between workability and production temperature. The most important applications in this period were as mortars and floor screeds. Carefully made gypsum plasters/binders were used for casting and moulding and for stucco works.

Fig. 1.1 Gypsum in Nature

In 20th century, gypsum increased its significance due to the development of prefabricated elements, starting with gypsum panels and variety of modern gypsum plasterboards and gypsum wallboards. Then came the development of machine-applied plaster mortars, self-levelling floor screeds based on anhydrite and alpha hemihydrate. The use of gypsum from flue gas desulphurization (FGD) is also receiving tremendous attention. *e.g.,* FGD gypsum from lignite fired power stations can be converted into alpha hemihydrate by autoclaving which can be used in applications where high strengths are required.

The natural gypsum is monoclinic agglomeration of gypsum plates or tabular shaped in habit (Fig. 1.1) generally found with lime or anhydrite deposits.

Over the centuries the gypsum industry has developed empirically out of the craft of gypsum plastering. The difference between gypsum plaster and lime, however, remained obscure up to the eighteenth century. Investigations into the principles of gypsum technology was begun in 1765 by Lavoisier, and has continued to this day.

1.2 The $CaSO_4.H_2O$ Crystal System

1.2.1 Phases

The $CaSO_4.H_2O$ system is characterized by five solid phases. Four exist at room temperature: calcium sulphate dihydrate, calcium sulphate hemihydrate, anhydrite III, and anhydrite II. The fifth phase, anhydrite I, exists at 1180°C and it has not proved possible to produce a stable form of anhydrite I below that temperature. Table 1.1 gives characteristics of the $CaSO_4.H_2O$ system. First four phases are of great interest to industry.

Table 1.1 Phases in the $CaSO_4.H_2O$ System

Characteristics	Calcium Sulphate Dehydrate	Calcium Sulphate Hemihydrate	Anhydrite III	Anhydrite II	Anhydrite I
Formula $CaSO_4$	$CaSO_4.2H_2O$	$CaSO_4.1/2H_2O$	$CaSO_4$	$CaSO_4$	$CaSO_4$
Molecular weight, M	172.17	145.15	136.14	136.14	136.14
Thermodynamic Stability, 0°C	< 40	metastable	metastable	40–180	>1180
Stages or Forms	—	two forms:	three forms:	three forms:	—
	—	Alpha β•	β-anhydrite III β-anhydrite III' α-anhydrite III	A-II, slowly soluble soluble AII-u, insoluble anhydrite AII-E, Estrichgips	—
Other Nomenclature based on USE	gypsum, raw gypsum chemical gypsum, by-product gypsum, hard-ened gypsum, synthetic gyp-sum, natural gypsum, min-eral gypsum	**α-form:** α-hemihydrate autoclaved plaster α-plaster **β-form:** β-hemihydrate stucco plaster β-plaster plaster of Paris gypsum plaster	soluble anhydrite	raw anhydrite natural anhydrite, synthetic anhydrite, chemical anhydrite, by-product anhydrite	high tem-perature anhydrite
Production Temperature, 0°C	< 40	α-form: 80–180 β-form: 120–180	β-anhydrite III & β-anhydrite III': 290 a-AIII: 1100	300–900 range; AII-s : <500-AII-U : 500–700, AIII-E: >700	>1180

Calcium sulphate dihydrate is starting material before dehydration and the final product after rehydration/casting. Gypsum dihydrate consists of a layer lattice in which double layers of water bonded by hydrogen bonds. The plane of layers is perpendicular to the *b axis* of the monoclinic structure, and this explains the perfect cleavage parallel to the (010) plane. Within the layers, the calcium and sulphate tetrahedral form chains parallel to the *c* axis as discussed by Florke[7]. A schematic illustration of the layer structure of gypsum is shown in Fig. 1.2.

Plaster is produced from the calcination of gypsum ($CaSO_4.2H_2O$) which partially dehydrates to produce a hemihydrate ($CaSO_4.1/2H_2O$) [$CaSO_4.2H_2O$. $CaSO_4.1/2H_2O + 3/2\ H_2O$].

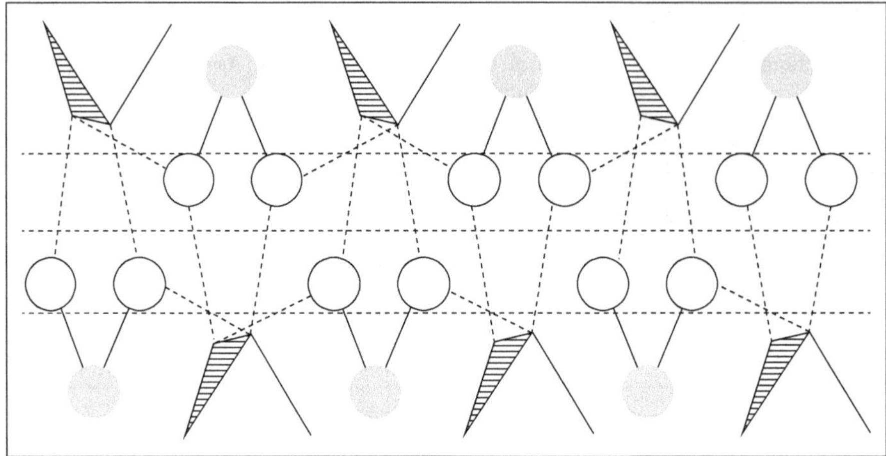

Fig. 1.2 Schematic illustration of the layer structure of gypsum showing a section parallel to the (001). The solid circles denote the position of Ca, the open circles those of water molecules, and the tetrahedrons, SO_4. (After Goto and Ridge) (Goto, M., and Ridge, M., J., J. Fac. Sci. Hokkaido Uni. Ser, 4, 13, 349, 1967.)

Calcium sulphate hemihydrate, $CaSO_4.1/2H_2O$, occurs in two different polymeric forms, alpha and beta, representing two limiting states[8]. They differ in their applications characteristics, their heats of hydration, and their methods of preparation. The alpha hemihydrate [Fig. 1.3 (*a*)] consists of compact, well-developed euhedral, large size crystals. The beta hemihydrate [Fig.1.3 (*b*)] forms flaky, anhedral to euhedral needle and prismatic crystals with presence of microcrystallanity.

Lehman, *et.al.*[9] gave three limiting stages for anhydrite III, also called as soluble anhydrite, beta anhydrite III', and alpha anhydrite II. The three phases of anhydrite III are characterized by X-ray analysis, differential thermal analysis, scanning electron microscopy, mercury porosimetry, and the specific surface area.

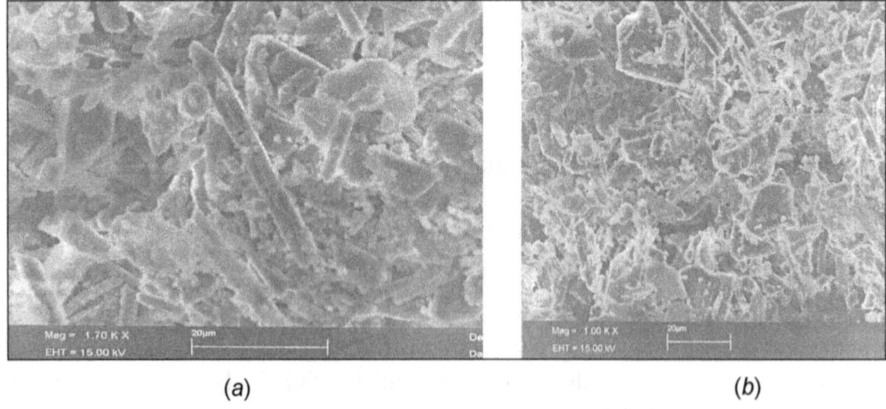

(*a*) (*b*)

Fig. 1.3 Microstructure of (*a*) Alpha Hemihydrate (*b*) Beta Hemihydrate

Anhydrite II is the naturally occurring form and also that produced by calcining the dihydrate, hemihydrate, and anhydrate III at elevated temperature. The important properties of the calcium sulphate phases are listed in Table 1.1.

1.2.2 Laboratory Synthesis

The thermodynamic stability ranges for the calcium sulphate phases are shown in Table 1.1 Below 40°C under normal atmospheric conditions, only calcium sulphate (gypsum dihydrate) is stable. The others are produced at elevated temperatures by progressive dehydration of the calcium Dihydrate in the following order:

Dihydrate > hemihydrate > anhydrite III > anhydrite II

The important physical properties of calcium sulphate are showin in Table 1.2.

Under normal atmospheric conditions hemihydrate and anhydrite III are metastable, and below 40°C in presence of water or water vapour they undergo conversion to dihydrate, as anhydrite performs. However, between 40°C and 1180°C anhydrite is stable.

To synthesize pure phases in the laboratory, β-hemihydrate is made from the dihydrate by heating at a low water-vapour partial pressure, *i.e.,* in dry air or vacuum, between 45°C and 200°C. Further careful heating at 50°C in a vacuum or up to ~ 200°C at atmospheric pressure produces β-anhydrite III.

Table 1.2 Physical Properties of the $CaSO_4.xH_2O$

Properties	Calcium Sulphate Dihydrate	Calcium Sulphate Hemihydrate α-form β-form	Anhydrite III	Anhydrite II	Anhydrite I
Water of Crystallization	20.92	6.22	0.00	0.00	0.00
Density, g/cc	2.31	2.7–2.8 2.62 – 2.64	2.57	2.93–2.98	2.9–2.93
Hardness, Moh's Scale	1.5	—	—	3 4	3.5–4.5
Solubility in water at 20°C, g/100 of Solution	0.21	0.67–0.88	hydrates to hemihydrate	0.27	—
Refractive indices	1.52–1.53	1.58–1.58	1.501–1.546	1.570–1.614	—
Optical character	+	—	+	+	—
Lattice symmetry	monoclinic	rhombohedral	hexagonal	rhombic	cubic
Lattice spacing, nm					
a	1.047	0.683	0.699	0.696	—
b	1.515	0.683	0.699	0.695	—
c	0.628	1.270	0.634	0.621	—

At very low water-vapour partial pressure, if water vapour is released rapidly and particle size is small, β-anhydrite III forms directly from the dihydrate, without formation of an intermediate hemihydrate. The specific surface area of such β-anhydrite III can be up to ten times that of β-anhydrite III.

α-Hemihydrate is obtained from the dihydrate at high water-vapour partial pressure, *e.g.,* above 45°C in acid or salt solutions, or above 97.2°C in water under pressure. Further careful release of water at 50°C in a vacuum or at 100°C under atmospheric pressure yields α-anhydrite III.

Anhydrite III is difficult to produce because anhydrite II begins to form above 100°C and anhydrite III reacts readily with water vapour to form hemihydrite.

The β-hemihydrates from β-anhydrite III and β-anhydrite III differ in their physical properties.[10] Therefore, hemihydrates from β-anhydrite III should be designated as β-hemihydrate. α-Anhydrite III absorbs water vapour to form α-hemihydrate. Likewise, the hemihydrates, in humid air, reversibly adsorb up to 2% of their weight in water without converting to dihydrate. This non-stoichiometric water in the hemi-hydrate can be completely removed by drying at 40°C. Anhydrite II is formed at temperatures between 200°C and 1180°C, above 1180°C it reverts to anhydrite II.

The dehydration kinetics of the dihydrate in contact with either aqueous solution[11] or a gas phase substantiate these reaction processes. However, all the hypotheses concerning reaction mechanisms, activation energies, and orders of reaction in the gaseous phase[12] have not yet been proved consistent. The problems are that the kinetics of the phase changes are often inhibited and that the test conditions have not been defined precisely.

1.2.3 Industrial Dehydration of Gypsum

Industrially, it is most important that dehydration is achieved in the shortest time with the lowest energy consumption, *i.e.,* that the costs be held to a minimum. Because of kinetic inhibitions calcination is carried out at much higher temperatures than those used in the laboratory (Table 1.1). Pure phases are rarely produced during manufacture; rather, mixtures of phases of the $CaSO_4$. H_2O system are produced. These types of calcined anhydrite II (anhydrous gypsum plaster of over burnt plaster) are manufactured, depending on burn temperature and time:

1. Anhydrite II-s (slowly soluble anhydrite), produced between 300 and 500°C.
2. Anhydrite II-u (insoluble anhydrite), produced between 500 and 700°C.
3. Anhydrite II-e (partially dissociated anhydrite, flow plaster, Estrichgips), produced above ~ 700°C.

In use the difference among those products lies in the rates of rehydration with water, which for anhydrite AII-s is fast, for anhydrite II-u slow, and for anhydrite II-u Transitions between these different stages of reaction are possible.

Anhydrite II-e consists of a solid mixture of anhydrite II and calcium oxide formed by the partial dissociation of anhydrite into sulfur trioxide and calcium oxide when raw gypsum is heated above 700°C. The pressure of impurities powers the normal dissociation temperature of anhydrite II ~ 1450°C.

1.2.4 Energy Aspects

Kelly *et al.* made a thorough study of the thermodynamic properties of the $CaSO_4$.H_2O system[13–14]. Tables 1.3 and 1.4 list the heats of hydration and dehydration of the various phase changes that are of industrial significance and backbone of gypsum industry.

1.2.5 Structure, Isomorphism, Mixed Compounds and Solubility Structure

The crystal structure of calcium sulfate consists of chains of alternate $Ca^{2+}$and tetrahedral SO_4^{2-} ions. For the most part these $CaSO_4$ chains remain intact during phase changes.

In calcium sulfate dihydrate the water of crystallization is embedded in between the layers forming a layer lattice and thus allowing easy cleavage along these planes. When calcium sulfate is dehydrated from dihydrate to hemihydrate, the volume decreases, and wide channels that run parallel to the $CaSO_4$ chain are formed. It is in definite positions in these channels that the water of crystallization is loosely bound. This water is able to escape relatively easily, which explains the facile conversion to anhydrite III. Anhydrite II exhibits the closest packing of ions, which makes it the densest and strongest of the calcium sulfates. However, lacking empty channels, it reacts only very slowly with water.

Table 1.3 Heats of hydration

Phase change				Heats of hydration per mole (gram) of dihydrate at 25°C, J		
β-$CaSO_4$.1/2H_2O	+	3/2 H_2O	→	$CaSO_4$.2H_2O	→	19300 ± 85
α-$CaSO_4$.1/2H_2O	+	3/2 H_2O	→	$CaSO_4$.2H_2O	→	17200 ± 85
β-$CaSO_4$ III	+	2H_2O	→	$CaSO_4$.2H_2O	→	10200 ± 85
α-$CaSO_4$ III	+	2H_2O	→	$CaSO_4$.2H_2O	→	25700 ± 85
$CaSO_4$	+	2H_2O	→	$CaSO_4$.2H_2O	→	6900 ± 85

Table 1.4 Heats of hydration

Phase change		Heats of hydration per mole (gram) of dihydrate at 25°C, J				
$CaSO_4$.2H_2O	→	β-$CaSO_4$.1/2H_2O	+	3/2 H_2O	→	86 700 ± 85
$CaSO_4$.2H_2O	→	α-$CaSO_4$.1/2H_2O	+	3/2 H_2O	→	84 600 ± 85
$CaSO_4$.2H_2O	→	β-$CaSO_4$ III	+	2H_2O	→	121 800 ± 85
$CaSO_4$.2H_2O	→	α-$CaSO_4$ III	+	2H_2O	→	117 400 ± 85
$CaSO_4$.2H_2O	→	$CaSO_4$	+	2H_2O	→	108 600 ± 85

Isomorphic incorporation of chemical compounds into the lattice of $CaSO_4$. H_2O phases is of interest in connection with by-product gypsum from flue-gas de-sulfurization and wet phosphoric acid processes. Isomorphic incorporation of calcium hydrogen phosphate dihydrate occurs because the $CaHPO_4$ $2H_2O$[15] mosodium phosphate, NaH_2PO_4, can also be incorporated into the gypsum lattice. Kitchen *et al.*[16] consider isomorphism possibility between the ions AlF^{2-} and PO_4^{2-} a possibility. Eipelatuer[17] reports on further isomorphism, incorporation of the anion FPO_4^{2-} as well as the incorporation of Na $< 0.2\%$ in the hemihydrate lattice, but only 0.02% in the dihydrate lattice. Chlorides are not incorporated.

1.3 Calcium Sulfate

There are triple sulfates of calcium with the divalent ions of the iron and zinc subgroups and of manganese, copper, and magnesium along with the univalent alkali metals, also including ammonium. A good example is the well known polyhalite, $2CaSO_4 - MgSO_4.2H_2O$ which also occurs in natural salt deposits.

Ettringite $3CaO - Al_2O_3 - 3CaSO_4 - 32H_2O$, is important in cement chemistry[18] as are syngenite[19] and thaumasite[2o].

1.4 Adducts

Gypsum when reacts with sulfuric acid forms adducts such as $CaSO_4.3H_2SO_4$ and $CaSO_4.H_2SO_4$. Calcium sulfate dihydrate can combine with four molecules of urea to form an addition compound.

1.5 Solubility

Gypsum is slightly soluble in water (Fig. 1.4). The solubility of the various forms of calcium sulfate are strongly affected by the presence of other solutes. The reference[21–22] can be consulted for particulars on the solubility of gypsum in acids, especially sulphuric acid, phosphoric acid, and nitric acid. Gypsum is readily soluble in glycerol and sugar solutions and in aqueous solutions of chelating agents, such as EDTA.

1.6 Occurrence and Raw Materials

1.6.1 Gypsum and Anhydrite Availability

Gypsum and anhydrite deposits are found in several countries. They originated from super saturated aqueous solutions in shallow seas, which evaporated and deposited first carbonates, then sulphates, and finally as chlorides, *i.e.,* in order of increasing solubility[23].

The various gypsum and anhydrite differ in purity, structure and colour. The major impurities are calcium carbonate (limestone), dolomite, marl, clay, silica, bitumen, glauberite, syngenite and polyhalite. The impurities which are present

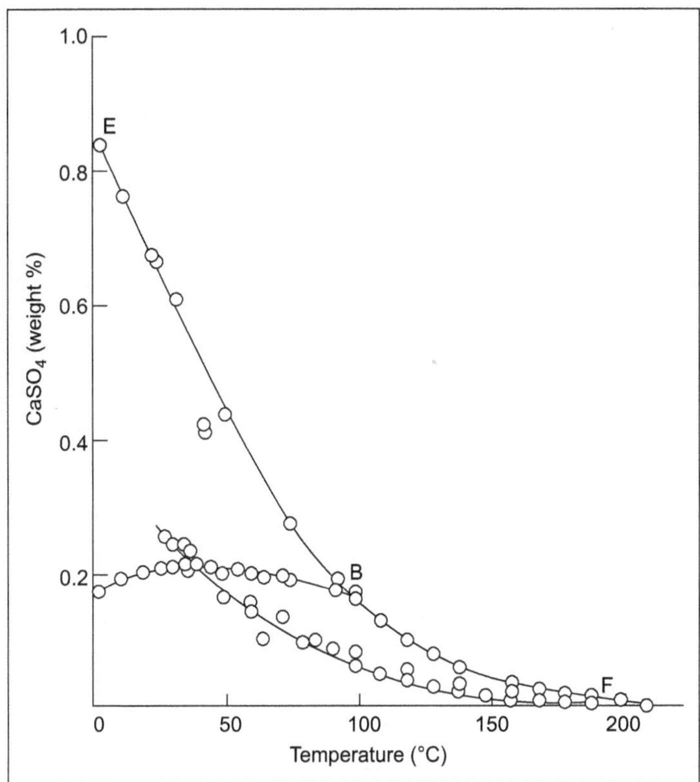

Fig. 1.4 Solubility diagram of systems $CaSO_4.H_2O$), Curves A-B., $CaSO_4.2H_2O$), C-D, Orthehombic $CaSO_4$, E-F $CaSO_4.H_2O$), Curves A-B $CaSO_4.2H_2O$ (Kelley, Southard and Anderson) (Kelley, K.K, Southard, J.C., Anderson, C.T., Tech. Pap. Bur. Mines, Wash, No. 625, 1941.)

from its formation, they are called primary impurities. Secondary impurities are formed during exposure to materials entrapped in to the cracks and leached cavities, but may also be introduced into the rock as waste material during mining. Gypsum is readily soluble in water and therefore rock is leached by surface water. Structurally, gypsum differs from anhydrite. The most important types of native gypsum are flaky gypsum (selenite), fibrous gypsum alabaster (grainy) gypsum, massive gypsum (rocky), porphyritic gypsum, earthly gypsum (gypsite) and gypsum sand *i.e.,* mixed with glauberite salt ($Na_2SO_4.10H_2O$). Anhydrite always crystalline, can either be sparry (anhydrite spar), coarse to close grained, or even rod shaped.

Gypsum can be pure white. If it contains iron oxide, it is reddish to yellowish. If clay and or bitumen is present, it is gray to black. Pure anhydrite is bluish white, but generally it is gray intermingled with blue tinge. The white colour veins sometimes are visible on the boundary between gypsum and anhydrite consist of glauberite ($CaSO_4.Na_2SO_4$) or Glauber's salt ($Na_2SO_4.10H_2O$).

1.6.2 Gypsum in India

1. Availability

As per UNFC, the total resources of mineral gypsum in India are estimated at 1,286 million tonnes (April 2010). Of these resources, 39 million tonnes have been placed under 'reserve' and 1,247 million tonnes under 'remaining resources'. Category wise, 62 million tonnes were proved reserves and 25 million tonnes probable reserves. Of theses total reserves, about 0.7 million tonnes were of surgical plaster grade, 28.3 million tonnes of fertilizer/pottery grade, 47.1 million tonnes of cement/paint grade, 2.4 million tonnes of soil reclamation grade and 8.3 million tonnes of unclassified grade.

2. Production, Stocks and Prices

The production of gypsum at 3.19 million tonnes in 2011–12 has been decreased by 35% from that in the previous year.

There were 35 reporting mines in 2011–12 as against 30 mines in the preceding year. Two main producers together accounted for about 99% production in 2011–12 (Table 1.5).

Rajasthan continued to be the leading producing state of gypsum contributing 99% out put. The remaining 1% out put comes from Gujarat, Jammu & Kashmir States (Table 1.6). The production of selenite (high purity gypsum) state-wise and their prices are shown in Tables 1.7–1.9 respectively.

Table 1.5 Principal Producers of Gypsum, 2011–12

Name & address of producer	Location of mine	
	State	District
Rajasthan State Mines & Minerals Ltd. C 89–90, Janpath Lal Kothi Scheme, Jaipur-302 015, Rajasthan.	Rajasthan	Bikaner, Sri Ganganagar, Hanumangarh, Jaisalmer, Jalore, Nagaur
FCI Aravali Gypsum & Minerals India Ltd., (formerly known as Fertilizer Corp. of India Ltd.) Mangu Singh Rajvi Marg, Paota 'B' Road, Jodhpur-342 010, Rajasthan.	Rajasthan	Bikaner, Sri Ganganagar, Jaisalmer

Table 1.6 Production of Gypsum, 2009–10 to 2011–12 (P)

(By States)

(Qty. in tonnes; value in ₹ 000)

States	2009–10		2010–11		2011–12 (P)	
	Quantity	Value	Quantity	Value	Quantity	Value
India	**3370322**	**1004631**	**4918170**	**1475454**	**3189229**	**1315174**
Gujarat	112	15	37	14	20	6
Jammu & Kashmir	33197	9959	38143	11443	29505	8852
Rajasthan	3337013	994657	4879990	1463997	3159704	1306316

Table 1.7 Production of Selenite, 2009–2010 to 2011–2012
(By State)
(Qty. in tonnes; value in ₹ 000)

States	2009–10		2010–11		2011–12 (P)	
	Quantity	Value	Quantity	Value	Quantity	Value
India	**14598**	**12408**	**6736**	**5726**	**12852**	**14547**
Rajasthan	14598	12408	6736	5726	12852	14547

Table 1.8 Production of Selenite, 2010–11 and 2011–12
(By Sector/State/Districts)
(Qty. in tonnes; value in ₹ 000)

State/District	2010–11			2011–12 (P)		
	No. of Mines	Quantity	Value	No. of Mines	Quantity	Value
India	**3**	**6736**	**5726**	**3**	**12852**	**14547**
Public sector	3	6736	5726	3	12852	14547
Rajasthan	**3**	**6736**	**5726**	**3**	**128528**	**14547**
Barmer	2	2170	1845	2	3475	3924
Bikaner	1	4566	3881	1	9377	10623

Table 1.9 Prices of Selenite, 2009–10 to 2011–12 (Domestic Market)
(in ₹ 000 per tonne)

Grade	Market	2009–10	2010–11	2011–12
Above 95% $CaSO_4.2H_2O$	Ex-pit Thob (Rajasthan)	850	850	1100–1150
Above 95% $CaSO_4.2H_2O$	Ex-pit Lunkaransar (Rajasthan)	850	850	1100–1150

3. Mining and Marketing

Gypsum is worked by opencast manual mining in Rajasthan except in a few semi-mechanised mines in Rajasthan. The deposits are found at shallow depths and scattered over large areas. Figure 1.5 shows mining in Bhutan. The production of gypsum is classified into four grades based on the calcium sulphate ($CaSO_4.2H_2O$) content: (*i*) above 90%; (*ii*) 85–90%; (*iii*) 80–85%: and (*iv*) less than 80%.

High grade gypsum is mined in Bikaner and Jaisalmer districts of Rajasthan. Selenite is also produced in Bikaner and dispatched to cements plants in Rajasthan, Gujarat, Madhya Pradesh, West Bengal, Uttar Pradesh, Bihar, etc. Sufficient quantity of gypsum containing 60–70% $CaSO_4.2H_2O$ is supplied to Punjab, Uttar Pradesh, Haryana, Delhi, etc. for reclaiming alkaline soil. Some quantity of gypsum is supplied from mines of Barmer, Bikaner, Sri Ganganagar and Nagaur districts of Rajasthan and Tehri Garhwal district of Uttarakhand to states of Rajasthan and Uttar Pradesh. Gypsum produced in Tamil Nadu and Jammu & Kashmir states is mainly for use in cement and plaster applications, hence

Fig. 1.5 Mining of Gypsum in Bhutan (Pemagatchal Mines, Bhutan)

sent to cement and gypsum industries in southern and northern India. Table 1.10 gives requirements of mineral gypsum in different industries as covered in Indian standard specification for Mineral Gypsum.

4. Consumption

About 8.12 million tonnes gypsum in all forms was consumed in organised industrial sector in 2011–12 as against 8.11 million tonnes in 2010–11. In addition, a substantial quantity of mineral gypsum as well as phosphogypsum was used in agricultural sector conditioning alkaline soil. The respective share of mineral gypsum, byproduct phospho and fluoro-gypsum and marine-gypsum and plaster of Paris moulds in total consumption in 2011–12 was about 54%, 41% and 5% respectively.

A major quantity of gypsum in 2011–12 was consumed in the manufacture of cement (99%). The remaining nominal consumption was in plaster of paris, asbestos products, ceramic, fertilizer, textile and chemical industries. The entire quantity of marine and gypsum moulds was consumed in cement and ceramic industries. Phosphogypsum was consumed mainly for manufacture of cement (99.9%) and a meager consumption was in ceramic industries in 2011–12 (Table 1.11).

5. Foreign Trade

Exports of gypsum and plaster at 51,732 tonnes in 2011–12 decreased by 49% from 100.918 tonnes in the preceding year. During the same period, export of alabaster was meagre at 1 tonne against 74 tonnes in the previous year. Gypsum & plaster were exported in bulk to neighbouring countries, *viz,* Nepal (84%) and Bangladesh (11%). Alabaster was exported to Nepal (Tables 1.12 and 1.13).

Table 1.10 Specifications of Mineral Gypsum in Different Industries

Constituent	Surgical Plaster	Ammonium sulphate fertilizer	Pottery	Cement	Reclamation of soil	Extender in paints
Free water	1.0% (max)	—	1.0% (max)	—	—	0.5% (max) when heated for 2 hr. at 45°C
CO_2	1.0% (max)	—	3.0% (max)	—	—	—
SiO_2 & Other insoluble matter	0.7% (max)	6.0% (max)	6.0% (max)	—	—	—
Iron & aluminium oxide	0.1% (max)	1.5% (max)	1.0% (max)	—	—	—
MgO	0.5% (max)	1.0% (max)	1.5% (max)	3.0 (max)	—	—
$CaSO_4.2H_2O$	96.0% (min)	85–90% (min)	85.0% (min)	70–75% (80–85% for export quality cement)	70% (min)	75% (min)
NaCl	0.01% (max)	0.003% (max)	0.1% (max)	0.5% (max)	—	—
Na_2O	—	—	—	—	0.75% (max) (Na)	—
Fineness	—	—	—	—	Residue on 2 mm sieve: Nil & on 0.25 mm sieve: 50% (max)	Residue on 240 mesh B.S. test sieve: 0.5%
Oil absorption	—	—	—	—	—	Within 5% of the approved sample
Colour	—	—	—	—	—	Close match to the approved sample
Lead & its compounds (calculated as metallic lead)	—	—	—	—	—	0.5% (max) when lead-free gypsum is required.
Physical form	—	—	—	—	—	In the form of dry powder.
Microscopic form	—	—	—	—	—	Material should match entirely with the characteristics of gypsum crystals.

Table 1.11 Reported Consumption of Gypsum, 2009–10 to 2011–12
(By Industry & Categorywise)

(In tonnes)

Category	Industry	2009–10	2010–11 (R)	2011–12 (P)
All Industries	**Grand Total**	**6984200**	**8114000**	**8121500**
Natural-Gypsum	**Total**	**3319800**	**4262800**	**4369700**
	Asbestos products	700(4)	700(4)	700(4)
	Cement	3305700(57)	4243700(70)	4330500(72)
	Ceramic	400(1)	400(1)	400(1)
	Fertilizer	100(1)	100(1)	100(1)
	Paint	++(2)	++(2)	++(2)
	Pharmaceutical	800(1)	900(2)	900(2)
	Plaster of Paris	12100(3)	17000(3)	37100(4)
	Refractories	—	++(1)	++(1)
	Textile	++(1)	++(1)	++(1)
By-Product Gypsum	**Total**	**3310400**	**3495300**	**3354800**
	Cement	3309700(69)	3494600(73)	3354200(74)
	Ceramic	600(1)	600(1)	600(1)
	Fertilizer	100(1)	100(1)	++(1)
Marine Gypsum	**Total**	**351300**	**353000**	**394100**
	Cement	351300(13)	353000(15)	394100(16)
Gypsum Moulds	**Total**	**2700**	**2900**	**2900**
	Ceramic	2700(5)	2900(5)	2900(5)

Table 1.12 Exports of Gypsum & Plaster (By Countries)

Country	2010–11		2011–12	
	Qty. (t)	Value (₹ '000)	Qty. (t)	Value (₹ '000)
All Countries	**100918**	**137642**	**51732**	**78485**
Bangladesh	51126	69141	5633	8281
Kenya	311	3866	363	4076
Italy	—	—	30	1556
Malaysia	57	346	400	8240
Nepal	44958	39453	43645	36473
Netherlands	25	2172	208	2688
Saudi Arabia	521	2939	43	1399
South Africa	203	2888	243	1812
Sri Lanka	137	1201	140	1500
Tanzania Rep	79	1446	110	1514
Other Countries	3501	14190	917	10946

Table 1.13 Exports of Alabaster (By Countries)

Country	2010–11		2011–12	
	Qty. (t)	Value (₹ '000)	Qty. (t)	Value (₹ '000)
All Countries	**74**	**329**	**01**	**49**
USA	—	—	01	49
Other Countries	74	329	—	—

Table 1.14 Imports of Gypsum & Plaster (By Countries)

Country	2010–11		2011–12	
	Qty. (t)	Value (₹ '000)	Qty. (t)	Value (₹ '000)
All Countries	**1697746**	**2212981**	**2776177**	**3979046**
Thailand	935292	1199952	1217598	1682179
Oman	101920	140394	589703	846513
Iran	496925	577849	452759	594900
Pakistan	14908	24820	351450	528509
Indonesia	40963	53823	69205	84134
USA	1485	42885	1547	45427
China	1495	20982	6401	45364
UAE	58	314	31955	43109
Afghanistan	29654	39620	25721	36472
South Africa	—	—	14609	16661
Other Countries	75046	112342	15229	55778

Table 1.15 Imports of Alabaster (By Countries)

Country	2010–11		2011–12	
	Qty. (t)	Value (₹ '000)	Qty. (t)	Value (₹ '000)
All Countries	**1237**	**19920**	**1138**	**21525**
Spain	398	6399	1089	20463
Italy	363	3631	49	1062
Other Countries	476	9890	—	—

6. Imports

Imports of gypsum & plaster at 27,76,177 tonnes in 2011–12 was substantially increased by 64% from 16,97,746 tonnes in 2010–11. Imports of alabaster marginally decreased to 1,138 tonnes in 2011–12 from 1,237 tonnes in 2010–11. Gypsum was imported mainly from Thailand (44%), Oman (21%) and Iran (16%). Alabaster was imported from Spain (96%) & Italy (4%). (Tables 1.14 and 1.15).

Table 1.16 Principal Producers of Phospho-gypsum

State	Unit
Andhra Pradesh	Coromandel International Ltd., Visakhapatnam.
Gujarat	(i) Gujarat State Fertilizers and Chemicals Ltd., Fertilizernagar, Vadodara district. (ii) Hindalco Industries Ltd., P.O. - Dahej.
Kerala	(i) Fertilizers & Chemicals Travancore Ltd., Udyogmandal, Ernakulam district. (ii) Fertilizers & Chemicals Travancore Ltd., Ambalamedu, Ernakulam district.
Maharashtra	Rashtriya Chemicals & Fertilizers, Chembur, Mumbai.
Odisha	(i) Paradeep Phosphates Ltd. (ii) IFFCO, Paradeep, district Jagatsinghpur.
Tamil Nadu	(i) Southern Petrochemical Industries Corporation Ltd., Thoothukudi. (ii) Coromandel International Ltd., Ennore, Thiruvallur. (iii) Sterlite Industries Ltd., Thoothukudi.
West Bengal	Tata Chemicals Ltd., Haldia.

1.7 By-Product Gypsum

In 2011–12, about 6–7 million tonnes of by-product gypsum has been produced in India. Generally, 4–6 tonnes of phosphogypsum is produced per tonne of phosphoric acid produced. The availability of phosphogypsum is shown in Table 1.16. The purity of phosphogypsum varies between 77% to 98% $CaSO_4.2H_2O$. It contains impurities of P_2O_5 (0.2–0.7%, fluoride (0.5–1.8%), organic matter, alkalies, radan, etc[24].

These impurities effect setting and strength properties of building products and cements produced from the unbeneficiated phosphogypsum. It is used in plaster and cement manufacture after proper washing and drying process[25].

1.8 Flue-Gas Gypsum

It is produced from the desulphrization of combustion gases of fossil fuels, such as anthracite, bituminous coal, lignite and oil in large combustion plants like power plants. Numerous processes have been developed and few have been put into use in the industrial applications. All these plants operate by countercurrent washing the flue gas with aqueous suspension of limestone or lime to remove sulphur dioxide[26–27]. This reaction which takes place at pH 7–8, gives practically insoluble calcium sulphite.

$$SO_2(g) + CaCO_3(s) + 1/2H_2O \longrightarrow CaSO_3.\tfrac{1}{2}H_2O(s) + CO_2(g) \text{ pH 7–8}$$

Calcium sulphite is further treated with SO_3 at reduced pH value to form soluble calcium bisulphate, $Ca(HSO_3)_2$.

$$2CaSO_3.1/2H_2O(s) + 2SO_3(g) + H_2O \longrightarrow 2Ca(HSO_3)_2\ (soln.)\ pH\ 5$$

The calcium bisulphate is easily oxidized by atmospheric oxygen to calcium sulphate dihydrate, calcium sulphate or - so called flue-gas gypsum

$$Ca(HSO_3)_2\ (soln.) + O_2 + 2H_2O \longrightarrow CaSO_4.2H_2O + H_2SO_4\ pH\ 5$$

The sulphuric acid produced reacts with the remaining limestone, thus forming additional gypsum

$$H_2SO_4 + CaCO_3(s) + H_2O \longrightarrow CaSO_4.2H_2O(s) + CO_2(g)$$

Large euhedral compact gypsum crystals are formed. These are separated from the aqueous gypsum suspension in hydro cyclones and vacuum drum filters or centrifuges. The product is moist, fine, pure powder with free water content less than 10%.

The overall process reaction is given below:

$$SO_2(g) + CaCO_3(s) + 1/2O_2(g) + 2H_2O \longrightarrow CaSO_4.2H_2O(s) + CO_2(g)$$

About 5.4 tonnes of gypsum are produced per tonne of sulphur in the fuel.

Japan, produced about 11.0 million tonnes of flue gas gypsum, the Federal Germany and United States 4.0 million tonnes each. These production may further increased due to increased environmental protection.

The FGD gypsum may be used as the raw material for gypsum industry in India when produced. The peculiar property of FGD gypsum is its high purity in terms of gypsum content and ultra white colour with high fineness. Fineness governs the usefulness of this waste. Cement and gypsum products are two major users of this value-added waste.

1.8.1 Phosphogypsum

Phosphogypsum is produced by acidulation of rock phosphate with sulphuric acid Disposal of phosphogypsum is of an important task. Normally, it is disposed off on land, in rivers or in sea. The dumping over the land makes the agricultural land barren, which contaminates ground water. The impurities of P_2O_5 and F are known to adversely affect the setting and strength development of plaster and cements. Consequently, the gypsum plaster sets fast and requires heavy retardation. Similarly, Portland cement or Portland slag cements suffer prolongation of setting and loss of strength at initial stage of hydration. Therefore, it is utmost essential to evolve beneficiation of phosphogpysum to reduce or remove the impurities to permissible level.

Large quantity of phosphogypsum are currently produced, and quantity may increase further due to establishment of new phosphate fertilizer Plants. The phosphogypsum is a moist powder with moisture content of 20–35% and lot of impurities, the exact impurities and their quantity depend upon the rock and the specific process of manufacturing adopted. For 1 tonne of phosphoric acid, about 4–5 tonnes of waste phosphogypsum are produced.

At present not more than 20% of the phosphogypsum produced is used. The major intricacies are their high free moisture and impuriries present in the gypsum.

1.8.2 Fluorogypsum/Fluoroanhydrite

Floroanhydrite or fluorogypsum is a by-product of hydrofluoric industry wherein fluorspar reacts with sulphuric acid. One tonne of fluorspar produces 1.75 tonnes of anhydrite. Fluoroanhydrite is used as the raw materials for gypsum industry only in Federal Republic of Germany and its neighbouring countries. Some researches have been done in India also[28–29].

1.8.3 Other Variety of By-product Gypsum

Calcium sulphate is formed in small amount in the production or treatment of organic acids such as tartaric acid, citric acid or oxalic acid or inorganic acids like boric acid. All these acids are produced by reaction of their calcium salts with sulphuric acid. Some gypsum is formed by treatment of waste water containing sulphate with lime as in zinc ore or production of titanium dioxide pigment. The utility of these variety of gypsum has yet to be discovered. Gypsum moulds from ceramic industry have been used so far as a raw material for the gypsum industry. Great efforts are needed to use this waste gypsum on mass scale.

1.9 Production

Several industrial processes have been developed to transform gypsum into cementitious calcium sulphate plasters over a long span of period, but a few plasters have been used extensively in the industry. The processes having potential of low capital and operating costs, simple design and operation, robust and long lasting equipment, and uniform high grade calcined gypsum further used to make different components are discussed in this book.

1.9.1 Natural Gypsum to Calcined Products

Natural gypsum is mined by the open-pit method[30] and by underground mining. In open-pit mining, gypsum is recovered by drilling and blasting. In deep mining, chamber blasting is used[31]. The blasting rock is used of large lumps, containing 0–35 free moisture. The amounts of explosives used are about 250 g/tonne for open-cast mining and 400g/tonne for underground mining.

The coarse rock is conveyed to the crushing plants by means of trackless loading and haulage gear. Impact crushers, jaw crushers, and single roll crushers with screen and oversize return are suitable for coarse size reduction. Impact pulverizes or roll mill are used for intermediate size reduction, and hammer mills, ball mills, etc. Or ring roll mills are used for fine grinding. The degree of size reduction by the calcining unit or the targeted use of gypsum:

Rotary Kiln	0–30 mm
Kettle	1–2 mm
Kettle with combined drying and grinding Unit	0.2–0.3 mm
Conveyor Kiln	4–50 mm
Gypsum and Anhydrite for Cement	10–50 mm

Generally, the mined and crushed gypsum is homogenized before being calcined. This is done in homogenizing plants with capacities of about one to two-week's production.

References

1. Schidegger, F, Aus der Geschichte der Bautechnic der, Basel, Boston, Berlin, Birkhauser-Verlag, 1990.
2. Benhamou, G; le Platre Paris, J.B. Bailliere, 1981.
3. Kruis, A; Gips. In: Ullmann Encyclopadie d. chem. Technik, 8, Band, Munchen, Berlin, Urban & Schwarzberg, 1957.
4. Steinbrecher, N. Gipsestrich und-mortel; Alite techniken wiederbeleben, Bausubstaz, 1992, Vol. 8, No. 10, pp. 59–61.
5. Reithmeir, C; We wird Stuckmarmor hergestellt, Stein, 1992, No. 10, pp. 21–25.
6. Fleischmann, B; Hochglanz-Marmor aus gips, Bautenschutz und Bausanierung, Vol. 14, 1991, No. 6, pp. 50–51.
7. Florke, O.W., Neues Jb., 1952, Miner. Abh., Vol. 84, 189.
8. Eipeltaur, E, Zem. Kal Gips, 1958, Vol. 11, pp. 264–273, 304–316.
9. Lehman, H, & Ricke,K, Tonind. Ztg. Keram Rundsch, 1973, Vol. 97, pp. 157–159.
10. Mehta, S. M, Dissertation, 1974, Techn. Universitat, Clausthal, Germany.
11. Satava, V, Zem. Kalk, Gips, 1971, Vol. 24, pp. 248–252.
12. Lehman, H and Rieke, K, Tonind. Ztg. Keram., 1974, Vol. 98, pp. 81–89.
13. Kelly, K. K., Southard, C.T. and Anderson, U.S., 1941, Bur. Mines. Tech. Paper, 625.
14. Kruis, A and Spath, H Tonind. Ztg. Keram, Rundsch, 1951, Vol. 75, pp. 341–351, 395–399.
15. Nippon Kokan Kabushiki, GB 1016007, 1962 (K. Araki).
16. Kitchen, D and Skinner, W.J., 1971, J. Appl. Chem. Biotechnol., Vol. 21, pp. 53–55, 56–60, 65–67.
17. Epiltaur, E, Tonind. Ztg. Keram. Rundsch, 1973, Vol. 97, pp. 4–8.
18. Ludwig, U, Zem. Kalk Gips, 1968, Vol. 21, pp. 81–90, pp. 109–119, pp. 175–180.
19. Sprung, S, Zem. Kalk Gips, 1974, Vol. 27, pp. 259–267.
20. Leifeld, G, Munchberg, W and Stegmaier, Zem. Kal. Gips, 1970, Vol. 23, pp. 174–177.
21. Gmelin, system No. 28, Calcium, Main B 3, 1961, pp. 675–785.

22. Slack, A.V., 1968, Phosphoric Acid, Marcel Dekker, New York, 1968, Part 1, Part II.
23. Grooves, A.W., Gypsum & Anhydrite, Overseas Geological Surveys, Her Majesty's Office, London. 1958.
24. Singh Manjit and Garg Mridul, Calcium sulphate hemihydrate activated low heat sulphate resistant cement, Construction & Building Materials (UK), 2002, Vol. 61, pp. 181–186.
25. Singh Manjit and Garg Mridul, Studies on waste gypsum from hydrofluoric acid industry, 6th NCB International Seminar on Cement and Building Materials, New Delhi, November, 1998, pp. 24–27, pp. X56–X64
26. Gunn, J.R., Gypsum, J, 1968, Vol. 49, pp. 14–18.
27. Kreuter, W, Zem Kal Gips,1974, Vol. 27, 222–225.
28. Singh Manjit and Garg Mridul, Durability of cementing binders from fly ash and other wastes, Construction & Building Materials (UK),Vol. 21, No. 11, November 2007, pp. 2012–2016.
29. Singh Manjit and Garg Mridul, Investigations on the use of fluorogypsum for making value-added building materials, 10th NCB International Seminar on Cement and Building Materials, 27–30 November 2007, Vol. 3, pp. VII-462-VII-471, New Delhi.
30. Gun, J. R., Gypsum J., 1968, Vol. 49, pp. 14–18.
31. Kreuter, W., Zem. Kalk. Gips., 1974, Vol. 27, pp. 222–225.

2

Production of Calcined Gypsum or Gypsum Plaster

2.1 Historic Background of Use of Gypsum Plaster

Gypsum plaster is viewed by many people in the conservation world as a modern building material. We know that it was used by the ancient Egyptians to plaster the pyramid at Cheops. In Britain, research indicate that considerable quantities of Plaster of Paris were imported from France during Henry VIII's reign for work on royal properties.

Our knowledge of the use of gypsum plaster prior to the 19th century is limited. However, Claire Gapper's research shows that it was being used in the 16th century with lime in floors, walls and ceilings, but decorative plaster work, which was previously assumed to contain gypsum, is proving to contain only minute traces; the sort of levels at which one would find it as an impurity in limestone. This contrasts with the use of gypsum over the last 200 years, when it was predominantly used for casting decorative elements and for gauging lime when running mouldings, whilst most flat work has been executed using plain lime plasters. Although further investigation is required, it would appear that gypsum was being used in these early gypsum/lime plasters very differently from the way we expected and there is no evidence, at the moment, that it was also used for mouldings or decorative work[1–2].

For small decorative artistic work such as scrolling leaves, fruit, figures and heraldic devices, cast decoration was allowed repetition. Gypsum or Plaster of Paris allowed crisp details to be produced as it was harder than lime and set before it was removed from the mould. Furthermore, casting in lime is more time consuming than using Plaster of Paris, because the lime has to be used very stiff and has to be punched into the mould. Nevertheless we find in 18th century work that casts were sometimes made in lime rather than gypsum. Its advantage is that, after being turned out of the mould, it can be adjusted whilst it is still soft, allowing minor variations in detail from one cast to another. The use of lime and gypsum in different cases may have been partly to do with different local traditions, but may have as much to do with the availability of materials.

We need to understand a lot more about the history of the production of gypsum plasters and also about their availability in different parts of the world, before we can fully understand why they were being used differently. In Derbyshire, for example, we know that alabaster was being burnt to make gypsum plaster in the 17th century, particularly for floors, whilst in other parts of the country gypsum was being imported from Paris. This may have been simply because the French plaster was purer, but it may also indicate that they were used for different purposes.

Plaster is the common name for calcium sulphate hemihydrate made by heating the mineral gypsum, the common name for sulphate of lime. Plaster was first made about 9000 years ago, and has been used by ancient Egyptian, Greek and Roman civilizations. However, it wasn't used on a large scale until 1700s, when it was required to be used in all construction in Paris. In 1666, a fire raged across London, destroying many parts of it. In its aftermath, the king of France ordered that all walls made of wood in Paris be immediately covered with plaster, as a protection against such fires. This resulted in large-scale mining of gypsum which was available around Paris in huge quantities. Thus, during the early 18th century, Paris (Montmartre) became the centre of plaster production, and hence the name, plaster of Paris.

Although the interest in the composition of old plaster may appear slightly academic it is an important part of repairing on a 'like for like' basis. If a repair is incompatible with the original, then the older material could end up being damaged. It is important, therefore, that original plaster is examined to establish its basic properties before repairs begin. With experience, it is often possible to tell whether an old plaster contains gypsum just by breaking a piece off and poking at it. In some cases dissolving a small piece in dilute hydrochloric acid can help, as a fine white residue is often left, along with the aggregate, if there is gypsum present. Occasionally, particularly with older plasters, it is worth carrying out a proper laboratory analysis to ascertain the proportion of gypsum used in the mix.

2.1.1 Conservation and Rehabilitation of Structures with Plaster

Gypsum is a naturally occurring crystal of calcium sulphate ($CaSO_4.2H_2O$). It can be quarried in different parts of the world in slightly different forms, but in this country the most commonly known form is alabaster. Plaster is made from gypsum by grinding it to powder and then gently heating it to drive off some, or all, of the water of crystallization.

If it is heated to about 150°C, then only some of the water is lost and the hemi-hydrate is formed ($CaSO_4.1/2H_2O$). Plaster of Paris, casting plaster, dental plaster and Helix, are all forms of hemihydrated gypsum. They all set rapidly (within 10–20 minutes) by recrystallising when mixed with water. Modern bagged plasters are also made from hemihydrated gypsum, but contain retarders which slow down the speed of the set, and also include fillers and other additives.

If gypsum is heated to higher temperatures, then all of the water can be driven off. Fully hydrated gypsum will not readily recombine with water and can only be used as a plaster when an accelerator is added to it. During the 19th century, and the first half of the 20th century, a number of patent plasters were produced using this method. Plasters like Sirapite, Keenes cement and Parian cement or Martin cement, as well as many others which appear in old plastering books, are made from fully hydrated gypsum. These plasters set in a few hours, and could be worked-over to achieve a highly-polished finish. They were particularly popular for use in hospitals and public buildings, because of their hardness and durability. We also sometimes find them used in domestic works to form skirting and door surrounds.

Lime and gypsum are obviously different materials with different properties. Lime sets slowly by absorbing carbon dioxide from the air, whereas gypsum plaster sets rapidly by crystallising (even fully hydrated gypsum plaster sets within about a day). Also, as a lime plaster dries it shrinks slightly, while a gypsum plaster expands slightly as it sets. Historically gypsum has generally tended to be used as an additive to lime. The speed at which a mix of lime and gypsum set (which is slightly slower than gypsum on its own) and the slight expansion that occurs, are particularly useful when running cornices and other mouldings *in situ.*

As we repair and reinstate historic plasterwork, it is important that we understand the materials and methods used in the first place. Many visually inappropriate and other physically damaging repairs have been executed in the last few decades because modern methods and materials have been used without thought.

The general knowledge and understanding of old plasterwork in the conservation world has increased enormously in the last five years or so. The reinstatement of fire damaged ceilings at Uppark (near Peters field, West Sussex) and Prior Park (Bath), have given people a chance to understand how they were put together in the first place. This has also led to the re-learning of some traditional skills, like the hand modelling of stucco. As we repair old plasterwork it is important not only to use compatible materials but also to use compatible skills and methods.

Over the next few years we need to investigate why, where and how gypsum was used historically in plastering. The challenge which then leads on from this is how we should train and encourage plasterers to match the methods and skills used originally in both lime and gypsum plasterwork. Research and the dissemination of information such as this are crucial to the future of the world rich heritage of historic plasterwork.

To understand exact behaviour of the calcined gypsum, it is necessary to know the gypsum calcination and its various applications in buildings. The gypsum crystal with two molecules of water on heating below 200°C produces wonderful material called calcined gypsum or plaster or popularly called as plaster of Paris. It exists in two polymeric forms *i.e.,* Beta (β) and Alpha (α) hemihydrates. Details of these plasters are described below:

2.2 β-Hemihydrate Plasters

Plaster of Paris or stucco or gypsum plaster, if natural/mineral gypsum is the starting material, are prepared by dry calcination between 120°–180°C, either in directly fired rotary kilns or else in indirectly heated kettles, to produce the β-hemihydrate. The kettles may be upright or horizontal.

2.2.1 Rotary Kiln Process

The rotary kiln up to capacity 100 to 600 tonnes/day is suitable for calcining granular (0–30 mm) gypsum rock. The gypsum (without predrying) is fed continuously to the kiln in concurrent flow with the hot gases with the weigh belt feeder and cellular wheel sluice (Fig. 2.1). The hot gases are produced in a brick – lined box combustion chamber ahead of the gypsum intake but consists of thick steel plate with steel inserters to ensure uniform distribution of gypsum particles across the furnace cross-section.

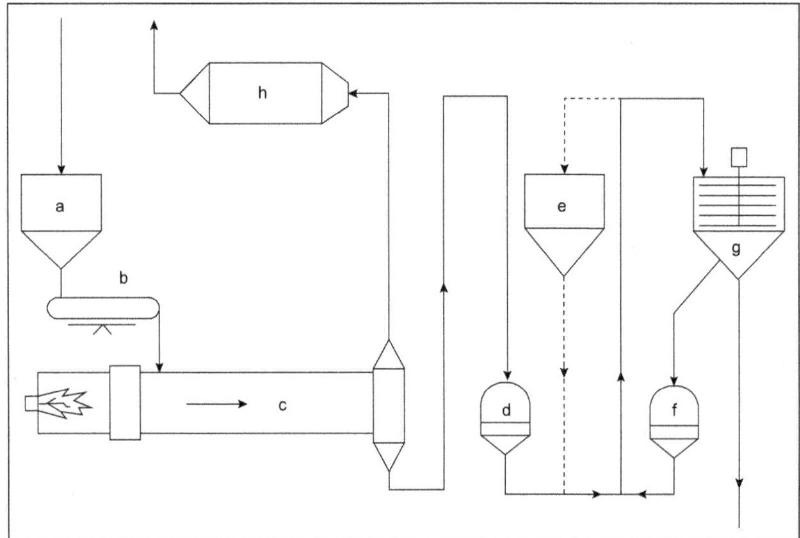

Fig. 2.1 Production of Beta (β)-Hemihydrate Plaster by Rotary Kiln Process (*a*) Silo for gypsum rock, (*b*) Weigh-belt feeder, (*c*) Rotary Kiln with Combustion Chamber, (*d*) Primary Mill, (*e*) Start up and Shut down Bin, (*f*) Fine mill, (*g*) Air Classifier, (*h*) Electrostatic precipitator, (*i*) β-hemihydrate (Courtesy-Ullman's Encyclopedia of Industrial Chemistry, Calcium Sulphate, Franz Wirsching, Germany, 1985)

(a)

(b)

Fig. 2.2 Rotary Kiln Calcination of Gypsum in (*a*) Germany and (*b*) France

High thermal efficiency is attained by direct transfer of heat from the hot gases to the gypsum particles. The residence time of gypsum rock in the kiln is self-regulating. It increases with the size of the grains, but complete calcination can be achieved irrespective of particle size, to produce β-hemihydrate to give uniform quality after fine attrition and air classifying. Any gypsum not calcined is added in small portions to the calcined gypsum. Rotary kiln plants are fully automatic and regulated by only one person. Rotary kiln calcination as adopted in Germany (*a*) and France (*b*) are shown in Fig. 2.2.

2.2.2 Kettle Process

The externally heated kettle, of capacity 20 tonnes or more, is the first of a series of similar calcining plants heated indirectly. The gypsum is pre dried and ground to < 1–2 mm. In modern plant the fine grinding, which used to follow the calcining process, is combined with the pre drying and carried out in drying – grinding units to produce particles < 0.2 mm. Generally, Raymond mill, Claudius Peters Mill, Attritor and others are used for grinding. The properties of the hemihydrate plasters are listed in Table 2.1.

A Kettle process or Gypsum Calcinator has also been developed and installed at Central Building Research Institute, Roorkee, India[3–4] for making plaster (Fig. 2.3). The Gypsum Calcinator has two main components namely, the mechanized churning system and the furnace. The mechanized churning system comprises of a mild steel pan, a vertical power shaft and a number of churning blades rigidly connected to the vertical power shaft at two levels and a sweeping chain attached

Table 2.1 Phase Composition of Calcined Plaster

	Mineral	Weir Product	Cyclone Product
Free water	0.3	—	—
Soluble Anhydrite	—	3.4	7.6
Gypsum	68.0	2.8	1.4
Hemihydrate	5.1	65.2	64.2
Insoluble Anhydrite	—	1.4	0.5

Fig. 2.3 Energy Efficient Gypsum Calcinator (Developed at Central Building Research Institute (CBRI), Roorkee, India

at the bottom of the lower set of the churning blades and a removable lid to cover the pan and allow churning of the gypsum charge without any dust loss, and finally a prime mover along with appropriate power transmission mechanism. The pan has a diameter of 1.25 m and depth 0.7 m. The bottom of the pan is typically bulging out at the centre towards inside of the pan. This design of the bottom of the pan helps in easy discharge of the calcined gypsum besides effective heating of the pan from below. The churning blades are made from the angle form section and two such blades are welded opposite to each other directly with the vertical shaft. A chain is attached to the lower set of the blades which sweeps the material on the bottom so that over calcination of gypsum may be avoided.

The vertical power shaft gives power to churning blades for rotation. It is maintained vertically and located centrally and held in position with the help of a bearing housing. The bearing housing is located centrally on top of the pan to provide support to the motor and other power transmission mechanism arranged for rotation of the churning blades. The bevel gear of the vertical power shaft is driven by a bevel pinion. A clutch is, however, arranged in between the bevel pinion and the output shaft of the speed reducer.

One set of calcination comprises of two pans. Two vertical power shafts with churning blades, one shaft for each pan are used. Both pans get power from a single prime mover and a speed reducer with two output shafts. A facility has been provided with which both the shafts may be actuated simultaneously or when required only one of the two sets of the churning blades can be kept in operation. To protect the gypsum charge against over-burning in situations when electric power may get suddenly off, a manually operated cranking system has been provided. Two persons can easily rotate the churning blades simultaneously in both the pans. Two small holes are provided in the lid in both the pans, one of which is used for putting a hopper for charging the pan with ground gypsum and the other for putting a small vent pipe through which water vapour may escape in atmosphere during calcination of gypsum.

Other important component of the Gypsum Calcinator is the furnace. The furnace used in this system is a vertical cylindrical shaft furnace. The pan is directly placed at the beveled - out top of the furnace. There are two such furnaces in one set. The general structure of the furnace is made out of ordinary burnt clay bricks, where the inner side lining of the furnace is claded with the fire bricks. A small flue ring is provided around each pan so that even the sides of the pan may be heated and maximum heat can be derived from the flue gases leading to the chimney. Both the furnaces are fired separately. The flue gases from both the pans meet at entry point to the chimney. A chimney of 6 m height helps in getting these flue gases out in atmosphere and also in securing the required draught for efficient burning of the coal/wood in the furnace.

To discharge hot calcined gypsum from the pan, a spout is provided with each pan. During calcination the spouts are kept closed to avoid release of gypsum through the spout. At the time of discharging calcined gypsum, the spout is opened.

1. Manufacture

Chemically, gypsum plaster is $CaSO_4.1/2H_2O$. It is manufactured by heating ground gypsum at 120–180°C in a suitable calcining plant when it loses one and-a-half molecules of water of crystallization in the from of steam. As a result gypsum plaster with half molecule of water is produced. It is also called the first settle plaster.

2. Preparation of Gypsum

Natural gypsum ground to passing 60 mesh or by-product phosphogypsum dried to 2 per cent moisture content either in the sun or in a dryer before calcination can be calcined in the kettle.

3. Fuel Preparation

Steam coal (with ash content 20%) is crushed to 50 mm size before feeding in the furnace. Alternatively, dried firewood may be used. Gypsum calcinator may be run with liquid fuels (diesel, LDO, furnace oil) as well as gaseous fuels.

4. Charging and Discharging of Pans

First of all both the pans are lighted. When the temperature inside the pans is about 120°C, the gypsum is fed through the charging hole in the hot pans with the churning mechanism already on, so that the material remains in agitation from the very beginning. The temperature rises to 130°C where the charge boils vigorously and gives out water of crystallization as steam. This temperature remains constant for some time depending upon the purity of gypsum. When the boiling subsides, the temperature starts rising quickly and reaches 160–170°C. Now the boiling ceases and the plaster starts settling indicating completion of the calcination process. The hot calcined gypsum is discharged immediately in the pit covered with a lid to avoid dusting. The system is again ready for next charge. The production of plaster is shown in Figs. 2.4 to 2.6.

The coal consumption in the gypsum calcinator is about 45 kg as compared to 150–200 kg in the open pan manual system for calcining one tonne of gypsum. The temperature of charge during calcination can be controlled by the rate of feeding of coal and use of damper in the furnace. The optimum and completion

Fig. 2.4 Crushing of Natural Gypsum before fine Grinding (Pemagatchal Plant in Bhutan)

Fig. 2.5 Gypsum Kettle (Pemagatchal Plant in Bhutan)

Fig. 2.6 Pulverization of Calcined Gypsum in Large Size Pulverizer (Pemagatchal Plant in Bhutan)

temperature of calcination can be noticed with the help of chromel-alumel thermocouples and indicators. Separate thermocouples and indicators are required for both pans. The optimum temperature of calcination is 130°C and temperature of completion of calcination is 160–170°C. The first charge of gypsum is calcined in 4.0 hours while the subsequent charge gets calcined in 3.5 hours, when coal is used as fuel. With LDO/Diesel fuels, the calcination time is 3.5 hrs initially and then on an average 2.5 hrs is sufficient to calcine one tonne of gypsum.

5. Quality of Plaster Produced

The calcined material after grinding to the desired fineness in a hammer mill or ball mill can be used for testing chemical and physical properties as per methods laid down in IS : 2542 (Part 1) –1978, Methods of tests for gypsum plaster, concrete and products.

6. Heat Required for Calcination of Gypsum

For calcining gypsum, the heat required as given below:

(*i*) Heat required for bringing gypsum to its calcination temperature.
(*ii*) Heat required to decompose gypsum for conversion into plaster.
(*iii*) Heat required to evaporate water molecules held in gypsum.
(*iv*) Heat required to evaporate the free moisture content in the gypsum charge.

On the average, the free moisture is taken as 10 per cent.

Assuming the ambient temperature of 25°C and average sp. heat of gypsum at the temperature range (25°C – 170°C) as 0.265 K Cal per kg per[5], heat required to raise temperature of gypsum from 25°C to 170°C would be:

$$= \text{mass} \times \text{sp. heat} \times \text{temp. difference}$$
$$= 1 \times 0.265 \times (170 - 25)$$
$$= 38.425 \text{ K Cal per kg} \qquad (i)$$

The heat absorbed in decomposing gypsum ($CaSO_4.2H_2O$) is 3921 calories per gram molecule[6] and is equivalent to 22.78 K Cal per kg for 100 per cent pure gypsum. Hence, heat of decomposition of gypsum into plaster

= 22.78 K Cal per kg *(ii)*

The gypsum plaster ($CaSO_4.½ H_2$) contains 6.2 per cent combined water and thus under average conditions, it amounts to a loss of 0.15 kg of water per kg of gypsum during calcination. Therefore, the heat required to evaporate 0.15 kg of water per kg of gypsum at mean average calcination temperature of 127°C (120°C to 130°C) would be:

= mass × latent heat of water evaporation
= 0.15 × 536
= 80.4 K Cal per kg of gypsum *(iii)*

Hence, the heat required to calcine 1 kg of gypsum (without and free moisture) would be:

= 38.425 + 22.78 + 80.4 [*add eq. (i), (ii) and (iii)*]

= 141.605 K Cal per kg of gypsum

For evaporating the free moisture contents (taken 10 per cent here) of the gypsum, heat required would be equal to, (*a*) + (*b*) as below.

(*a*) Heat required to raise temperature of free moisture, per kg from 25°C (ambient) to 100°C,

= mass × sp. heat × temp. difference

= 1 × 1 × (100 – 25)

= 75 K Cal per kg of moisture

(*b*) Heat required to evaporate 1 kg of water = 536 K Cal

Hence, total heat required to evaporate moisture,

= (*a*) + (*b*)

= 75 + 536

= 611 K Cal per kg of water

Hence, heat required for calcining 1 kg of gypsum with 10 per cent moisture,

= weight (dry) of gypsum, kg × heat required per kg of gypsum + weight of free moisture, kg × heat required to evaporate per kg of moisture.

= 0.9 × 141.605 + 0.1 × 611

= 188.54 K Cal

Now, the heat (Q) required to produce 1 kg of plaster of Paris when 1 kg of gypsum (with 10 per cent free moisture) gives 0.765 kg of plaster of Paris, can be calculated as below:

$$\text{Heat (Q)} = \frac{188.54}{0.765} = 246.5 \text{ K Cal per kg of plaster of Paris}$$

7. Rate of Coal Burning and Grate Area

The rate of coal burning in any furnace very much depends upon the grate area of the furnace. Grate is a place where coal is fed and burnt inside the furnace. The grate cross-sectional area depends upon the quantity and quality of coal, type of coal, size of lumps, design of fire box and draft requirements and also on the consideration for protection of brick work of the furnace. Coal with higher ash contacts will however, need little larger grate area, expressed in different terms, the required grate surface may be calculated from the quantity of fuel burnt in a unit time. Although it is possible to burn 350 kg of good coal per hour sp. metre of grate surface with a draft of 25 mm of water, the temperature of the fuel bed resulting from so high a combustion rate, may burn out the roof over the grate within a few days. The coal required to calcine 1000 kg of gypsum, assuming calorific value of coal as 5500 K Cal/kg of coal and a furnace efficiency of say 65 per cent, can be calculated in the following way.

1000 kg of raw gypsum with 10 per cent free moisture gives 765 kg plaster of Paris. Assuming heat required for 1 kg of plaster of Paris as 246.5 K, Cal, the quantity of coal required for 765 kg of plaster would be:

$$\begin{aligned}
&= \text{wt. of plaster} \times \text{heat required for 1 kg of plaster calorific} \\
&\quad \text{value of coal} \times \text{efficiency of furnace} \\
&= 765 \times 246.5 \\
&= 5500 \times 0.65 \\
&= 50 \text{ kg (approx.)}
\end{aligned}$$

Hence, for calcining 1000 kg of raw gypsum, coal required should be theoretically equal to 50 kg. Since two pans, each of 500 kg capacity, are used in the Basic Unit of Calcinator, for each furnace (one furnace for one pan) the coal required would be 25 kg. Taking average calcination time as 3.5 hours.

$$\begin{aligned}
\text{Rate of coal burning} &= \frac{25 \times 100}{3.5} \\
&= 7.14 \text{ kg per hour}
\end{aligned}$$

The low rate of combustion (as above) has an additional advantage. The alternate streams or threads of combustible gases and excess air rise slowly in the fire box without any turbulence. These gases are later mixed and a slow, sustained secondary combustion is initiated which heats the hearth (here pan bottom) uniformly.

This low rate of heating thus fulfils the most important requirement of uniform heating of the gypsum (placed above the hearth and agitated regularly) during calcination thereby ensuring uniformity in the quality of plaster produced. Also low rate heating gives a smaller furnace and consequently reduced that losses from the furnace.

8. Thermal Efficiency of Calcinator

The calcination of 1000 kg of gypsum with assumed normal percentage of free moisture as 10% results in the production of 765 kg of plaster of Paris. The preparation of plaster of Paris consumes about 246.6 kilo calories per kg of plaster. Since the coal of calorific value 5500 K Cal per kg was consumed in calcinations at the rate of 45 kg of coal for one tonne of gypsum, the thermal efficiency of the calcinator can be calculated as below:

Heat required for calcining 1000 kg of gypsum with 10 % free moisture

$$= \text{Weight of plaster} \times \text{heat required per kg of plaster}$$
$$= 765 \times 246.5$$
$$= 188572 \text{ K Cal}$$

Heat supplied by burn = wt. of coal × calorific value of coaling 45 kg of coal

$$= 45 \times 5500$$
$$= 247500 \text{ K Cal}$$

$$\text{Thermal Efficiency} = \frac{\text{Heat cons. in calcination}}{\text{Heat supplied}}$$
$$= \frac{188572 \times 100}{247500}$$
$$= 76.1\%$$

Hence, Thermal Efficiency = 76.1%

9. Calculation of Fuel Saving in Innovative Gypsum

Calcinator - A Practical Example

The raw gypsum is first ground into powder form usually passing IS : 60 Sieve. The powdered gypsum is filled in bags and invariably stored temporarily before it is taken for calcination. During storing, the powdered gypsum picks up some moisture because of its hygroscope nature. The amount of moisture picked up may depend upon several factors, *e.g.*, fineness of the powder, weather, humidity and method and period of storing. The free moisture thus picked up by the powdered gypsum usually vary between 3 to 7 per cent but it may go up to 10 per cent also in certain cases. In addition to the free moisture, the gypsum ($CaSO_4.2H_2O$) contains two molecules of water in the combined form. In converting gypsum into plaster, gypsum loses 1.5 molecules of water (which amounts to 15% loss in the weight of gypsum) besides the free moisture present in the powdered gypsum. The heat thus consumed by gypsum during calcination comprises of following:

(*i*) Heat required for evaporating free moisture from the gypsum charge.
(*ii*) Heat required for bringing gypsum to its calcination temperature.
(*iii*) Heat required to decompose gypsum for conversion into plaster of Paris.
(*iv*) Heat required to evaporate 1.5 molecules of water held in gypsum (thus leaving only 0.5 molecules of water finally present in the plaster).

The above four heats (*i*) to (*iv*) when added constitute what is known as Theoretical Heat required for the calcination of gypsum.

Heat Calculations for Industrial Unit of Innovative Calcinator at Silliguri (West Bengal, India)

Samples of gypsum, plaster and coal used at the Silliguri plant were collected and analyzed. Gypsum samples were found to possess about 3 per cent free moisture whereas the calorific value of the coal samples was found to be 3220 K/kg. The coal was consumed at the plant with the rate of 60 kg of coal per tonne of gypsum .

Theoretical Heat for Calcination

(*i*) For evaporating free moisture contents (here 3%) in the gypsum, heat required is calculated as below:

(*a*) Heat required to raise the temperature of free moisture (per kg) from ambient temperature (with an average value of 25°C in this case to 100°C.

$$= \text{mass} \times \text{sp. heat} \times \text{temp. difference}$$
$$= 1 \times 1 \times (100 - 25)$$
$$= 75 \text{ K Cal}$$

(*b*) Heat required to evaporate 1 kg of water

$$= 536 \text{ K Cal}$$

Hence, total heat required to evaporate per kg of moisture,

$$= (a) + (b)$$
$$= 75 + 536$$
$$= 611 \text{ K Cal per kg of water} \qquad (i)$$

(*ii*) For bringing gypsum to its calcination temperature, the heat required is calculated as below:

With average value of ambient temperature as 25°C and average specific heat of gypsum at the temperature range (25°C – 170°C) as 0.265 K Cal per kg °C, heat required to raise temperature of gypsum 25°C – 170°C would be:

$$= \text{mass} \times \text{sp. heat} \times \text{temp. difference}$$
$$= 1 \times 0.265 \times (170\text{–}25)$$
$$= 38.425 \text{ K Cal per kg of gypsum} \qquad (ii)$$

(*iii*) The heat absorbed in decomposing gypsum ($CaSO_4.2H_2O$) is 3921 Calories per gramme molecule and is equivalent to 22.78 K Cal per kg, 100 per cent pure gypsum.

Hence, heat of decomposition of gypsum into plaster

= 22.78 K Cal/kg (*iii*)

(*iv*) The heat required to evaporate 1.5 molecules of water (*i.e.*, 15% water by weight combined with gypsum) during the conversion of the gypsum into plaster of Paris at mean average calcination temperature, 127°C (120°C to 130°C):

= mass × latent heat of water evaporation

= 0.15 × 536

= 80.4 K Cal/kg of gypsum (*iv*)

Hence, the total heat required to calcine 1 kg of dry gypsum (*i.e.*, without free moisture) would be:

= 38.425 + 22.78 + 80.4 [*add eq. (ii), (iii) and (iv)*]

= 141.605 K Cal per kg of dry gypsum (*v*)

The theoretical heat required for calcining 1 kg of gypsum with 3 per cent moisture would be:

= wt. of gypsum, kg × Heat required per kg of dry gyp sum [from Eqn. No. (*v*)] + wt. of free moisture, kg × heat required to evaporate per kg of moisture [from Eqn. No. (*i*)].

= 0.97 × 141.605 + 0.03 × 611

= 155.68 K Cal per kg of gypsum (*vi*)

The theoretical heat required per kg of plaster can be calculated with the help of Eqn. (*vi*). One kg of gypsum charge will finally give 0.8245 kg of plaster of Paris after losing 3 per cent free moisture and 15 per cent chemically combined water. Hence, theoretical heat per kg of plaster.

$$= \frac{\text{Theoretical Heat per kg of gypsum}}{\text{Wt. of plaster produced per kg of gypsum}}$$

$$= \frac{155.68}{0.8245}$$

= 188.8 K Cal per kg of plaster (*vii*)

Hence, the Theoretical Heat in respect of Silliguri Calcinator is 188.8 K Cal per kg of plaster.

From above, the actual heat consumed per kg of plaster produced in Silliguri plant:

$$= \frac{\text{Coal(kg) per tonne of gypsum} \times \text{C.V. of coal (K Cal/kg)}}{\text{Wt. of plaster produced per tonne of gypsum}}$$

$$= \frac{60 \times 3220}{0.8245 \times 1000}$$

$$= 234.32 \text{ K Cal/kg of plaster} \qquad (viii)$$

Hence, the Actual Heat Consumed is 234.32 K Cal/Kg of Plaster. Thermal Efficiency of Silliguri Calcinator, %age

$$= \frac{\text{Theoretical Heat regd. per kg of plaster}}{\text{Actual Heat consumed (through coal) per kg of plaster}} \times 100$$

$$= \frac{(vii)}{(viii)} \times 100$$

$$= \frac{188.8}{234.32} \times 100$$

$$= 80.5$$

Hence, the Thermal Efficiency is 80.5%

Heat Required for Calcination of Gypsum in Open Pan System

The theoretical heat of calcination for the gypsum is 171 K Cal per kg of plaster. Since nothing is mentioned about the contents of free moisture present in the gypsum, this figure may be safely considered as the value of theoretical heat of calcination of gypsum devoid of free moisture. That way it is quite comparable with the value of theoretical heat calculated in foregoing pages in Equation (*v*), from where the theoretical heat of 141.605 K Cal per kg of dry gypsum is equal to 166.6 K Cal per kg of plaster. Besides free moisture, the theoretical heat may also vary because of variation in purity of gypsum.

In case of open pan, taking the theoretical heat as 171 K Cal per kg of plaster and the actual heat consumed (through fuel) as 1587 K Cal per kg of plaster, the thermal efficiency of open pan can be calculated as below:

Thermal Efficiency, %age

$$= \frac{\text{Theoretical heat per kg of plaster}}{\text{Actual heat consumed per kg of plaster}}$$

$$= \frac{171 \times 100}{1587}$$

$$= 10.77 = 11 \text{ per cent (say)}$$

This value of Thermal Efficiency is very close to 12% which is reported in Reference[7].

Saving in Fuel Through Calcinator

From description of heat calculations given in the foregoing pages in respect of innovative calcinator and the open pan system, the following emerges:

Actual heat consumed in calcinator per kg of plaster produced = 234.32 K Cal

Actual heat consumed in open pan system per kg of plaster produced = 1587 K Cal.

Hence, % Saving in fuel in calcinator w.r.t. open pan

$$= \frac{\text{Actual heat in open pan} - \text{Actual heat in calcinator}}{\text{Actual heat in open pan}} \times 100$$

$$= \frac{1587 - 234.321}{587} \times 100$$

$$= 85.23\%$$

Therefore, the innovative calcinator brings savings in fuel to the tune of over 80% in comparison to Open Pan System.

2.2.3 Continuous Type Gypsum Kettles (Horizontal/Conical Kettle Types)

1. Background

The development of the conical kettle was a logical consequence of the improvements which had been made thermal efficiencies by the use of the submerged combustion technique. This technique (Fig. 2.7) utilizes a normal continuous

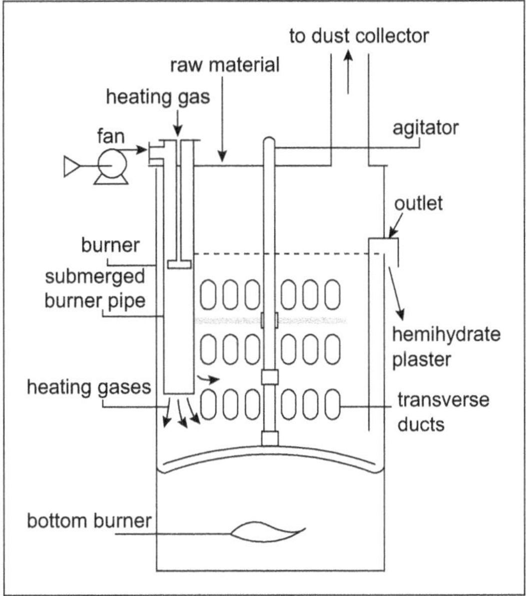

Fig. 2.7 Kettle with submerged combustion burner for continuous production of β-hemihydrate (plaster of Paris)

kettle with heat input via the bottom and cross-tubes. Additionally a submerged combustion gases are discharged into the calcining mass of the kettle. The gases are rapidly and intimately mixed with the gypsum in the kettle and thus achieve very efficient heat transfer. Thus, a typical stack gas temperature for combustion gases leaving a standard kettle would be 350°C to 400°C, whereas the combustion gases from a submerged combustion unit will leave the calcining mass at a temperature slightly above the calcining temperature, typically 155°C. This temperature difference is an indication of the extra heat which has been used where it is wanted, in the calcination process. Thus, the installation of a submerged combustion unit to a standard kettle will provide an increase in the overall efficiency and in the kettle output – in some cases doubling the stucco production rate.

Recently, the submerged combustion method – indirect heat transformation of the externally heated boiler casing combined with direct heat transformation – has been introduced. The advantages are high thermal efficiency and low energy consumption. Kettles with submerged burners can be used up to 500 t beta plaster per day. Existing kettles can also be fitted with submerged combustion, which reduces energy requirements and improves flow rates[8]. Batch horizontal kettles, with capacities of 5–10 tonnes/hour, are famous in France under the name of Beau. Almost the entire global production of hemihydrate is carried out in plants of the kind described above. Most of the beta-hemihydrate or stucco produced is used in gypsum building components and to the lesser extent, it is used in special building plaster. A few plants produce hemihydrate and multiphase plaster in a fluidized bed [9–10].

The effectiveness of the submerged combustion technique extended to the technique to enable a kettle to be fired solely by a submerged combustion system.

2. Development (Prototype of the Pilot Plant)

Initially, attempts were made to run a submerged combustion kettle with the bottom burner off, but it was found that it was necessary to provide a heat input to the kettle bottom in order to fluidize the contents. It was shown that it was possible to reduce the heat input to the kettle and thus increase the ratio of submerged to bottom heat but it was felt that a greater improvement could be made by a total redesign of the kettle.

A basic requirement was for a kettle which could be fluidized by a submerged combustion unit without the need for an external heat source. It was considered that a conically shaped vessel would give this property and a temporary pilot unit was designed and built. A schematic diagram of this unit is shown in Fig. 2.8.

As may be seen the kettle consisted of a simple conic (approximate angle 30°) with a submerged combustion tube placed centrally. The cone has a maximum diameter of approximately 1.2 meters and height of 1.8 meters. The submerged combustion tube contained a nozzle mix burner with a spark igniter and a flame detector probe.

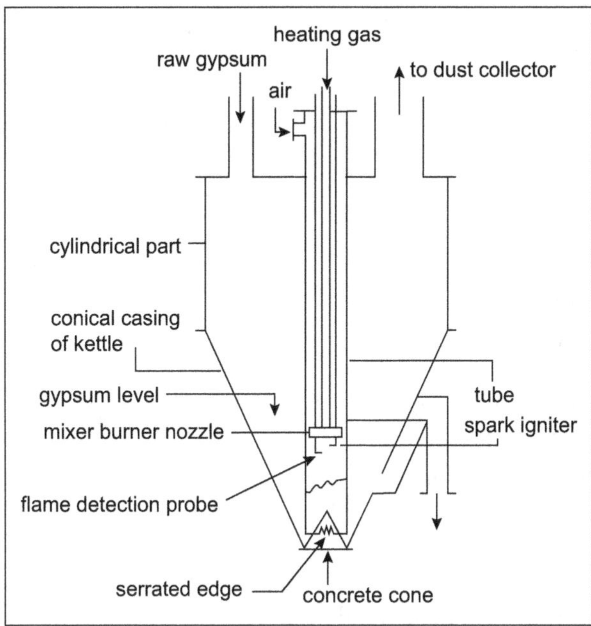

Fig. 2.8 Conical Kettle with Sub

The combustion gases pass through the bottom opening of a tube and though holes spaced around the circumference near the bottom of the tube. Gypsum is fed into the kettle from a normal mineral bin via a rotary valve which is controlled by a thermocouple located in the plaster bed, and plaster product leaves via the overflow weir. The combustion gases and associated dust pass to a dust collector.

The unit was operated for a short period of time to prove the viability of the project by producing plaster at a rate of about 1 tph on a continuous basis solely by the use of submerged combustion. Having proved the principle it was decided to refine the design of the kettle by examining few patterns in Perspex models.

Two models were produced with cone angles of 30° and 60° respectively, and these were tested for suitability by the following criteria:

(*a*) Evenness and regularity of observed flow patterns in both solids and gases.
(*b*) Minimization of dead areas and also surging or bridging of solids.

A number of test runs were carried out at a differing air flows and using various solids, such as polystyrene, vermiculite and coarsely ground mineral. In all cases it was found that the performance of the 60° cone was superior to the 30° cone for the criteria listed above. The design was then slightly modified in line with space requirements and a new pilot unit with a 50° cone was constructed. This had a slightly larger diameter of about 1.5 meters with the height of the conical section being 1.2 meters and a 0.3 meter cylindrical section on top. A further modification of this unit was the addition of a conical insert at the submerged tube outlet to assist mixing at this point and prevent material remaining in the area. The unit had a production capacity of about 2 tph, the main limitation being the volume rating of the fabric dust collector.

3. Gas Fired Conical Kettle

A schematic diagram of the first production unit is shown in Fig. 2.9. It can be seen that there were no major changes necessary in the scale up exercise and that the only significant difference is in the longer cylindrical section which was added to reduce the amount of particle carry-over. The cone has a maximum diameter of 3.0 metres and its height is 2.8 meters, while the cylindrical section is 1.8 meters in height. Both sections are constructed of 10 mm carbon steel and are insulated with 100 mm of mineral wool. The kettle is mounted on compression load cells and full instrumentation is provided to monitor such parameters as weight of material in the kettle, temperatures within the kettle and air and gas flow rates. The kettle was installed at Roberts bridge plant in 1981.

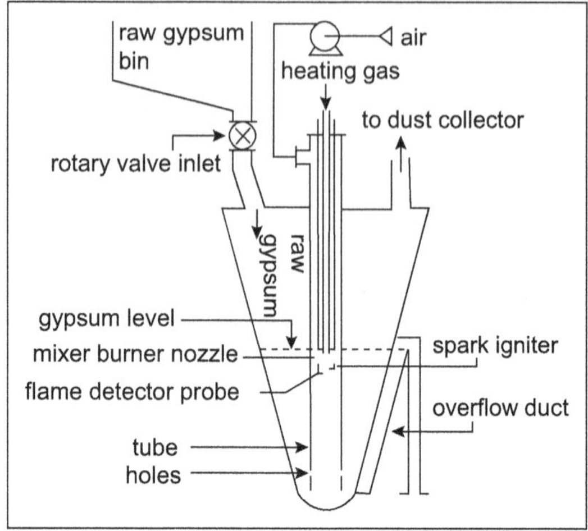

Fig. 2.9 Gas Fired Conical Kettle

Since the commissioning of the kettle, there have been two main modifications carried out. They are:

(*a*) The replacement of the refractory cone with a radiator column in the bottom section of the submerged combustion tube.

(*b*) The installation of a booster fan in the gas supply line to increase the available pressure of the gas and thus the potential capacity of the kettle from 20 to 25 tph.

The radiator column was installed in the kettle soon after commissioning as it was found that the product was containing unacceptably high levels of insoluble anhydrite. It was felt that this was due to the high volumes of high temperature gases leaving the bottom of the tube and over calcining the plaster at this point. Discussions with the burner manufacturer suggested that a significant reduction in this temperature could be achieved by means of a radiator which would increase the heat transfer through the tube. Various sizes of radiator were tested before the present design, a 0.8 meter diameter, 1.6 meter high spun steel cylinder with a cast steel dome was installed. This has been in position for over 12 months, and although showing signs of wear due to the extreme conditions it has had to withstand, it does not yet need replacing. The radiator was successful in reducing the amount of insoluble anhydrite found in the kettle from 5 – 10 % to around 4%.

The installation of the gas booster was carried out in April, 1983 in order to increase the kettle production. No problems were encountered with either quality or reliability following the increase in throughput.

The kettle has been run as a production unit since January, 1982 and has been run virtually continually to supply the board plant. Figures obtained over this period indicate that the kettle is consistently using approximately 6.5 gross therms/tonne stucco (690 kJ/kg of stucco), for calcining a mineral of approximately 78% gypsum content.

In use on the board plant the stucco behaves the same as continuous or submerged kettle stucco apart from a faster set. This normally requires the accelerator addition to be cut to about one half of normal but all other settings can remain the same.

4. Oil Fired

Many plants do not have natural gas available and use heavy oil as the primary fuel source. In view of the experience with the gas fired unit it was decided that a heavy oil fired unit should be designed and installed at an appropriate plant. A schematic diagram of the unit is shown in Fig. 2.10. The basic design of the kettle shell is identical with the gas fired unit, the only difference being an increase in diameter from 3 meters to 3.6 meters. The overall height is 4.5 meters and the cylindrical disengagement zone is 1.65 meters in height. The increased diameter is designed to give a similar gas velocity in the disengagement zone when allowance has been made for the increased size of the combustion chamber. The kettle

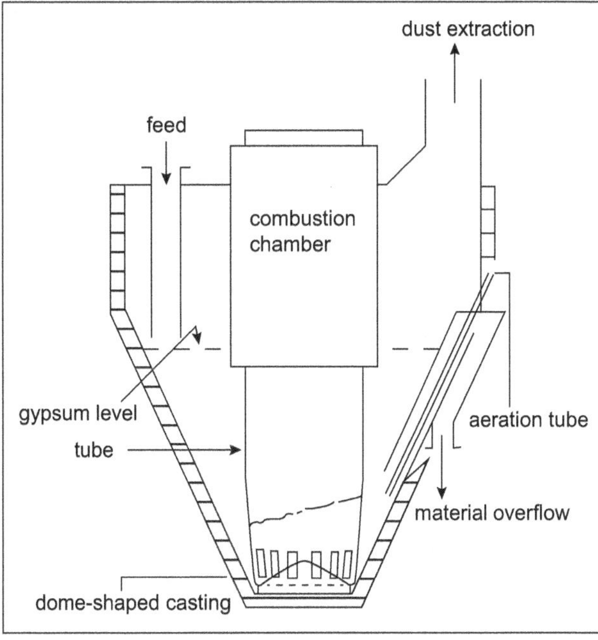

Fig. 2.10 Oil Fired Conical Kettle

is again manufactured from 10 mm carbon steel and is insulated externally with 100 mm mineral wool. It will be noted that there is no radiator in the tube, only a small dome shaped splitter at the tube outlet. It was felt that the different properties of the oil flame would make a radiator unnecessary, so the kettle was designed to start without one, but to be easily modified should the product quality require it.

The kettle was designed to replace an existing heavy oil fired continuous kettle which had reached the end of its operating life. By replacing only the kettle it enabled the existing mineral and plaster handling equipment to be used with only limited modifications for increasing throughputs. The installation was carried out at Kirby Thore Works in 1983.

The unit was commissioned in the summer of 1983 and has run continually since September of that year. No major modifications were necessary and the kettle has continued to produce about 21 tph of stucco which is being used to feed a board plant. When introduced onto the board plant in place of the normal continuous plaster, it was found that the only parameter which needed to be changed was the accelerator addition rate.

Once again this had to be reduced to about one half of its previous rate but all other conditions remained constant.

Plaster quality was found to be acceptable without the need for a radiator in the submerged combustion tube. Initial runs which were done at the normal continuous kettle calcination temperature of 160°C were found to give high levels of soluble anhydrite. However, by gradually reducing the calcination temperature to its present 147°C it was found that this could be reduced without leaving residual gypsum. As well as improving the quality of the stucco, this also reduced the energy input and increased the thermal efficiency.

Energy input under present conditions has been measured at 6.2 therms/ tonne of stucco (650 kJ/kg stucco) for calcining a mineral of approximately 72% gypsum content.

Following the results obtained from the first oil fired unit a further two have been installed at another plant and are now being commissioned.

4. Advantages

The main advantages of a conical kettle as compared with a standard continuous kettle are:

(*a*) Thermal efficiency
(*b*) Construction
(*c*) Operation

(*a*) Thermal Efficiency

As explained earlier the thermal efficiency of the kettle is very high due to the high rates of heat transfer in the kettle, ensuring that the temperature of the exhaust gases leaving the kettle is very close to the calcining plaster temperature. Tests carried out at the operating plants have consistently given fuel usage figures of around 6.2 therms/tonne stucco (650 kJ/kg stucco) for a mineral purity of around 72% gypsum.

(*b*) **Construction**

Another major advantage of the conical kettle is the simplicity of its design, with no moving parts in the kettle and a minimum of refractory. As well as reducing the maintenance requirements of the kettle this also gives a relatively low capital cost.

A further point which was borne in mind when the kettle was designed was ability to fit into the space at present occupied by a continuous kettle. A totally new plant is a rare occurrence but kettle replacements are a continuing requirement.

(*c*) **Operation**

In many ways the most attractive feature of the kettle is its ease of operation. This is mainly due to the relatively small hold-up of gypsum in the kettle, normally around 5 tonnes, and to the direct contact between gypsum and hot gases. Thus, it can be seen that the time from flame ignition to calcination temperature being attained in the kettle is approximately 15 to 20 minutes. Furthermore, as there is no mass of refractory and brickwork to heat up to operating temperature, the kettle is virtually immediately producing its desired output rate.

The output which is obtained from the kettle is dependent upon the capability of the combustion unit to provide heat and of the dust extraction unit to handle the combustion gases. As mentioned previously this has recently been demonstrated on the gas fired unit whereby the production was increased from 20 to 25 tph simply by installing a booster to increase the available gas pressure.

Another feature that has assisted in enabling the kettle to be used as a replacement for existing continuous kettle is the ability to process minerals with a wide range of particle size. This has meant that the existing grinding mills can be used to supply the kettle without any change being made in their operation, and that a conical kettle and a continuous kettle can operate side by side without problems.

5. Future Developments

The main development being carried out at present is to extend the availability of fuel to include a coal-fired unit. A fluidized bed coal combustor has been chosen for this unit as it is a relatively simple and efficient means of burning coal.

The combustion takes place within a bed of sand particles which are fluidized by the combustion air. The sand bed is pre-heated to a temperature of 650°C at which point combustion will take place and coal is fed into the bed. The combustion of the coal takes the temperature of the bed up to 950°C and the coal feed is then controlled to maintain this temperature. It is not allowed to go higher than this as it then approaches the softening point of the coal ash which can lead to fusing of the ash.

The hot gases from the bed are then fed via insulated ductwork into a standard design of conical kettle. The kettle is 1.8 meters in diameter with an overall height of 2.4 meters and a disengagement zone of 0.9 meters. The pilot unit has been operated at throughputs of up to 1.8 tph and with heat consumption of 7.5 therms/tonne of stucco (90 kJ/kg of stucco).

Because the temperature of the gases entering the gypsum bed, is lower than for gas or oil firing there is very little tendency for the formation of insoluble anhydrite. Typical results obtained are given in Table 2.1. Designs are being prepared for a full scale unit and it is hoped that a production unit will be operated soon.

Figure 2.11 shows a flow diagram for such a plant. Such type of kettles up to capacities of 300 tonnes of plaster per day have been claimed. The Saint-Gobain Gyproc India Ltd. at Jind in Haryana (India) has installed such type of kettle for making β-hemihydrate plaster for paper coated gypsum plaster boards.

6. Fluidized Bed Calciner

Background

Within the United Kingdom the price of coal to produce one net therm is cheaper by far than any other conventional fuel. Typical prices as at April, 1984 are:

Coal	£ 0.205/gross therm
Natural gas	£ 0.303/gross therm
Heavy oil	£ 0.397/gross therm

Fig. 2.11 Calcination of Gypsum in Continuous Type of Kettle as Installed by Saint-Gobain Gyproc India Ltd. Jind, India

In addition the known reserves of coal are very high, thus guaranteeing long term availability.

With these two points in mind it was decided to design calciner specifically to burn coal and to ensure that it had a very high thermal efficiency. A joint exercise was carried out by M/s Engineering Department in conjunction with the Coal Research Establishment of the National coal Board, and a successful design was produced.

Once again a fluidized bed is used to provide an efficient and adaptable combustor with the added advantage that the design uses heat transfer through the wall of the combustor to reduce the temperature of the bed and hence the amount of excess air needed, and thus increase the thermal efficiency of the unit.

Description

A schematic diagram of the calciner is shown in Fig. 2.12. The calciner is based on the use of two fluidized beds – an inner one of sand particles for burning the coal, and an outer one of gypsum for calcining. The combustor contains a sand bed which is supported by a distributor with a number of stand-pipes for fluidizing air. The combustion gases leave the combustor through ducting which connects to manifolds and thus to a number of sparge pipes. In order to reduce heat loses from the system the ducting and manifolds are insulated with ceramic fibre.

The sparge pipes are located near the bottom of the outer vessel which contains the material being calcined. The combustion gases leaving the spurge pipes fluidizes the gypsum thus ensuring efficient calcination. After passing through the gypsum bed the combustion gases leave the calciner and pass to a dust collection system. Dust carryover from the system is normally quite low but a cyclone is used to collect dry product and it is passed to the hot

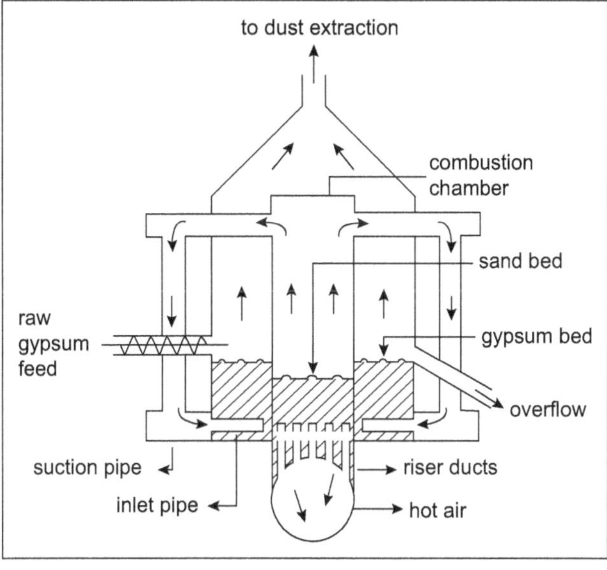

Fig. 2.12 Fluidized Bed Furnace for Making Gypsum Plaster

pit. The outer calcining section performs in a manner similar to a continuous kettle with the mineral feed being controlled by the calcining gypsum and a continuous overflow of product via a weir to a hot pit.

A major element of the design which ensures the high thermal efficiency is the location of the combustor within the calcining bed. This enables heat to transfer directly from the high temperature sand bed to the lower temperature gypsum bed with very good heat transfer coefficients being obtained in both fluidized beds. As stated previously it is necessary to ensure that the sand bed temperature does not rise above 950°C otherwise the ash from the coal will soften and fuse together, preventing the bed from fluidizing. In a free standing combustor this is normally done by using a relatively high level of excess air (around 100%) and this reduces the overall efficiency of the system. Direct cooling of the combustor bed, however, means that much lower levels of excess air may be used (around 30%) thus reducing heat losses from the exhaust.

Operating Experience

In order to check the performance of the calciner a 5 tph pilot plant unit was built and installed at out East Leake Works. For this output, the calciner, which is roughly cubic in shape, had dimensions of 3 × 3 × 3 meters. After commissioning by design staff the unit was operated by normal production personnel from October, 1981. During this proving period the calciner produced approximately 23,000 tonnes of plaster which was mixed with product from continuous kettles and used for plasterboard production. The average production rate was 4½ tph with a fuel usage of 6.6 therms/tonne of plaster (700 kJ/kg of plaster) for a mineral with a gypsum content of 81.5%.

The relatively low temperature of the combustion gases entering the gypsum bed (normally around 900°C meant that the unit did not tend to form insoluble anhydrite). The product has been found to be very similar to that obtained from a continuous kettle but with a faster setting time. A typical analysis of the product is given below:

Soluble anhydrite	—	—	8%
Gypsum	—	—	2%
Hemihydrate	—	—	70%
Insoluble anhydrate	—	—	0%

During the continuous operating period it was noted that there was a gradual, very small carryover of sand particles from the combustor into the gypsum bed. This was insufficient to affect the quality of the plaster but by lowering the sand bed level, reduced the amount of coal which could be burned, and thus the output of the calciner. Initially, this sand could only be replaced by stopping the unit for a short period, but an automatic injection system was developed which enabled the unit to run at maximum output on a continuous basis. Also an automatic bed cleaning system could be incorporated if needed for certain grades of coal.

Following the successful operation of the pilot unit, a full scale unit having a capacity of 18 to 20 tonnes per hour has been built and is presently awaiting installation. After proving of the full scale unit in continuous production conditions it is proposed to build a number of further units. Limited exploitation jointly by British Gypsum and the National Coal Board is envisaged. Also, opportunities for use in calcining or drying other minerals will be investigated.

Another development is the indirectly heated, continuously operated horizontal kettle. Externally, it resmbles a rotary kiln. In this hot gases first passes through a central tube in concurrent flow and then through further heating tubes in counter-current flow. Similar is the continuously operated Holoflite calcining plants/units, using hot oil or superheated steam as the heat exchange medium. In this case the gypsum is moved along on a screw conveyor that has hot oil passing through it[11].

2.3 Anhydrous and Multiphase Plasters

Anhydrous or over burnt plasters are produced in dry calcining process at temperatures between 300–900°C. There are three reaction stages:

1. AII-s, slowly soluble anhydrite, < 500°C
2. AII-u, insoluble anhydrite, 500 – 700°C
3. AII-E, Estrichgips, >700°C

The properties and phase composition are shown in Tables 2.2 – 2.3.

In over burnt plaster and also in multiphase plaster, the above three stages of reaction have definite proportions which governed by the raw materials and the calcination process. In some of the techniques, gypsum is calcined in such a way, that the anhydrite phase is obtained along with hemihydrate phase. In other set of processes, other over burnt plaster is produced separately and then blended with the hemihydrates plaster afterwards. These plaster are called as multiphase plasters, putzgips in Germany.

The latest process for making overburnt plaster is the conveyor kiln developed by KNAUF[12]. Today a plant of capacity 1200 tonnes per day are available.

Table 2.2 Phase Composition of, in % of Over burnt Plaster, Plaster of Paris and Multiphase

	Plaster of Paris (Rotary kiln)	Over burnt (Conveyor plaster)	Multiphase Plaster (Mixture of *POP & Over burnt plastert)
Dihydrate $CaSO_4.2H_2O$	0–0.6	0–2.1	0–2
β -Hemihydrate $CaSO_4.1/2\ H_2O$	75	6	26
β-Anhydrite III β-$CaSO_4$ III	19	18	14
Anhydrite II $CaSO_4$ II**	5	74	58

Table 2.3 Properties of Calcined Gypsum

Method of Production	Type of Plaster	Fineness Residue on 0.2 mm sieve	Combined water (%)	W/p ratio	Setting time (Minutes) Initial Final		Strength +FS **CS *H (MPa)			Density (kg/m^3)	Applications
Natural gypsum											
Rotary kiln	β-plaster	1.0	4.5	0.70	12	26	4.5	11.5	19.1	1070	Building components Special building plasters
Kettle	β-plaster	3.2	5.4	0.63	10	22	5.1	14.0	26.3	1130	Building components Special building plasters
Conveyor Kiln	• MPP	35.0	0.80	0.58	6	35	5.2	15.4	26.0	1235	Machine applied plaster Multiphase plaster
Autoclave	α-plaster	1.0	6.0	0.38	11	24	12.4	40.4	92.0	1600	Molding plaster Dental plaster
Flue-gas gypsum											
Rotary kiln (Knauf Process)	β-plaster	3.0	3.5	0.74	9	22	4.7	11.1	12.1	1060	Premixed plaster-Jointing plaster Gypsum building Components
Kettle (Knauf Process)	β-plaster	2.5	5.1	0.71	15	29	5.3	13.0	88.5	1130	Premixed plaster Jointing plaster Gypsum building components
Conveyor Kiln (Knauf Process)	MPP	25.0	1.88	0.56	6	20	6.8	19.9	36.0	1335	Machine applied plaster Multiphase plaster
Autoclave (Nitto Gypsum plaster)	α-plaster	0.8	6.0	0.37	13	28	12.0	44.5	90.3	1580	Molding plaster Dental plaster Industrial plaster

*Tested as per DIN 1168, + Flexural strength, **compressive strength, *Hardness, *MPP = Multiphase plaster

Before feeding in to the kiln, gypsum rock is crushed to size 4–60 mm and split into 3 or 4 sieves fractions – 7–25 mm, 25–40 mm, 40–60 mm or 4–11 mm, 11–25 mm, 25–40 mm and 40–60 mm. The fractions are put on the continuous grate, the smallest on the bottom. The speed of grate is 20–35 m/h, passes through a calcining hood. The hot gases are drawn through gypsum bed by exhaust fans. The top layer can reach to the temperature of 700°C, the bottom up to 300°C. The temperature of the heat resistant plates does not exceed 270°C. The gypsum is not mixed during calcinations, hence little gypsum dust is produced. Nearly half the gases are discharged into the atmosphere as waste gas through chimney at 100°C and remaining gas goes to combustion chamber. In Germany and elsewhere conveyor kilns with capacity of 2 Mt/day or more are successfully operating.

Multiphase plaster are used to finish interior walls and ceilings. Since 1965 machine applied plaster has been produced in large quantities by adding chemical additives to such multiphase plaster. Table 2.4 shows energy requirements in manufacturing calcined gypsum.

2.3.1 Alpha Hemihydrate

Alpha-hemihydrate can be produced by wet calcining processes, either under elevated pressure in autoclaves or at atmospheric pressure in acids or aqueous salt solutions between 80 and 150°C. However, only the autoclave processes have so far achieved industrial importance for the small amounts produced.

Alpha-hemihydrate from natural gypsum is always a batch process. For instance, gypsum rock (particle size 150–300 mm, >95% $CaSO_4.2H_2O$ rocklike) is put in the wire baskets, and either stacked in upright autoclaves or wheeled into horizontal autoclaves of capacities of 0.5–10 m^3. The autoclaves are heated directly or indirectly with steam at 130–135°C. The heating is carried out so that after about 4–5 hours a pressure of 4–5 bars has built up. The autoclave is then empted, the alpha-hemi hydrate formed is immediately transferred in the baskets to a chamber to be dried at 105°C under atmospheric pressure. It is then ground to 120 mesh or even more. The variation in temperature and pressure during dehydration and drying can be used to effect the properties of the products.

Alpha-hemihydrate is often mixed with beta-hemihydrate. For that reason processes have been developed that produce a mixture of α- and β-hemihydrates in a single operation.[13] In one such process the α-hemihydrate is dried in a rotary kiln, which at some time serves to both calcine the β-hemihydrate and blend the two types of material.

2.3.2 Process Control Parameters

Dry gypsum calcining processes are usually equipped with dry dust collection systems. The dust formed depends on the type of calcining unit. Generally dust collectors, electrostatic filters are used. The dust collected is added to the calcined gypsum. The water liberated in the process is discharged through stacks into the atmosphere. There are no wastes or by-products are produced during the process. No environmental problems are encountered in the normal processing of the plaster. The energy consumed in the process is the sum of the fuel used in the calcination of the gypsum and the electric power required to operate the plant. Table 2.4 shows energy comparison of various types of processes.

2.4 Calcined Products from By-product Gypsum

2.4.1 β-Hemihydrate and Multiphase Plasters

1. Flue-Gas Gypsum

Gypsum obtained from the desulphurization of combustion gases from power plants, flue-gas gypsum, is a moist fine powder [Fig. 2.13, *a* (FGD), *b* (Natural anhydrite)] with free moisture content of below 10% and contains only minor impurities. These impurities are soluble salts of Na^+, Mg^+, Cl^- and calcium sulphite, $CaSO_3.1/2H_2O$.[14–15] Hence, this variety of gypsum can be used by cement and gypsum industry with out beneficiation. In most cases, FGD gypsum require drying before calcination Concurrent drying (Hazemag and Babcock Driers) are most suitable for this purpose. The gypsum is then calcined to beta plaster and converted into building materials with out any grinding, etc.

In case making building plaster and multiphase plaster, FD gypsum has to be treated further. Its particle structure ranging from cubic (bulk density 1.2 t/m^3) to lath or rod shaped (bulk density 0.5 t/m^3) and the small particle size like 20–60 micron to maximum 200 micron, is unsuitable for multiphase plaster production and also gives thixotropic behaviour. Briquetting (2 cm thick × 6 cm long) of FGD gypsum is suitable for making multiphase plaster and to obviate thixotropic problem of the FGD gypsum. The briquetting plants such as Knauf and Saarberg in Germany are in operation. In future the increased availability of FGD gypsum may be a problem for use. However, the combination of fly ash with FGD gypsum to be used in agriculture and land fill may be the profitable preposition in future.

In India, the production and use of FGD gypsum has not started yet. Some industries are now producing mixture of such wastes from oil refinery or combustion plants.

2. Phosphogypsum

Finely divided phosphogypsum (95% < 0.2 mm) is a by-product of the production of wet phosphoric acid. It is available as a filter cake with a free water

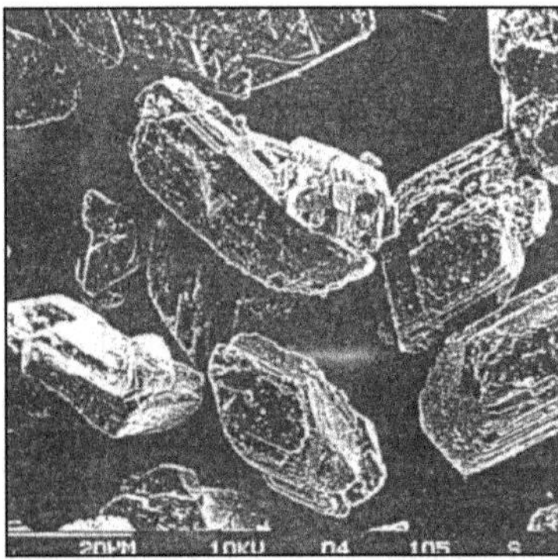

Fig. 2.13 Microstructure of FGD Gypsum showing Prismatic and Tabular Crystals with Twinning (*a*)

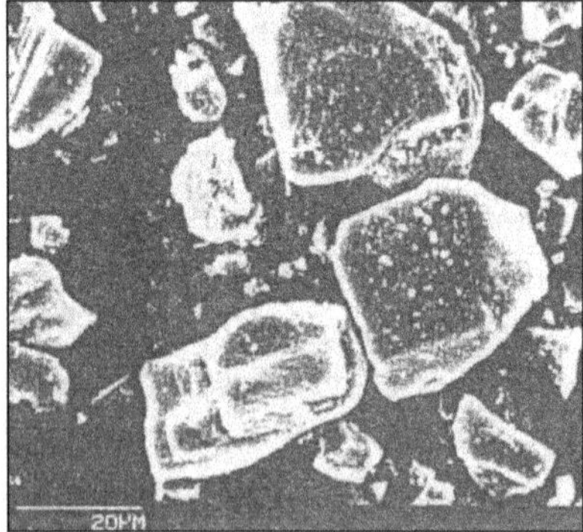

Fig. 2.13 Microstructure of Natural Anhydrite showing broadened Prismatic Crystals (*b*)

content of 20–30%. The proportion of impurities is high, comprising organic and inorganic substances, water soluble and water-insoluble substances, some adsorbed at the surface, some incorporated into the crystal lattice. When phosphogypsum is used by the gypsum industry, these impurities have to be removed by washing, flotation, and recrystallization.

Phosphogypsum produced from magmatic (*e.g.*, Kola) or fused raw phosphates contains no organic impurities. Sedimentary phosphate (*e.g.*, Morocco) contains organic substances and small amounts of radioactive radium, at least some appearing in the gypsum. All phosphogypsums contain inorganic impurities, *e.g.*, phosphates, silco fluorides, and sodium. Unfortunately, they affect the properties of the gypsum.

The one- and two-stage processes that have been developed for the manufacture of wet phosphoric acid produce different qualities of phosphogypsum. In the one-stage processes, *e.g.*, that of Prayon, altogether representing 84% of all phosphogypsum produced as dihyhrate) or hemihydrate (hemihydrate process, *e.g.*, that of VEBA[16] or Fisons[17], but representing less than 1%.

In the two-stage phosphoric acid processes, either gypsum or hemihydrate is produced in the first stage. In the second stage it is converted to another state of hydration before being removed by filtration. These processes include the hemihydrate-dihydrate process (*e.g.*, Nissan[18]; 15% of all phosphogypsum produced) and the dihydrate-hemihydrate process (*e.g.*, Central Prayon [70%], less than 1%). Generally, gypsum obtained from the two-stage processes is of better quality in regard to in-organic impurities such as phosphate, and sodium.

The gypsum industry is faced with the need for a considerable amount of extra treatment if it attempts to use phosphogypsum[19–20]. Removal of organic impurities, which discolour the gypsum, and of water-soluble inorganic contaminants, which cause efflorescene, involves first remashing the phosphogypsum with water, then subjecting it to flotation, classification (hydrocyclone), thorough washing, and filtering up to 5 t of water per tonne of phosphogypsum is consumed. Gypsum purified in this way is obtained as filter cake with a free water content of 20–30%.

For production of β-hemihydrate intended for gypsum building components the filter cake is dried (*e.g.*, rapid dryer from Hozemag, or the contact dryer from Serapic) and then calcined to β-hemihydrate in the same way as finely ground natural gypsum. Today this method is still the principal one used by the gypsum industry in Japan, where this method was developed in 1940 by Yoshino.

For use in gypsum building plaster and multiphase plaster, phosphogypsum is made unsuitable by its particle shape, fineness, and isomorphous acid phosphate impurities. Its particle shape and fineness seriously impair the workability of the multiphase plaster, *i.e.*, the plaster is thixotropic. The cocrystallized acid phosphates cause calcined products to develop lime sensitivity, which interferes with setting and development of the strength.[21]

In order to overcome these deficiencies the phosphogypsum is dried, calcined, and after addition of aqueous calcium hydroxide suspension simultaneously agglomerated and recrystalized in a pelletizer. Such alkaline-recrystallized pellets are used as starting material for gypsum building plaster and multiphase plaster.

In one process phosphohemihydrate is the starting material. If the process is carried out with adequate care, the phosphohemihydrate is sufficiently pure that the purification phase can be dispensed with. The fine phosphohemihydrate, having a residual water content of 20–25%, is mixed immediately with a calcium hydroxide suspension or calcium hydroxide powder so that the calcium sulfate crystallizes into coarse-grained dihydrate. Part of the moisture and all of the acid phosphate are bound chemically, and the particle size and structure are satisfactory.

The phosphogypsum lumps thus produced can then be calcined and further processed in rotary kilns or, after grinding, in kettles to produce a β-hemihydrate plaster similar in composition to plaster of Paris. They can be converted on a conveyor grate into over burnt plaster, which consists of coarse particles and is comparable to multiphase plaster. These plasters can be processed into all kinds of machine-applied and premixed plasters.

Such processing methods, still in use today, were developed and put into production by Knauf in 1962 and 1970[22]. Most other processes designed to use phosphogypsum *(e.g.,* those of Rhone-Progil, Charbonnages de France, Chimie-Air Industrie, Imperial Chemical Industry, Buell) have not resulted in viable commercial operation.

The extra treatment required by phosphogypsum involves additional capital expenditure and operating costs, jeopardizing its competitiveness with natural gypsum and generally rendering phosphogypsum uneconomical for commercial use. The repercussions felt after the first energy crisis, in 1973, because the extra treatment is energy intensive, were an additional set-back.

Only Japan has so far managed to continuously use phosphogypsum, an accomplishment favoured by its total lack of natural gypsum resources and the prohibitive transport costs from the exceedingly distant sources of supply in North Africa and Australia. But even in Japan there is a growing tendency to use flue-gas gypsum. South Korea also uses phosphogypsum as a source of gypsum.

2.4.2 Methods of α-Hemihydrate Plaster

Continuous autoclave processes were developed in the Germany by Giulini in 1962[23] and in Japan by Nitto Gypsum in 1970 for the manufacture of α -hemihydrate plaster. These facilitate wet preparation without intermediate drying. However, the use of phosphogypsum as a starting material does necessitate prior flotation and countercurrent washing. Flue-gas gypsum obtained from power plants is of sufficient purity to be processed directly to α-hemihydrate by this method. In the Nitto Gypsum process (licensee Knauf), flue-gas-gypsum having a free water content of 10% is slurried with water (one part gypsum, two parts water) and pumped continuously into the autoclave where it is dehydrated to α-hemihydrate under controlled conditions (135°C, 2h). Additives in the suspension change the crystal habit of the α-plaster and yield a product of defined, consistent properties. The α-hemihydrate produced is withdrawn as an aqueous suspension and dewatered in a vacuum filter. The product that has a free water content of 10–15% is immediately dried at 150°C in an indirectly heated dryer and ground. This dry α-hemihydrate can then be used for all types of α-hemihydrate plasters and products.

Today the continuous Nitto Gypsum autoclave process is also used industrially for the processing of finely divided natural gypsum and other types of finely divided byproduct gypsum.

Other continuous autoclave processes have ceased operations within a few years (Giuilni), never got beyond the pilot stage (ICI), or remained dormant in patents. Hochst even shut down their phosphogypsum plant at Knapsack for the manufacture of α-hemihydrate and its further "wet-in-wet" processing into gypsum partition panels.

Researches have been reported by Singh & Rai[24] and others that an autoclaved plaster can be produced by heating the naturally occurring selenite and by-product phosphogypsum under steam pressure in an autoclave. Various factors effecting the calcination of gypsum such as pressure, time of autoclaving, process and the particle size of gypsum lumps were investigated. The formation of hemihydrate was examined by differential thermal analysis and weight loss determinations. The mineralogical studies of alpha plaster were determined using microscopy and X-ray diffraction (XRD). Figure. 2.14 showed that complete inversion of gypsum into hemihydrate takes place by heating gypsum particles (size 15 to 20 mm^3) at 1,75 kg cm^{-2} steam pressure for 7 h (Fig. 2.15). The microscopy of plaster revealed formation of well developed euhedral to tabular shaped crystals (Fig. 2.16). Figure 2.17 shows XRD of alpha plaster.

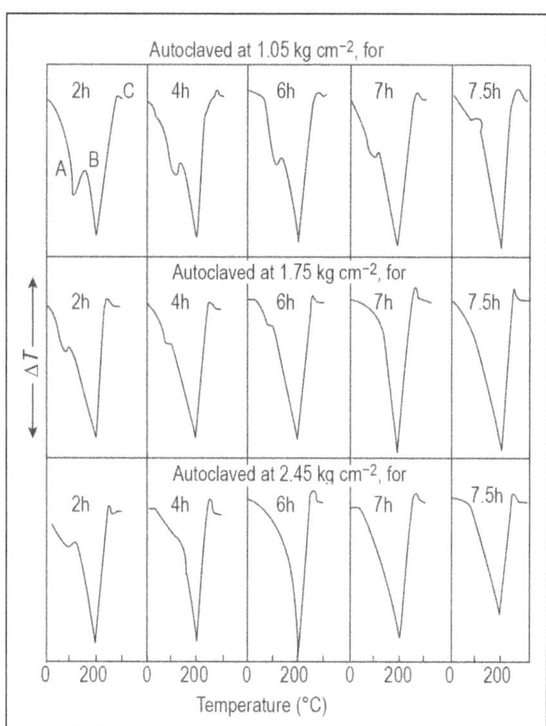

Fig. 2.14 DTA of Gypsum Samples Autoclaved under Different Conditions

XRD showed alpha hemihydrate of stronger reflections at similar *d*-spacing than the beta hemihydrate plaster (Fig. 2.17). The

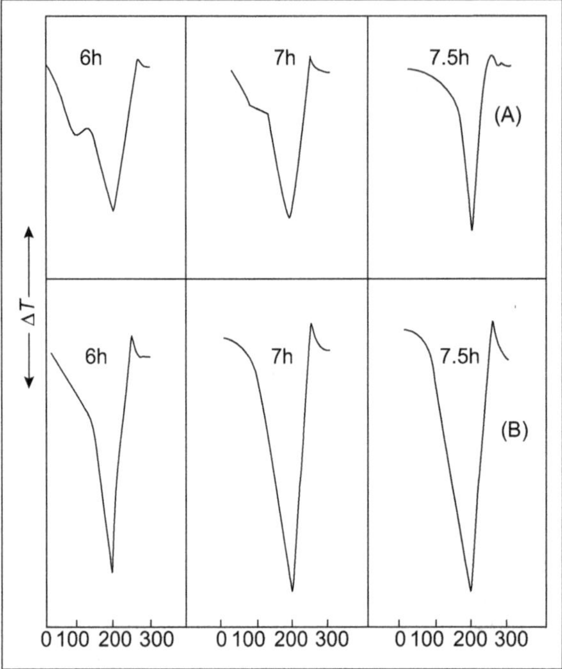

Fig. 2.15 DTA of Unprocessed (*a*) and Processed Phosphogypsum Samples (*b*) Autoclaved under Different Conditions

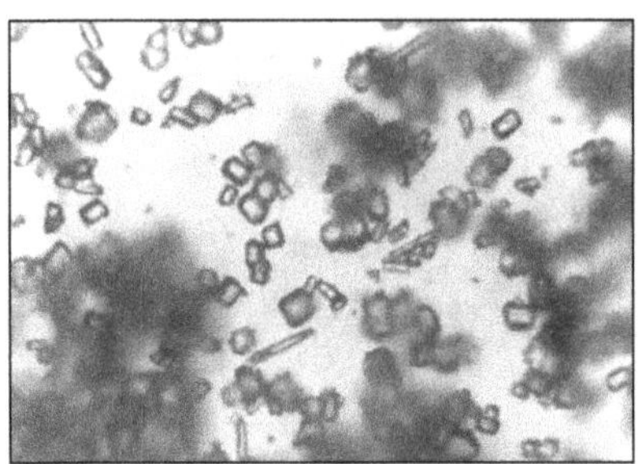

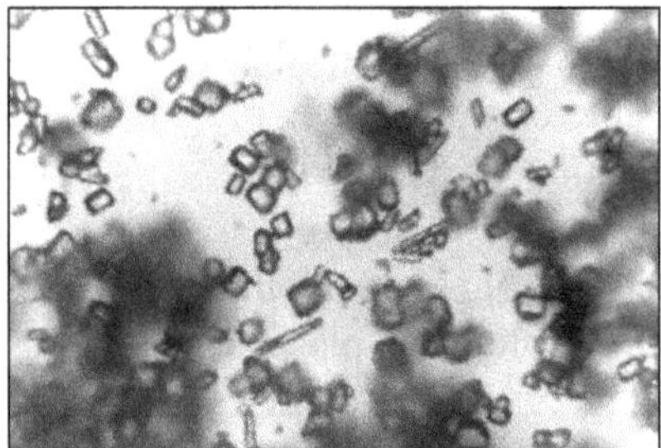

Fig. 2.16 Photomicrograph of of Alpha Plaster produced from Selenite and Phosphogypsum (x350)

hydraulic properties of plaster were studied and found to be of high strength of 22–23 MPa. The alpha plaster is suitable for newer applications like masonry mortars (1:4 plaster-sand) and fibrous boards (density 1200–1240 kg/m^3, breaking load 750 N) for constructional works.

Alpha plaster is also produced by slurry process wherein gypsum slurry is autoclaved in presence of suitable activators cum crystal modifier at short duration[25]. Thus, selenite gypsum (Passing 600 micron) and beneficiated phosphogypsum is mixed with equal volume of water to form uniform slurry. A small quantity (0.15–0.25%) of chemicals like sodium succinate, sodium citrate, or

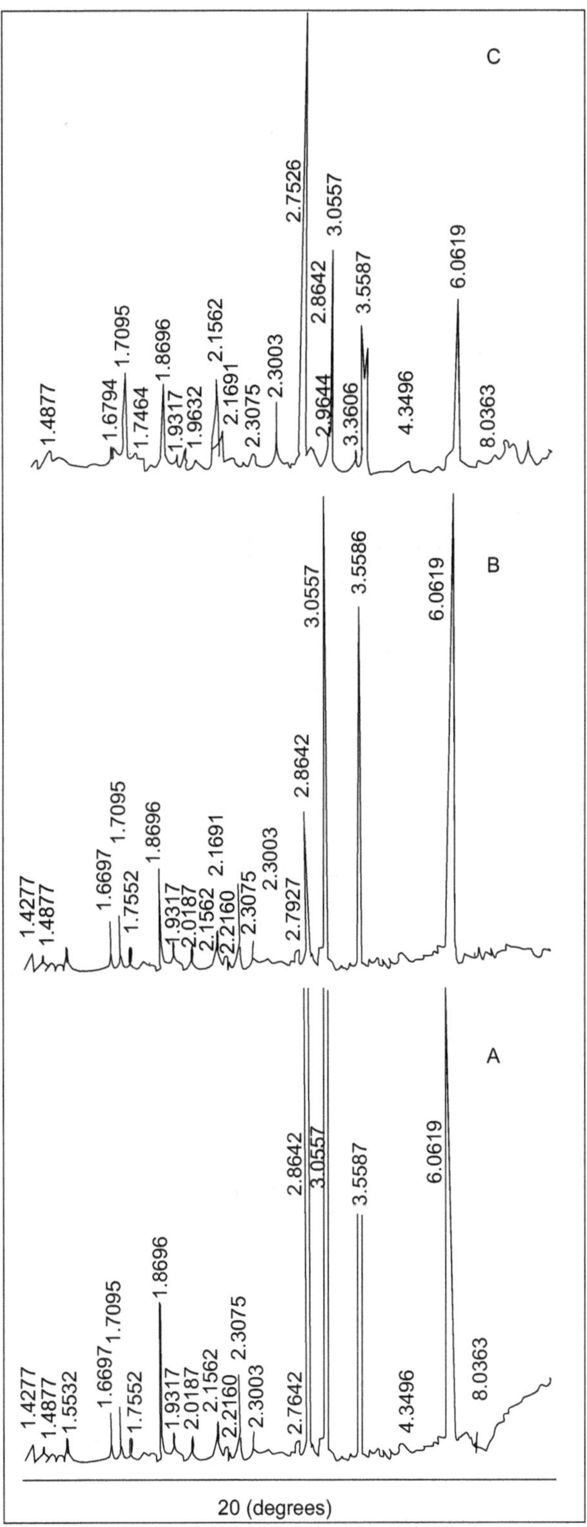

Fig. 2.17 XRD of Gypsum Plaster Samples (*a*) Alpha Hemihydrate from Selenite Gypsum and (*b*) Alpha hemihydrate from Phosphogypsum and (*c*) β-Hemihydrate

Table 2.5 Physical Properties of Alpha Plaster

Material Characteristics			Properties		
α-plaster	**Additives Sod. Succinate (%)**	**Consistency (%)**	**Setting Times (Minutes)**	**Bulk density (kg/m^3)**	**Compressive strength (Nmm^{-2})**
Made at 35 psi ($2.45\ kgcm^{-2}$) steam pressure and dried at 130°C	0.15 0.20 0.25	50.0 40.0 38.0	18.0 19.0 19.0	1515 1540 1688	17.00 20.20 25.50
Made at 60 psi ($4.2\ kgcm^{-2}$) steam pressure and dried at 130°C	0.15 0.20 0.25	52.0 50.0 47.0	20.0 20.0 19.0	1430 1456 1480	15.73 17.82 22.32

magnesium sulphate are added to gypsum slurry and then heated at 30 and 60 psi steam pressure for a period of 1.5–2.0 hours. After autoclaving, gypsum slurry is filtered and died at 130°C and ground to pass 150 micron sieve to form alpha plaster. The properties of plaster tested as per IS : 2542 (Part 1)-1978 are given in Table 2.5. According to results, 0.25% of sod: succinate is an optimum dose for making alpha plaster as it gives maximum strength values.[26–34]

2.5 Z-Gypsum Plaster

CERACEM (A United States Corporation based in New York) has developed a water resistant, external quality gypsum to replace outlawed asbestos facing materials in most construction applications, including external walls. Z-gypsum can be used with non-ferrous reinforcement to produce structural elements to replace timber, and concrete structures which will not corrode. This is now possible because the basic crushing strength of Z-gypsum is equal or in excess of that of concrete and it sets in less than 24 hours. No sand or aggregates are required as needed in concrete. Once the material is made it does not dissolve in water and its impermeability is reduced to that of concrete or lower. The Z-gypsum possesses high compressive strength (40–80 N/mm^2). They showed structures consisting of ribbon like prismatic, well shaped often grow across one another to form a mesh. These crystals penetrate into voids and form a structure which makes a network of intersecting interstitial spaces. The Z-gypsum is a mixture of calcium silicates and calcium sulphate as the crystalline phase. Z-gypsum is targeted for those geographical areas where wood or clean sand is scarce and where gypsum is available in abundance.

2.6 Anhydrite Plaster

Anhydrite plaster is produced by grinding anhydrite rock in tube mills or impact pulverizes to a particle size below 0.2 mm. Activators to promote setting are added together with the gauging water. However, the very fine grinding is expensive.

The activators are mixtures of alkali-metal or heavy-metal salts and calcium hydroxide, up to ~ 2 wt% of the anhydrite. Acid activators, *e.g.*, potassium hydrogen-sulfate or iron (II) sulfate, can also be used [35].

Fluoro anhydrite, a dry fine powder, is neutralized with calcium hydroxide and ground very finely for use as an anhydrite plaster. Sulfates, *e.g.*, potassium sulfate and zinc sulfate, and calcium hydroxide or Portland cement are activators, which are usually added and mixed with the anhydride powder in the factory.

Natural anhydrite and fluoroanhydrite differ from each other in crystal structures. Fluoro anhydrite consists of very small primary crystals that have been agglomerated to secondary particles with a high specific surface area and high reactivity, whereas natural anhydrite consists of large primary particles, which are rendered reactive by fine grinding.

Lot of research and developmental work was accomplished in Central Building Research Institute, (C.B.R.I.), Roorkee, India. Phosphogypsum and fluorogypsum wastes were evaluated for use in developing tiles, binders and in cement manufacture. Phosphogypsum does not occur as anhydrite but fluorogypsum is produced as anhydrite from the hydrofluoric acid industry. The anhydrite does not set at all or sets very slowly of poor strength. For building applications, plaster of medium strength and plasticity is required. However, for producing flooring tiles, high strength plaster is needed. A process has been developed for making high strength plaster from these industrial calcium sulphates which consists of converting phosphogypsum into anhydrite by heating phosphogypsum at elevated temperature (900–1000°C) whereby insoluble property of $CaHPO_4$ is transformed into inert calcium pyrophosphate. The microstructure of anhydrite are shown in Fig. 2.18. The anhydrite after fine grinding is blended with chemical activators (2–3% by mass of anhydrite) like alkali or alkaline earth sulphates, chlorides, carbonates, hydroxides with or without cement for quick dissolution of anhydrite to strong dihydrate gypsum.

2.7 Medical Plasters

2.7.1 Dental Plaster

Plaster for dental purposes should as a rule, carry not less than 93% of calcined gypsum and it should all pass a 30 mesh sieve and 95% pass 100 mesh. Only the whitest and purest of gypsum can be employed for dental plaster. IS : 6556–1972 covers Dental Impression Plaster.

2.7.2 Bandage Plaster

Plaster for bandage purpose should have setting time 4–7 minutes and tensile strength 0.15 N/mm^2. The compressive strength and loss of ignition of plaster is 10.0 Mpa and 4.5–8.0%. The purity of plaster should not be less than 90% ($CaSO_4.1/2H_2O$). IS : 4738–1993 governs, Plaster of Bandage, Plaster of Paris-Specification.

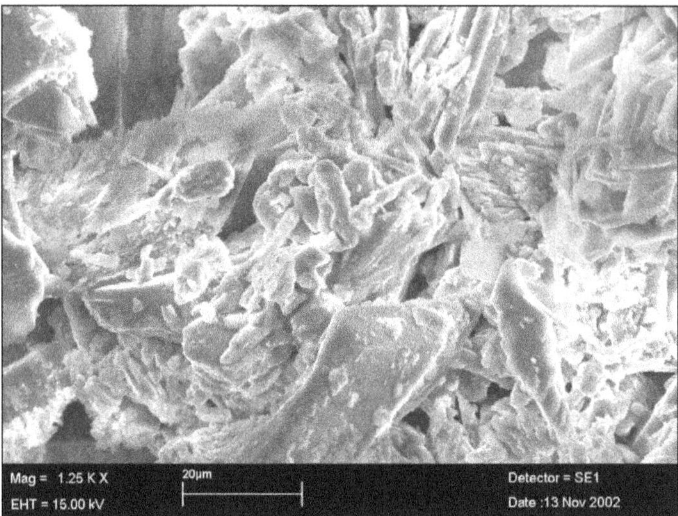

Fig. 2.18 SEM of Fluorogypsum Anhydrite

2.7.3 Surgical Plaster

Surgical plaster is used in treating the broken/cracked bones under surgical procedures. The plaster has special properties and requirements have to be as per Pharmocopoia of India for the surgical plaster. Researches have been carried out at CBRI, Roorkee to make surgical plaster from selenite gypsum of chemical composition SiO_2 2.1%, R_2O_3 0.80%, CaO 31.72%, MgO 0.41%, Alkali 0.20%, SO_3 44.20%, Cl 0.22%, Loss on ignition 20.05%, $CaSO_4.2H_2O$ 95.03%. On washing gypsum with tap water, the earthly impurities like SiO_2, R_2O_3 are reduced and the purity ($CaSO_4.2H_2O$) of gypsum is increased to 96.88%. The selenite gypsum was crushed and ground to pass 150 micron sieve and then calcined at 140–145°C in the Gypsum Calcinator to form beta plaster. The plaster on grinding to 300 m^2/kg (Blaine's) was evaluated for different properties (Table 2.6) as per Pharmacopoeia of India (The Indian Pharmacopoeia) Vol. 1, p. 399, 3rd Edition, 1985.

2.8 Plaster of Paris for Ceramic Industries

This type of plaster is used for making pottery moulds in which the potteries are cast. The pottery or ceramic slip is prepared by different clays with additives. The ceramic plaster is produced by calcining high purity gypsum ($CaSO_4.2H_2O$ above 85%) in kettles, autoclaves or rotary kilns at 150–160°C.

IS : 2333–1992 prescribes requirements and test for Plaster of Paris for use in ceramic and optics industries. There are four types of material as follows:

Type 1	Suitable for moulds for slip casting,
Type 2	Suitable for moulds for jiggering and case and block making,
Type 3	Suitable for mounting optical glass items.
Type 4	Suitable for automatic machine jiggering and for roller head.

Table 2.6 Properties of Surgical Plasters

Properties	Results		Limits
	Unwashed	Washed	
Consistency	55.0	50.0	N.S.
Setting times (a) Vicat (minutes) (b) Crumbling test	 9.0 Passes	 4.0 Passes	 N.S. Should notcrumble under thumb impression after 2 hrs. of casting
Bulk density (kg/m^3)	1500	1530	N.S.
Compressive Strength (N/mm^2)	21.0	23.0	N.S.
Physical State	Grayish White Odourless	Grayish Odourless	Should be white Odourless
pH	7.05	7.0	6.5–9.0
Loss on Ignition (%)	8.0	8.0	6.0–9.0
Acid Insoluble (%)	2.6	2.1	Should not exceed 5% (Max.)

The material should be in the form of a fine white powder of smooth texture, free from foreign matter and lumps. It should be calcined gypsum and shall correspond essentially to the formula $CaSO_4.1/2H_2O$.

The ceramic plaster is generally, made from the mineral gypsum of different purities. As the phosphogypsum retains higher purity, this material has great potential for use in ceramic plaster which may increase the use of phosphogypsum mani-fold in near future.

IS : 2333–1992 covers the requirements of ceramic grade plaster. This standard was earlier a combined standard of dental laboratory standard, dental impression plaster and the gypsum building plasters up to 1972. Later, the standard was revised and since 1981 new standard *i.e.,* IS : 2333 was made as the standard for Plaster of Paris for Ceramic Industry.

The physical and chemical requirements for Plaster of Paris are given in Table 2.7.

2.9 Properties of Gypsum Plasters and Products and Anhydrite Plasters

2.9.1 Hydration, Setting and Hardening

Calcium sulfate hemihydrate, anhydrite III, and anhydrite II undergo hydration under ambient conditions, converting into calcium sulfate dihydrate. If hydration is carried out with just enough water to produce a homogeneous, fluid, stable, non sedimenting slurry, then this mixture sets and hardens because the calcium sulfate dihydrate forms needles that intergrow and interlock.

Table 2.7 Specifications for Plaster of Paris for Use in Ceramics (IS : 2333–1992)

(a) Physical Requirements for Plaster of Paris

Characteristic	Requirement			
	Type-1	Type-2	Type-3	Type-4
(i) Fineness				
(*a*) Material retained on 150 micron IS sieve (max.)	Nil	Nil	Nil	Nil
(*b*) Material retained on 75 micron IS sieve (max.)	7	Nil	Nil	Nil
(ii) Normal consistency	60 to 80	45 to 60	55 to 65	40 to 55
(iii) Setting time (minutes)	15 to 30	10 to 30	10 to 15	10 to 15
(iv) Temperature in °C (min.)	12	12	12	12
(v) Modulus of rupture MPa (min.)	4.0	5.0	5.0	7.0
(vi) Dry compressive strength MPa (min.)	9	15	17	20

(b) Chemical Requirements for Plaster of Paris

Characteristic	Requirement			
	Type-1	Type-2	Type-3	Type-4
(i) Free moisture (max.)	2.0	2.0	0.5	0.5
(ii) Carbonates as $CaCO_3$ (max.)	3.0	3.0	1.0	0.5
(iii) Matter insoluble in Hydrochloric acid (max.)	7.0	7.0	2.0	2.0
(iv) Calcium sulphate as $CaSO_4$	85.0	85.0	90.0	90.0

Extensive research has been done into the mechanism of hydration. Around 1900, Le Chateler[36–37] established the theory of crystallization, which gained universal acceptance[38–39]. According to this theory the calcium sulfate hemihydrate in water first forms a saturated solution, about 8 g/litre at 20°C. However, this solution is actually supersaturated, because at 20°C calcium sulfate dihydrate has a solubility of only 2 g/litre, thus, $CaSO_4.2H_2O$ precipitates.

Formation of dihydrate crystals conforms to the laws of nuclei formation and crystal growth (Fig. 2.19). Mixing and wetting of the hemihydrate powder, which causes disintegration of the hemihydrate particles, is followed by a short induction period, after which nuclei begin to form from the supersaturated solution. The accumulation of very small dihydrate crystals with much excess water has been described by Kronert *et al.* as clustering[40]. Subsequently, after this nucleation, crystal growth begins, which at least initially is accompanied by continuous recrystallization[41]. The rate of nuclei formation is proportional to the relative super saturation (Von Weimarn's theory), and the rate of crystal growth is proportional to the absolute super saturation (Nernst-Noyes equation). Hemihydrate is converted directly into dihydrate, there are no intermediate stages. Anhydrite III is converted via the hemihydrate, and anhydrite II is converted directly into

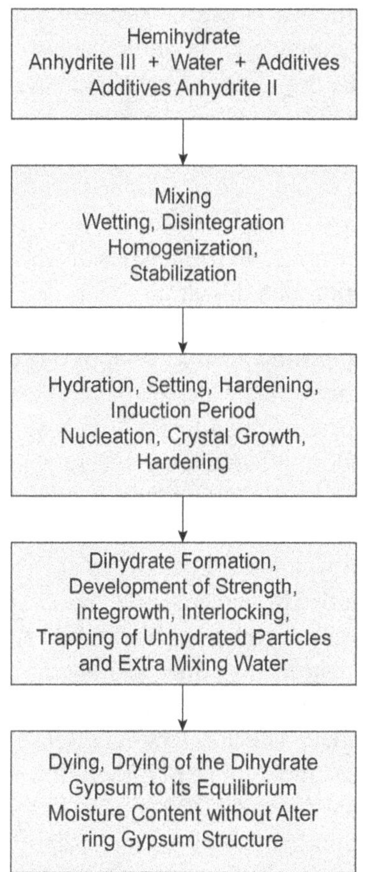

Fig. 2.19 Stages of Hydration, Setting and Hardening of Calcium Sulphates

dihydrate, without anhydrite III or hemihydrate intermediates. If the proportion of water is correct for setting and hardening, the slurry hardens by forming a dihydrate structure, a final crystallization, which according to Ludwig *et al.*[42] consists of intergrown, overgrown, and interlocking dihydrate crystals and inclusion of unhydrated components. Excess water can be removed by drying.

The rate of hydration of β-hemihydrate as shown in Fig. 2.19 is demonstrated by combined water content, intensity of the X-ray diffraction water content and rise in temperature due to the heat of hydration[43].

There are many ways in which these processes of hydration, setting, and hardening can be applied in practice. Parameters for characterizing these processes are the water-to-plaster ratio on mixing, the consistency of the mixture, the initial and final set, the rate of strength development, and the strength and density of the final dry gypsum product.

The method of manufacture of the plaster influences the gypsum technology to a very large extent. For instance, β-hemihydrate from a rotary kiln requires more water to produce fluid slurry of uniform consistency than does plaster from a kettle. The latter, in turn, requires more water than multiphase plaster, which, in turn, requires more than autoclave plaster. This water-to-plaster ratio (water capacity of the gypsum plaster), an inverse of the quantity of gypsum plaster in grams per 100 g of water, is related to the strength and density of the set and hardened gypsum product.

α-plasters, which are workable with little water, can be simply turned into gypsum products of high strength and density.

β-plasters and multiphase plaster require more water than α-plaster to obtain fluid consistency. Generally, very fine plaster require more water than coarse-grained plaster. β-hemihydrate particle may disintegrate on first contact with water, breaking up into a multitude of very fine loose particles, thus changing their particle size distribution[44]. The particle size distribution of multiphase plasters determines their workability to a great extent. When stored, calcined gypsums are subject to changes in their properties, called aging. This aging is caused to some extent by the uptake of water vapour from the air. The degree of aging affects the water needed for given consistency; more water is needed for fresh calcined gypsum than for aged. If a considerable amount of water vapour is adsorbed, dihydrate nuclei may form, accelerating the hydration process. However, the reactions taking place on curing have not yet been entirely explained[45–46]. Natural aging of calcined gypsum produces gradual changes in the properties of the plaster over a period of months. To avoid this, methods have been developed to bring about aging artificially, so that the plaster undergoes no significant changes during storage. One process is called aridization; calcium chloride or similar salts are added in quantities up to about 0.2 wt%. to the raw gypsum before calcinations. Aging of calcined gypsum is also achieved by injecting small quantities of water containing a wetting agent to prevent the formation of a dihydrate[47–48].

The mixing and gauging of calcined gypsum with water to form a slurry of specific consistency can be affected by various wetting agents. Most of these, called plasticizers or water-reducing agents, lower the water demand. They include alkal aryl sulfonates, lignosulfonates, or melamine resins[49]. It is also possible to increase the water requirement by adding flocculating agents, *e.g.*, polyethylene oxide[50]. Chemicals that thicken, *e.g.*, cellulose and starch ethers, can be added to stabilize the water plaster slurry or prevent sedimentation and segregation; however, these have little effect on water requirement[51].

Setting and hardening can be accelerated or retarded by numerous additives[52–53]. Many inorganic acids and their salts are useful as accelerators, especially sulfuric acid and its salts. Calcium sulfate dihydrate is regarded as a special additive in this respect. Finely divided, it acts as a strong accelerator and therefore must be completely removed when raw gypsum is calcined. The accelerating effect of these substances is due to an increase in solubility and the rate of dissolution of the calcined gypsum and to an increase in the rate of nuclei formation.

Retarders are usually organic acids[54] and their salts and organic colloids that are the decomposition and hydrolysis product of bio-polymers such as proteins as well as salts of phosphoric acid or boric acid. The mechanism of retardation varies; high molecular mass colloids prolong the induction period because they are nuclei poisons. Other retarders slow down the rate of dissolution of the hemihydrate or the growth of the dihydrate crystals. The hydration of anhydrite II usually does not have to be retarded since it is slow enough and almost always requires acceleration.

In every case, temperature affects the rate of hydration of plaster, the rate increasing up to ~ 30°C, after which it decreases[55].

The strength of set dried gypsum is directly proportional to its density, therefore dependent only on its porosity or, less directly, on the water-to-plaster ratio and the size and structure of the pores[56–57]. The strength is affected by moisture or additives without a change in density. The strength of gypsum products with a moisture content exceeding 5% is only about one half that of air-dried gypsum products. When a gypsum product dries, the strength begins to increase below 5% moisture content, becomes evident around 1% moisture content, and reaches its final value in the region of its equilibrium moisture content[58].

Continuous moist conditions mitigate strength, because crystalline and textural changes, especially recrystallization, take place as a result of the solubility of gypsum in water[59]. The deformation or creep of moist gypsum products under mechanical stress is likewise the result of structural change. Addition induces a change in structure by changing the crystal habit of the dihydrate so that without a change in density strength is changed even in the dry state.

An extreme example of reducing strength is the effect of citric acid, commonly used as a retarder. Used sparingly, less than 0.1% it does have a retarding effect and lessens the strength only slightly. More, say, above 0.2%, changes the crystal habit of the dihydrate to such an extent that hardening of the gypsum is no longer possible because the crystals no longer interlock and inter grow[60].

Murat[61] has studied the morphology of natural and synthetic calcium sulfate dihyrates with up-to-date techniques and the effects of additive upon the crystal habit of the dihydrate.

With the increase in industrial activities, the amount of wastes generated will increase manifold. Industrial and mineral wastes from mineral processing industries, such as metallurgy, petrochemicals, chemicals, paper and pulp account for nearly 1,000 million tonnes per annum in India. The more important wastes from these industries from the view-point of building materials are fly ash, slags, phosphogypsum, press mud (lime sludge), red mud, mine tailings, etc.

In India, over 6.0 million tonnes of phosphogypsum is being produced annually as a by-product from the wet process phosphoric acid manufacture at present. For every tonne of P_2O_5 produced as phosphoric acid, about 4–5 tonnes of gypsum is thrown out as waste product containing about 15–30% free moisture. Phosphogypsum contains impurities of phosphates, fluorides, organic matter, alkali, etc. which interfere with the normal setting and hardening of the plasters and cements produced out of it. There is a serious problem of its disposal, particularly in the context of rising cost of disposal and growing awareness of pollution hazards.

For a long time, the use of phosphogypsum in building industry was considered prohibitive due to adverse effect of impurities associated with it. It was Japan which started utilizing phosphogypsum as early as 1955 in the manufacture of cement and plaster products (boards, blocks, cement additive) to overcome the problem of non-availability of mineral gypsum in the country. Various processes were developed in France, Romania, USA, Germany and several other countries for the beneficiation of phosphogypsum and its use as building material. In India, about 20% of phosphogypsum is being utilized for the manufacture of ammonium sulphate fertilizer, cement and soil reclamation purposes annually. Bulk of the material produced remains stockpiled adjacent to the phosphatic fertilizer plants for future use.

2.9.2 The Assessment of Plaster

1. Chemical Method

Five tests are normally carried out an estimation of moisture content, an ignition loss, an estimation of the hemihydrate content, an estimation of the γ-$CaSO_4$, and a total sulphate content. To these an extraction with NH_4Cl is sometimes added to determine insoluble impurities.

2. Moisture Content

A weighed sample (4g) is dried overnight at 45°C and reweighed, a loss in weight indicating the presence of moisture, a gain in weight the presence and amount of γ-$CaSO_4$.

3. Ignition Loss

A weighed sample (2g) is ignited at 700°C for 1 h, if known to be pure, or at 400°C for 2h if regarded as impure. The loss in weight will indicate the amount of water of hydration.

4. Hemihydrate Content

Determined by adding an equal amount of water to a weighed amount of plaster (3g), shaking to mix, allowing to stand for Ih, then drying overnight at 45°C, cooling and weighing.

$$\% \text{ Hemihydrate} = \frac{5.37 \times \text{gain in wt.} \times 100}{\text{Wt. of sample}}$$

5. Soluble Anhydrite (γ-$CaSO_4$)

A similar test is carried out, but the mix is allowed to stand for 24 h before drying at 45°C overnight, cooling, and weighing.

$$\% \text{ Soluble anhydrite} = 15.1 \times \% \text{ gain in weight on drying}$$

6. Total SO_3

This is determined by standard chemical procedure given in IS : 1288 or ASTM C 471 (As given below):

Dissolve 0.5 g of gypsum sample in 50 mL HCl (1:5 by Vol.), boil and add 100 mL of boiling distilled water and again boil for 5 min. Filter (through Whatman 41), wash with hot water. Boil and add 20 mL of 10% BaC_{12} slowly. Digest hot (Over water bath) for 1 hr or until the precipitates settle. Filter (through Whatman 42), dry, char (gently) and ignite (red heat) the precipitate, cool and weigh.

Calculate SO_3 to the % of sample as received. (Wt. of ppt. × 0.343 × 100/ Wt. of sample).

7. Gypsum ($CaSO_4.2H_2O$)

$$\% \text{ Gypsum} = 4.78\ \%\ (\% \text{ compound water} - \% \text{ Hemihydrate} \times 0.0621).$$

8. Insoluble anhydrite

$$\% \text{ Insoluble Anhydrite} = (\% \text{ calcium sulphate} + \% \text{ combined water}) - (\% \text{ hemihydrate} + \% \text{ gypsum})$$

9. Impurity Content

5 g of plaster stirred with a litre of 20% NH_4C1 and the residue is collected on a tared sintered-glass filter, dried and weighed.

References

1. Stark, J, Wicht., B, The History of gypsum and gypsum anhydrite.
2. Zem. Kalk. Gips. International, 1999, Vol. 52, No. 10, pp. 527–533.
3. Kaushik, J.P., Dass Bhagwan, Singh Manjit, Saini, S.K., Sharma Shailendra & Kumar Narendra, Improved Calcination of Gypsum through Mechanized Pan System, Building Research No. e No. 46, 1985, CBRI, Roorkee.
4. Singh, Manjit, Kaushish, J.P., and Bhagwan Dass and Saini, S.K., Properties of Gypsum Plaster Produced in the Mechanised Pan System, Chemical Engineering World, 1985, Vol. 20, No. 5, pp. 143–145.
5. Riddle, Wallace. C, Kettle Process of Calcination, Rock Products, 1945, Vol. 48, No. 8, pp. 88, 89 and 152.
6. Ward, A.G.T., Methods of Reducing Energy Requirements of Gypsum Processing in the Calcining Kettles, Zem. Kalk. Gips., 1980, Vol. 33, No. 11, pp. 594–597.
7. Kaushik, J.P., Dass Bhagwan, Singh Manjit, Saini, S.K., Sharma Shailendra & Kumar Narendra, Mechanised Pan Calcination System for Gypsum Plaster and Plaster Boards, Project Proposal No. 61., 1986, CBRI, Roorkee.
8. Lewis, R, Improved Calcining Processes for Gypsum, Zem. Kalk.Gips., 1985, Vol. 38, No. 5, pp. 250–255.
9. J. Stein Kuhl, O. Wiech Mann, K, Moldan, Zem. Kalk.Gips, 1972, Vol. 25, pp. 383–386.
10. Sorgel, P, Bergmann, J, Fietsch,G, Silikattechnik, 1971, Vol. 22, pp. 225–230.
11 Skinner, S.D., National Gypsum Co., 1954, US Pat. 2788960.
12. Gebruder Knauf, 1961, DE 1143 430, (A.N. Knauf).
13. Roddenrig & Co.,1964, CH 445 359, (H. Roddewing Sen.).
14. Knauf, A.N., Zem. Kalk.Gips, Ed.B, 1983, Vol. 36, pp. 271–274.
15. Wirsching, F, Hamm.H, Hutter, R, Kraftwerk, Umwelt, VGB Knot. 1981, pp. 96–101.
16.. Kurandt, ISMA Tech. Conf. 1974.
17. Fishon GB 1135951, 1966 (N. Robinson).
18. Nissan Kakasu KKK, US 3653826, 1968, (T. Ishihara).
19. Getting Rid of Phosphogypsum I-IV, Phosphorus Potassium, Vol. 87, 1977, pp. 37; Vol. 89, 1977, p. 36; Vol. 94, 1978, p. 24; Vol. 96, 1978, p. 30.
20 International Symposium on Phosphogypsum, 5–7 No. 1980, Florida Institute of Phosphate Research (47 Papers).
21 Beretka, T. Brown, J. Chem. Technol Bio.Technol, Vol. 32, 1982, pp. 607–613; Vol. 33 A, 1983, pp. 299–308.
22 Wirsching, TIZ, Vol. 105, 1981, pp. 285–299.
23 J. Forster, Chem. Ing. Tech. Vol. 44, 1972, pp. 969–972.
24. Singh Manjit & Rai Mohan, Autoclaved Gypsum Plaster from Selenite and Byproduct Phosphogypsum, J. Chem. Tech. Biotechnol, 1988, Vol. 43, pp. 1–12.
25. S. Nagai and M. Sekiya, Preparation of Gypsum Plaster by the Special Hydrothermal Method, J. Japan.Ceram. Assoch., 1948, Vol. 56, pp. 43–47.
26. Peredrii, I.S., High Strength Gypsum, Chem. Abst., 1940, 34, 8210n.
27. Engert, H.J., Koslowski,T, Zem. Kalk. Gips, 1998, Vol. 51, No. 4, pp. 229–237.
28. Offutt, J.S. amnd Lambe, C.M., Plasters and Gypsum Cement for the Ceramic Industry, Bull. Amer. Cerm. Soc., 1947, Vol. 26, pp. 29–36.
29. Combe, E and Smith, D.C., Studies on the Preparation of Calcium Sulphate Hemihydrate by an Autoclave Process, J. Appl. Chem., 1968, Vol. 18, pp. 307–12.

30. Khalil, A.A. and Hussein, A.T., Weight Loss as an Method of Assessment of Gypsum Plaster, Trans. Brit. Ceram. Soc.,1972, Vol. 71, pp. 67–70.
31. Powell, D.A., The α and β forms of Calcium Sulphate Hemihydrate, Nature, 1960, Vol. 185, pp. 375–60.
32. Forster, H.J., Production of α-Hemihydrate from By-product Gypsum, Chem.Ing. Tech.,1972, Vol. 44, pp. 969–72.
33. Saito, T., Observation on the Process of Dehydration and Rehydration of Gypsum by means of Propton Magnetic Resonance, Bull. Chem. Soc., Japan, 1961, 34, 1454–7.
34. Hummel, H.U., Freyer, O., Schneider, J, and Voiger, W, The Effect of Additives on the Crystalline Morphology of Alpha Calcium Sulphate Hemihydrate –Experimental Findings and Molecular Simulations, Zem. Kal.Gips Intl., 2003, Vol. 56, No. 10, pp. 61–69.
35. G. Grimme, Zem. Kalk. Gips. 1962, Vol. 15, pp. 285–299.
36. Le Chatellier, M.H., Hebd, C.R., Seances Acad. Sci, Vol. 96, 1983, pp. 1668–1671.
37. Vanit Hoff, J.H., Armstron, E.F., Hinrescher, W., Weigert, W., Weigert, G.and Z. Just, Phys. Chem. Stoechiom, Werwandt Schaftsl, 1903,Vol. 45, pp. 257–306.
38. Chiffon, J. R., Report, 1973, V135, TN 755, pp. 1–28.
39. Ridge, M.J., Beretka, J, Rev. Pure. Appl. Chem., 1969, Vol. 19, pp. 17–44.
40. Kronert, W., Haubert, P. Unpublished Results, RWTH Aachen, Germany.
41. Kronert, W., Haubert, P., Zem. Kalk. Gips, 1972, Vol. 25, pp. 553–558.
42. Ludwig, U., Kuhlmann, J. Tonind. Ztg. Keram, Rundsch, 1974, Vol. 98, pp. 1–4.
43. Koslowski, Th., and Ludwig, U, The Effect of Admixtures in the Production and Application of Building Plasters, Zem. Kalk. Gips., 1999, Vol. 52, No. 5, pp. 274–286.
44. Lane, M. K., Rock Prod. 1968,Vol. 71, No. 3, pp. 60–63, p. 108; 1968, Vol. 71, No. 4, pp. 73–75, p. 16, p. 17.
45. Lehmann, H., Mathiak, H., Kurpiers, R., Bor. Dtsch. Keram. Ges., 1973, Vol. 50, pp. 201–204.
46. Certain-Teed Products Corp., 1934, US 2067762, (G.A. Hoggatt).
47. Lelong, B., Zem. Kalk. Gips., 1984, Ed. B 37, pp. 205–218.
48. Gerbrider Knauf. DE-AS 2023853, 1970 (A.N. Knauf).
49. Aignesberger, A., Krieger, H., 1968, Zem. Kalk. Gips, Vol. 21, pp. 415–419.
50. Imperial Chem. Ind., 1963, GB 1049184, (K.G. Cunningham).
51. Imperial Chem. Ind, 1959., DE 1126792, (K.G. Cunningham).
52. G. Benz, Stuckgewerbe, 1969, No. 12, pp. 533–544.
53. Knauf, A.N., Kronert, W, Haubert, P., Zem. Klk. Gips., 1972,Vol. 25, pp. 546–552.
54. Koslowki, T., Ludwig, U., Zitrronensdure, Institute fur Gesteinshitten Kunde der RWTH, Aachen, 1983.
55. Aichinger, K., Wandser, B., Zem. Kalk. Gips., 1948, Vol. 1, pp. 33–37, pp. 50–51.
56. Satava, V., Tonind. Ztg. Keram. Rundsch, 1967,Vol. 91, pp. 4–5.
57. Robler, M., Dissertation, 1983, Techn. Universitat Clausthal.
58. Albrecht, W., Stuckgips und Putzgips, Fortscharitte und Forchungen in Bauwesen, 1953, No. 15, Franckh Sche Vertag Shandlung, Stutgart.
59. Wandser, B., Zem. Kalk. Gips 15, 1962, pp. 437–438.
60. Ridge, M.J., Surkevicius, H.J., 1961, Appl. Chem. Vol. 11, pp. 420–434.
61. Murat, M., Tonind, Ztg. Keram. Rundsch., 1973, Vol. 97, pp. 160–164, 1974, Vol. 98, pp. 33–37, pp. 73–78.

3

By-product Gypsum

3.1 By-product Gypsum

3.1.1 Waste Phosphogypsum

Among waste gypsum, marine gypsum is recovered as a by-product while manufacturing common salt from solar evaporation of sea water, and phosphogypsum as a by-product from phosphatic fertilizer plants. Marine gypsum comes from the coastal areas of Gujarat, Maharashtra, Karnataka and Tamil Nadu. While phosphogypsum is the by-product of phosphatic fertilizer plants located in A.P., Gujarat, Maharashtra, Rajasthan, Kochi, Visakhpatnum, etc.

Phosphogypsum is produced by the interaction of rock phosphate with sulphuric acid by two processes *viz.,* (1) Single step and (2) Two step process (Fig. 3.1)[1–3]. The single step process involves treatment of ground rock phosphate with sulphuric acid at 98–100°C. The reaction is violent and the phosphogypsum with large amount of impurities (P_2O_5, F) is produced whereas in the two step process (Hemihydrate-dihydrate) first of all hemihydrate is formed at 110–120°C which is then cooled to 55–60°C to form dihydrate gypsum crystals with comparatively lesser amount of impurities than the dihydrate gypsum. Overall reaction of rock phosphate with sulphuric acid may be given below:

$$Ca_5\,(PO_3)_3F + 5H_2SO_4 + 10H_2O \longrightarrow 5CaSO_4.2H_2O + 3H_3PO_4 + HF$$

$$Ca_3(PO_4)_2 + 3H_2SO_4 + 3H_2O = 3CaSO_4.1/2H_2O + 2H_3PO_4 + 3/2H_2O$$

Dihydrate Process

$$CaSO_4.1/2H_2O + 3/2\,H_2O = CaSO_4.2H_2O$$

Hemihydrate-dihydrate Process

Disposal of phosphogypsum is of paramount concern. Normally, it is disposed off on land, in rivers or in sea (Figs. 3.2–3.3). The dumping over the land makes the agricultural land barren, which contaminates ground water. Disposal of phosphogypsum in river or sea adversely affects the aquatic biota.

The impurities of P_2O_5 and F are known to adversely affect the setting and strength development of plaster and cements. Consequently, the gypsum plaster

Fig. 3.1 Phosphatic Fertilizer Plant

Fig. 3.2 Phosphogypsum Disposal in the Open Yard

Fig. 3.3 Phosphogypsum Disposal Yard

sets fast and requires heavy retardation. Similarly, Portland cement or Portland slag cements suffer prolongation of setting and loss of strength at initial stage of hydration. Therefore, it is utmost essential to evolve beneficiation of phosphogpysum to reduce or remove the impurities to the permissible level.

World wide, about 150–160 million tonnes of phosphogypsum are currently produced, and the quantity may increase further due to establishment of new phosphate fertilizer Plants. The phosphogypsum is a moist powder with moisture content of 20–35% and lot of impurities, the exact impurities and their quantity depend upon the rock and the specific process of manufacturing adopted. About 1.7 tonnes of gypsum are produced per tonne of raw phosphate giving rise to 5 tonnes of gypsum per tonne of P_2O_5 produced.

At present not more than 20% of the phosphogypsum produced, is used. The major intricacies are their high free moisture and impurities present in the gypsum. It is fact that quantity of phosphogypsum produced every year surpasses more than the natural gypsum and anhydrite.

Phosphogypsum, is, indeed a futuristic source of gypsum for most of the industry requiring sulphate as the major component. The high or low purity mineral gypsum has to be replaced as the mineral gypsum deposits will soon be over in near future. Then phosphogypsum will be a promising substitute of the mineral gypsum. This gypsum is a rich source of SO_3 and thus serve the purpose of plaster and cement industry. Hence, proper characterization and beneficiation of the waste gypsum have to be actuated in right perspective.

3.1.2 Fluoroanhydrite/Fluorogypsum

Floroanhydrite or fluorogypsum is a by-product of hydrofluoric industry wherein fluorspar reacts with sulphuric acid as per following chemical reaction :

$$CaF_2 + H_2SO_4 \longrightarrow CaSO_4 + 2HF$$

$$CaSO_4 + 2H_2O \longrightarrow CaSO_4.2H_2O$$

One tonne of fluorspar produces 1.75 tonnes of anhydrite. Fluoro anhydrite is used as the raw material for gypsum industry in many countries including India also.

3.1.3 By-product Gypsum from Different Sources

Calcium sulphate is formed in small amount in the production or treatment of organic acids such as tartaric acid, citric acid or oxalic acid or inorganic acids like boric acid. All these acids are produced by reaction of their calcium salts with sulphuric acid. Some gypsum is formed by treatment of waste water containing sulphate with lime as in zinc ore or production of titanium dioxide pigment. The utility of these variety of gypsum has yet to be discovered.

Saltgypsum

It is produced by reaction of calcium compound with sulphate in making NaCl from sea water.

$MgSO_4 + CaCl_2 = MgCl_2 + CaSO_4.2H_2O$ Salt gypsum Titangypsum

It is produced by neutralization reaction of waste sulphuric acid in making TiO_2 from ilemnite.

$$FeTiO_2 + 2H_2SO_4 = TiOSO_4 + FeSO_4 + 2H_2O$$
$$TiOSO_4 + H_2SO_4 = TiO(OH)_4 + H_2SO_4$$
$$FeSO_4 \text{ or } H_2SO_4 + CaCO_3 \text{ or } Ca(OH)_2 = CaSO_4.2H_2O \text{ Titangypsum}$$

Sodagypsum

It is produced by reaction of $CaCl_2$ with Na_2SO_4, from soda ash industry and rayon industry.

$$CaCl_2 + Na_2SO_4 + 2H_2O = 2CaSO_4.2H_2O$$

Sodagypsum

3.1.4 Pottery Gypsum/ Plaster Board Waste

Gypsum moulds from ceramic industry or waste gypsum boards have not been used so far as a raw material for the gypsum industry. Although R&D efforts are being continuously being made to use them. The availability of pottery or ceramic gypsum is small and requires some treatment before putting it into use.

Among the by-product gypsum, phosphogypsum is available to the greatest amount, hence, and maximum R&D work has been made for its utilization, hence, it was considered worthwhile to describe this waste in detail. Other wastes shall also be discussed in this chapter later.

3. 2 Details of Phosphogypsum

3.2.1 Characterization of Phosphogypsum

In this chapter, characterization of phosphogypsum by chemical analysis and instrumentation (DTA & SEM) has been reported.

1. Chemical Composition

The samples of phosphogypsum was analysed for various constituents like SiO_2, R_2O_3 ($Al_2O_3 + Fe_2O_3$), CaO, MgO, SO_3, loss on ignition (LOI), P_2O_5, F, organic matter, Cl, alkalies ($Na_2O + K_2O$) and pH as per IS : 1727–1967, Methods of tests for pozzolanic materials, IS : 1288–1983, Methods of test for mineral gypsum and as per standard test procedures [4-5]. The results of chemical analysis are shown in Table 3.1.

The data indicate that the materials contain impurities of P_2O, F, Organic matter, alkalies ($Na_2O + K_2O$) and Cl. The low pH of phosphogypsum shows its acidic character.

2. Differential Thermal Analysis (DTA)

The procedure as reproted by Machanzie was adopted for DTA studies of the phosphogypsum. The thermograms are plotted in Fig. 3.4. It can be seen that the phosphogypsum designated as 'A' shows two endotherms of 140° and 220°C

Table 3.1 Chemical Analysis of Phosphogypsum Samples Collected from Different Sources in India

Constituents	Percentage					
	*PPL Paradeep	RCF Mumbai	SPIC Tuticorin	AMD Ambernath	CFL Vizag	FACT Udyog mondal
P_2O_5	0.47	0.52	1.28	0.92	0.65	1.02
F	0.52	0.25	1.80	1.03	0.28	0.51
Organic matter	0.31	0.06	0.58	0.13	0.11	0.26
SiO_2 + insoluble in HCl	2.40	0.90	1.05	6.60	5.15	1.64
R_2O_3 (Al_2O_3 + Fe_2O_3)	0.51	0.06	0.50	5.86	0.84	1.82
CaO	32.14	31.50	31.20	27.76	31.47	32.02
MgO	0.09	0.05	0.06	0.25	0.43	0.40
SO_3	41.17	45.10	43.00	39.65	41.98	43.60
Na_2O + K_2O	0.51	0.06	0.50	0.46	0.32	0.36
Combined water	19.20	19.80	19.20	17.84	19.76	18.01

*PPL: Paradeep Phosphate Ltd., Paradeep (Orissa), RCF: Rashtriya Chemicals & Fertilizers, Co., Mumbai, SPIC: Southern Petrochemical Industries Corpn., Tamilnadu, AMD: Albright Morarji Dharam Dassji Ltd., Maharastara, CFL: Coromondal Fertlizer Ltd., (A.P.), FACT: Fertlizer & Chemicals, Travancore, (Kochi)

due to double dehydration of gypsum forming hemihydrate ($CaSO_4.1/2H_2O$) and soluble anhydrite ($CaSO_4$ III) respectively. The appearance of exotherm at 450°C indicates inversion of $CaSO_4$ (III) → $CaSO_4$ (II). The formation of exotherm at elevated temperature of 450°C instead of 360°C normally obtained for the mineral gypsum can be attributed to the presence of impurities of P_2O_5 and F in phosphogypsum.

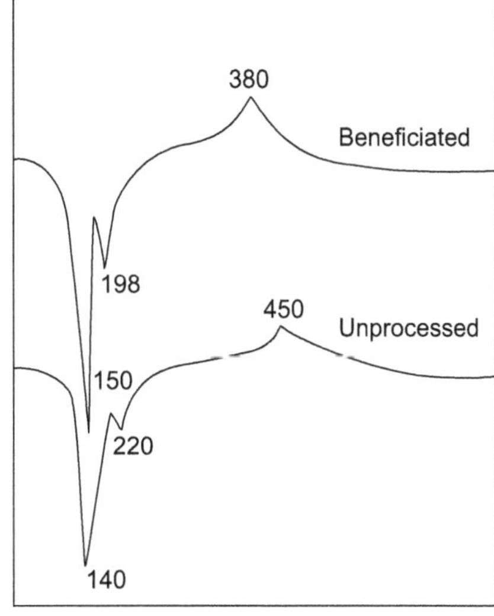

Fig. 3.4 Differential Thermograms of Phosphogypsum

It can be seen that phosphogypsum samples PPL, RCF and CFL contains lower level of P_2O_5, F and organic impurities than the other samples. Probably these gypsum samples have been produced by two step hemihydrate-dihydrate process. The phosphogypsum samples SPIC and FACT contain higher level of P_2O_5 and F which may be attributed to their manufacturing by the single step dihydrate process. It is claimed that two step hemihydrate-dihydrate process yield lesser amount of impurities than the single step dihydrate process.

As is the case with all chemical processes using natural raw materials, by-product gypsum is always contaminated with impurities like P_2O_5 in the form of mono and dicalcium phosphates, undecomposed phosphate rocks, residual acid, fluorides and traces of radioactive materials like radium. Large portions of the impurities are found on the surface of gypsum crystals and in the interstices of agglomerated crystals[6]. Some portions of phosphates also enter into solid solutions with gypsum by substitution of HPO_4^{2-} for SO_4^{2-} because of similar crystal lattice parameters and crystal habits[7–8]. While quantitative estimation of the impurities by methods other than chemical analysis is difficult, identification of the nature of the impurities is possible by techniques such as differential thermal analysis (DTA), X-ray diffraction (XRD) and infra-red (IR) spectroscopy. Using these techniques, Murakami in his principal paper in the V international Symposium on Chemistry of Cement observed that all phosphogypsum produced from the dihydrate process were agglomerated crystals in which a large quantity of impurities like P_2O_5 and fluoride were included and that these impurities were difficult to remove. Employing IR spectroscopy, Maki and Suzukawa[9–12] identified the presence of HPO_4^{2-}, manifesting characteristic weak bands of HPO_4^{2-} at 1015 cm^{-1} and 837 cm^{-1}, Dehlgren[13] from his independent studies concluded that the amount of HPO_4^{2-} substituted in the crystal lattice of phosphogypsum varied with the changes in concentration of phosphoric acid and H_2SO_4 in the wet process. Singh, *et al.*

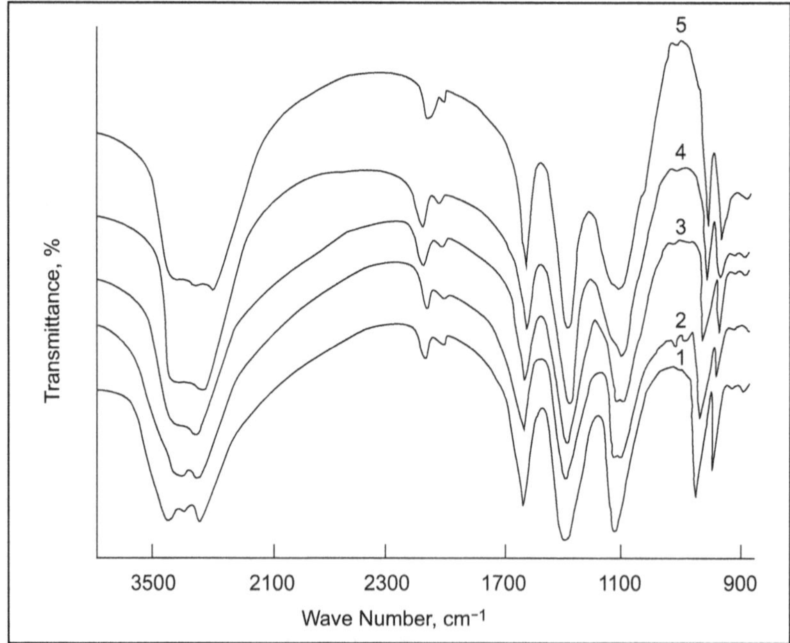

Fig. 3.5 IR Spectra of Unprocessed (2 and 4) and Processed Phosphogypsum (3 and 5)

Fig. 3.6 SEM Photograph of Typical Unprocessed Phosphogypsum

also identified these impurities by IR spectroscopy[14–15]. Data (Fig. 3.5) showed that intensity of absorption bands due to HPO_4^{2-} at 840 cm^{-1} is more pronounced in the unprocessed phosphogypsum than the processed phosphogypsum samples designated as 3 and 5. No HPO_4^{2-} group is present in the mineral gypsum.

3. Scanning Electron Microscopy (SEM)

The scanning electron microscopy of the phosphogypsum sample was carried out on SEM apparatus Model LEO 438 VP (U.K.) by sprinkling process in which the waste gypsum powder were uniformly sprinkled on the carbon tab. The samples were scanned at different magnifications to achieve optimum results.

The SEM microphotograph of unprocessed phosphogypsum is shown in Fig. 3.6. The SEM of phosphogypsum shows formation of euhedral to subhedral and anhedral prismatic, tabular and lath shaped crystals of variable sizes. Some crystals can be seen agglomerated and coated with small crystals occasionally showing twinning. Most of gypsum crystals show some defects like cavities and irregular boundaries. However, the prismatic crystals were well stacked. The agglomeration and uneven surface and boundaries of the crystals may be ascribed to the presence of impurities of P_2O_5, F alkalies, etc.

4. X-Ray Analysis

X-Ray powder diffraction analysis of phosphogypsum showed the presence of impurity of $CaHPO_4.2H_2O$ in $CaSO_4.2H_2O$ only. No fluoride compound could be detected in the diffraction pattern. Presumably it had gone into solid solution. The X-Ray diffraction pattern of different phosphogypsum samples were found to be identified. A typical diffraction pattern powder lines for AMD PG sample is given in Table 3.2. The solid solution of $CaHPO_4.2H_2O$ in $CaSO_4.2H_2O$ is

Table 3.2 X-Ray Powder Data of Phosphogypsum Sample AMD

d	I	Mineral Identified	
		$CaSO_4.2H_2O$	$CaHPO_4.2H_2O$
7.56	VS	"	"
4.225	VS	"	"
3.25	W	—	"
3.52	W	"	"
3.08	VVS	"	"
3.885	VVS	"	"
3.698	VVS	"	"
2.50	VS	"	"
2.243	MS	—	"
2.04	S	"	"
1.899	MS	"	"
1.807	MS	"	"
1.669	W	"	"
1.621	W	-	"
1.439	VW	-	"
1.37	VW	-	"
1.328	VVW	-	"
1.25	VVS	-	"
1.207	VVW	-	"
1.142	VVW	-	"

Table 3.3 Lattice Constants and Space Group of Gypsum and Dicalcium Phosphate Dihydrate

Lattice Constants	$CaHPO_4.2H_20$	$CaSO_4.2H_2O$
a	5.68 Å	5.812 Å
b	15.18 Å	15.180 Å
c	6.58 Å	6.239 Å
d	118.23′	116.26′

due to their similar lattice parameters or constants and space groups[14,9,13] as shown in Table 3.3. This P_2O_5 in lattice dissolves with difficulty in water and is known to effect the hydration and development of strength adversely. Besides P_2O_5, fluorides may also be present as AlF_5^{2-} ion in crystal lattice.

3.2.2 Beneficiation of Phosphogypsum

As we know phosphogypsum contains impurities of P_2O_5, F, organic matter and alkalies which normally effect the normal setting and strength of plaster and cements adversely, it is, therefore, essential to beneficiate phosphogypsum to get best quality of plaster and plaster products.

Investigations have been reported abroad to reduce the impurities in phosphogypsum[16–19]. Detailed studies were carried at the (CBRI), Roorkee regarding beneficiation and utilization of phosphogypsum.

Different techniques of beneficiation were applied on phosphogypsum which are described below:

Simple Water Washing

Phosphogypsum samples were washed several times, with (*i*) ordinary tap water and (*ii*) hot water at 80°C and dried to constant weight at 42 ± 2°C. The washed phosphogypsum samples were analyzed for residual impurities.

Wet Sieve Analysis

The coarser fraction in phosphogypsum is known to be rich in undecomposed phosphate rock, alkali silicates and larger particles of organic matter. To determine the extent to which these impurities can be removed through the elimination of coarser fraction, the phosphogypsum samples were wet sieved through different IS sieves[20]. The fractions retained over each sieve were further washed and dried at 42 ± 2°C. These were analyzed for residual impurities.

Thermal Treatment

Neutralization of calcined phosphogypsum is reported to inactivate the impurities of P_2O_5 and F[21]. Following this approach, phosphogypsum was calcined at 160°C and neutralized with milk of lime [$Ca(OH)_2$]. The treated phosphogypsum samples were analysed for residual impurities.

Calcination at Higher Temperature

The presence of dicalcium phosphate ($CaHPO_4$, $2H_2O$) occluded in the gypsum crystal-lattice adversely affects the hydration of cement. In order to render dicalcium phosphate insoluble to water and inert, the phosphogypsum samples were calcined at 750°C for 4 hours[22].

Chemical Treatment

Treatment with hot Ammonium Sulphate Solution $(NH_4)_2SO_4$.

Phosphogypsum samples were thoroughly shaken with variable concentrations of hot aqueous $(NH_4)_2SO_4$ solution in mechanical shaker for 6 hours, filtered and washed with 0.5% $(NH_4)_2SO_4$ solution, followed by plain water washing[23]. The resultant gypsum was dried at 42°C and analyzed for residual impurities. The results are reported elsewhere.

Treatment with Aqueous Citric Acid

The phosphogypsum samples were thoroughly shaken with 2 to 5 % aqueous citric acid solution in a mechanical shaker for 15–25 minutes at 30°C, filtered through a Buchner funnel and washed with 0.5–1.0 % aqueous citric acid solution then washed with plain water 2 to 3 times. The purified gypsum samples complied to respective Indian Standards, 12679, Specification for by-product gypsum for use in plaster block and boards. A patent of the process has been claimed by the author[24].

Treatment with Sulphuric Acid and Reactive Silica (H_2SO_4 + SiO_2)

Phosphogypsum samples were treated with H_2SO_4 acid diluted with water to concentrations ranging from 40 to 65% (50 mL for 100 g phosphogypsum) and ground silica gel (–45 micron IS sieve) in the range of 2 to 6% on the weight of phosphogypsum samples. The temperature and period of reaction was kept at 70°C and 3 hours respectively. The treated phosphogypsum samples were analysed for residual impurities.

Table 3.4 Impurity Contents after Cold and Hot Water Washing

Sample Designation	After Cold Water Washing (%)			After Hot Water Washing (%)		
	P_2O_5	F	C	P_2O_5	F	C
AMD	0.46	0.38	0.04	0.45	0.36	0.03
CFL	0.37	0.32	0.09	0.36	0.30	0.08
FACT	0.49	0.46	0.10	0.43	0.44	0.09
SPIC	0.22	0.83	0.46	0.30	0.81	0.41

Table 3.5 Impurity Contents after Wet Sieve Analysis

Sample Designation	+ 300 micron (%)			+ 150 micron (%)			+ 90 micron (%)			Passing 90 micron (%)		
	P_2O_5	F	C	P_2O_5	F	C	P_2O_5	F	C	P_2O_5	F	C
AMD	0.69	0.49	0.07	0.62	0.43	0.06	0.57	0.39	0.04	0.45	0.36	0.03
CFL	0.56	0.39	0.08	0.48	0.31	0.08	0.36	0.21	0.07	0.35	0.20	0.06
FACT	0.48	0.51	0.16	0.41	0.46	0.13	0.33	0.44	0.11	0.31	0.43	0.10
SPIC	0.26	0.90	0.48	0.23	0.86	0.40	0.19	0.77	0.32	0.16	0.71	0.29

Treatment of phosphogypsum with ammonium hydroxide (15–20%) has been reported to beneficiate the waste for various applications in building sector[25].

1. Washing Treatment

The data obtained showed that cold water washing of phosphogypsum samples was effective in removing impurities partially (Table 3.4). As can be seen substitution of cold water with hot water washing was not of much advantage. On wet sieving of phosphogypsum, it was found that total combined material retained on 300 and 150 micron sieves for samples AMD, CFL, FACT and SPIC was 11.6, 12.0, 11.0 and 15.0 per cent respectively. The content of impurities was comparatively more in fractions retained and passing 300 and 150 micron IS sieve (Table 3.5). Considering that fractions retained on 300 and 150 micron IS sieve form only 11.0 to 15.0 per cent of the bulk, it would be proper to reject these fractions to reduce the overall content of impurities in phosphogypsum. Any undecomposed phosphate rock is also eliminated through the rejection of coarse fraction.

2. Thermal Treatment

Data obtained on the content of impurities in phosphogypsum samples after calcination and neutralization with milk of lime, *i.e.*, $Ca(OH)_2$ showed only slight reduction in the content of total P_2O_5 F and organic matter in the phosphogypsum samples. The optimum percentage of $Ca(OH)_2$ was found to be 4% for samples AMD, CFL & FACT and 3% for sample SPIC. The data showed that level of impurities of P_2O_5, F and organic matters were almost the same before and after the treatment except the organic matter which showed slight reduction. This may be ascribed to the fact that impurities remain intact in phosphogypsum even after the treatment with $Ca(OH)_2$ and it only the form of compounds [$Ca_3(PO_4)_2$ and CaF_2] in which they are present after treatment gets changed.

On calcining phosphogypsum at 750°C to anlydrite, the impurities of dicalcium phosphate ($CaHPO_4.2H_2O$) gets converted into calcium pyrophosphate ($Ca_2P_2O_3$). It is insoluble in water and thus harmless. Data obtained before and after calcination showed no loss of P_2O_5 on heating but reduction in fluoride content through volatilization.

3. Chemical Treatment

The residual impurities left after the treatment of phosphogypsum with aqueous $(NH_4)_2SO_4$ are P_2O_5 0.33, 0.29, 0.40 and 0.15; F 0.60, 0.33, 0.49 and 0.80 and organic matter 0.80, 0.9, 0.14, 0.31 for phosphogypsum samples AMD, CFL, FACT and SPIC respectively. The optimum concentration of $(NH_4)_2SO_4$ was obtained as 5% for AMD, CFL and FACT and 2.5% for HDS samples. The reason for lower quantity of $(NH_4)_2SO_4$ needed for SPIC sample is considered to be due to its having initially lower content of impurities.

On treating phosphogypsum with $H_2SO_4.SiO_2$ mixture at 70°C, the impurities of $CaHPO_4.2HO_3$ and $Ca_3(PO_4)_2$ are converted into water soluble H_3PO_4 and

Table 3.6 Impurity Contents after $H_2SO_4 - SiO_2$ Treatment

Sample Designation	Optimum Percentage of H_2SO_4	Optimum Percentage of SiO_2	Constituents in %		
			P_2O_5	F	C
AMD	55.0	2.0	0.38	0.46	0.05
		3.0	0.36	0.34	0.05
		5.0	0.32	0.26	0.03
		6.0	0.35	0.26	0.04
CFL	55.0	2.0	0.30	0.29	0.05
		3.0	0.30	0.22	0.05
		5.0	0.27	0.20	0.04
		6.0	0.29	0.22	0.04
FACT	55.0	2.0	0.38	0.37	0.11
		3.0	0.30	0.32	0.10
		5.0	0.29	0.29	0.08
		6.0	0.29	0.29	0.09
SPIC	40.0	2.0	0.21	0.70	0.33
		3.0	0.16	0.51	0.31
		4.0	0.11	0.30	0.22
		5.0	0.13	0.31	0.23

Ca $(H_2PO_4).H_2O$ compounds[26]. At the same time reactive SiO_2 combines with fluoride and come out in the soluble form as H_2SiF_4. The residual impurities left after the treatment of phosphogypsum with H_2SO_4–SiO_2 mixture are reported in Table 3.6. The results show reduction in P_2O_5, F and organic matter with an increase in the concentration of H_2SO_4 and reactive SiO_2 has been found to be 5% for AMD, CFL and FACT & 4% for SPIC samples respectively.

4. Relative Efficacy of Different Treatments

The efficacy of various beneficiation techniques can be rated as varying. Each treatment has it own limitations. Although, all processes are effective in separating the impurities from phosphogypsum, best results have been obtained with H_2SO_4 and reactive SiO_2 process. However, this process is not feasible on economic grounds. Similarly other chemical treatment, though feasible on technological ground is not financially viable. The wet sieve analysis process of phosphogypsum appears to be feasible both on technological as well as economical grounds. Hence, this treatment was chosen for further beneficiation work.

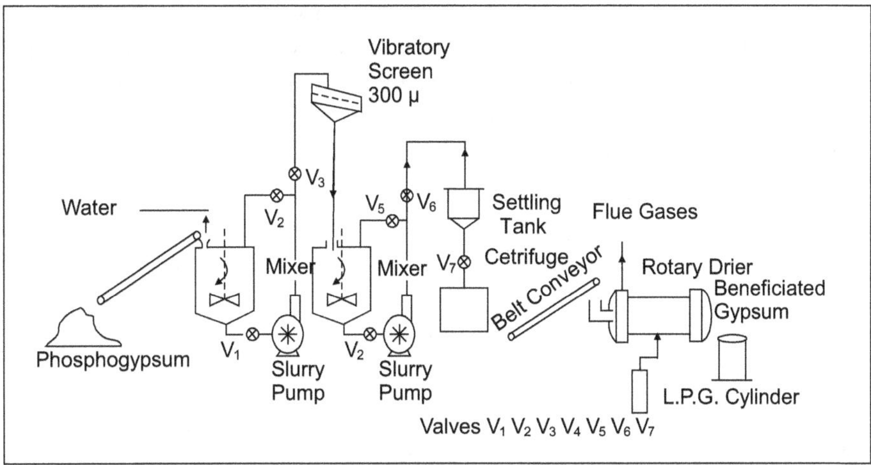

Fig. 3.7 Process Flow Diagram for Beneficiation of Phosphogypsum Plant

5. Pilot Plant for Beneficiation

Based on wet sieving of phosphogypsum through 300 micron sieve[27–28], a pilot plant of capacity 3 tonnes per day has been designed and installed at CBRI for the beneficiation of phosphogypsum. The pilot plant is comprised of major equipment such as mixer, vibrating screen, rotary drier, ball valves and centrifugal pumps. The process flow diagram of the pilot plant is shown in Fig. 3.7. Various steps involved in the beneficiation of phosphogypsum are listed below:

1. Churning and Mixing
2. Pumping
3. Vibro Screening
4. Centrifuging/Vacuum filtering
5. Rotary drying of wet phosphogypsum

The beneficiation of phosphogypsum requires extraction of impurities present in it. During churning and mixing of phosphogypsum, the water-soluble impurities are solubilized. The phosphogypsum water slurry in the proportion 13 by volume was used in the beneficiation trials.

During the mixing process in the first mixer, all the ball valves were kept closed. The valves V_1 and V_2 were used to get uniformly mixed gypsum slurry. After five minutes of agitation, the slurry valve V_3 was opened and the valve V_2 was shut to deliver the slurry to vibrating screen to filter gypsum slurry. The coarse fraction rich in impurities of undecomposed phosphate rock, quartz, organic matter, etc. retained over 300 micron sieve fitted in the vibrating screen was rejected. The partially cleaned gypsum slurry was further mixed with water in the second mixer to solubilize any remaining water-soluble and water-removable impurities. The gypsum slurry was pumped to settling tank through valve V_6. From the tank, the gypsum slurry was discharged to centrifuge by opening the valve V_7 to remove extra water to form gypsum cake. The gypsum cake (with 25–26% moisture) was finally dried in the rotary drier at 110–120°C to get the dried beneficiated material with moisture content below 5.0 per cent. The recovery of the centrifuged material was 85%.

The flow rate of gypsum slurry was 125 kg/hour. The phosphogypsum contains some undecomposed phosphate rock and pseudo lumps. These materials have to be removed from the gypsum by screening through a sieve of size 3/8" or 3/16" so that gypsum slurry should not create any mixing or pumping problem

Fig. 3.8 Pilot Plant for the Beneficiation of Phosphogypsum Set up at CBRI Roorkee

Fig. 3.9 Process Flow Diagram for Beneficiation of Phosphogypsum Plant Set up at M/S Orissa Gypsum Ltd., Cuttack

during operation. The pseudo lumps are hand breakable. The raw materials may be fed either by buckets or belt conveyor. The gypsum after drying in the rotary drier may be cooled to ambient temperature and packed in the polythene lined bags to obviate any moisture.

The dimensions of vibratory screen are: 2 m length × 1 m width × 1/4 m height with screen aperture-300 micron (stainless steel). Figure 3.8 shows the pilot plant for the beneficiation of phosphogypsum. Later a commercial unit of 21 tonnes/day of beneficiation of phosphogypsum has been setup at M/s Orissa Gypsum Pvt. Ltd., Cuttack (Orissa) (Fig. 3.9). The plant is running successfully. The capacity of the plant has been increased to 84 tonnes per day after installation of four units of 21 tonnes/day each. There is a tremendous scope of putting up such more units at various places in the country where phosphatic fertilizer plants are available. M/s Rashtriya Chemicals and Fertilizers, Mumbai has shown great interest to put a beneficiation plant of higher capacity than the Cuttack unit in near future.

Evaluation of Beneficiated Phosphogypsum

The samples of phosphogypsum beneficiated by wet sieving process were analyzed for various constituents as per standard test procedures. The results are listed in Table 3.7. Data show that impurities of P_2O_5, F, organic matter and alkalies are reduced considerably in the Fraction passing 300 micron sieve than the fraction retained over the sieve. An increase in SO_3 content and the pH values also indicates removal/reduction of impurities to a greater extent than the unprocessed phosphogypsum samples.

The extent of removal of impurities from the phosphogypsum can be supplemented by DTA and SEM of the beneficiated samples.

Table 3.7 Impurities in Phosphogypsum Beneficiated by Wet Sieving Process

	Phosphogypsum Fractions (%)					
	PPL (PG)		RCF(PG)		SPIC (PG)	
	Fraction	Fraction	Fraction	Fraction	Fraction	Fraction
Impurity	Retained over	Passing	Retained over	Passing	Retained	Passing
Content	300 micron	300 micron	300 micron	300 micron	300 micron	300 micron
(%)	sieve	sieve	sieve	sieve	sieve	sieve
P_2O_5	0.25	0.21	0.28	0.24	0.868	0.41
F	0.29	0.23	0.14	0.11	1.23	0.57
Organic matter	0.16	0.15	0.034	0.026	1.239	0.34
$Na_2O + K_2O$	0.12	0.10	0.06	0.03	0.18	0.15

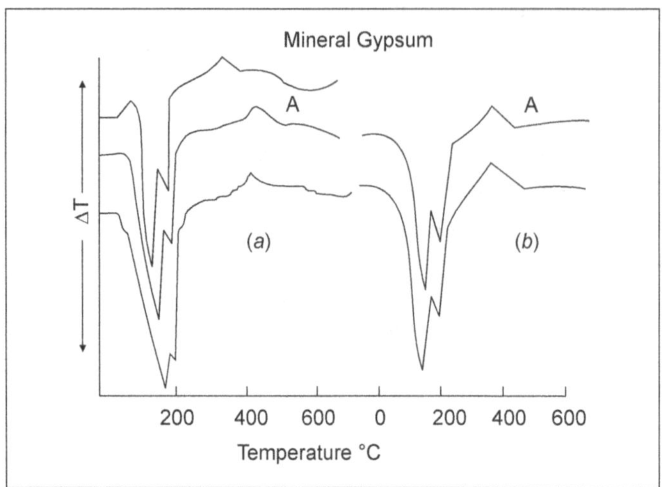

Fig. 3.10 Differential Thermograms of Unbeneficiated (*a*) and Beneficiated Phosphogypsum (*b*) in the Pilot Plant

DTA

The DTA of the phosphogypsum (samples A and B) is shown in Fig. 3.10. It clearly shows reduction in the inversion peak temperatures for $CaSO_4$ (III) to $CaSO_4$ (II) from 410–440°C for unprocessed PG to 360–370°C for the beneficiated PG. The peak of double dehydration are akin to mineral gypsum. The shifting of inversion temperature to 410–440°C may be ascribed to the formation of solid solution of $CaSO_4.2H_2O$ with the $CaHPO_4.2H_2O$ due to their similar lattice parameters. The unprocessed PG samples A and B contained impurities of P_2O_5, F, organic matter (%) in the range 0.32 – 0.36, 0.8–0.9, 0.30 – 0.45 which were reduced after beneficiation to P_2O_5, F, organic matter (%) in the range 0.23 – 0.27, 0.40 – 0.43, 0.11 – 0.15 respectively.

SEM

The SEM photographs of the beneficiated phosphogypsum is shown in Fig. 3.11. It shows formation of well developed euhedral prismatic and tabular crystals.

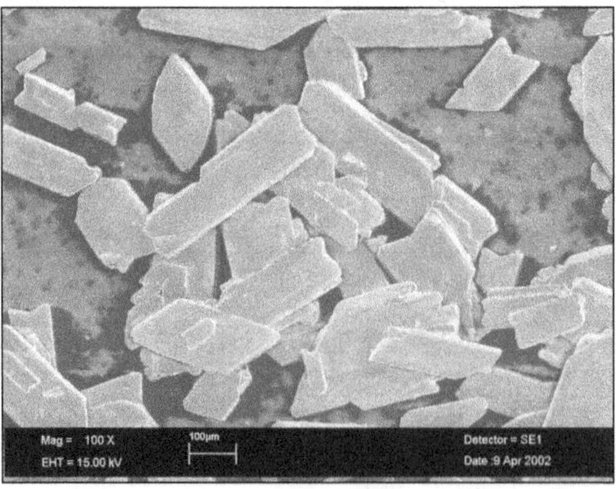

Fig. 3.11 SEM of Beneficiated Phosphogypsum

The crystals are individual and without any agglomeration and coating. The depth of crystals can be seen with sharp boundaries. No twinning or fusion of boundaries in the gypsum crystals were noted. Generally, the crystal with broader surface having length 390.45 micro meter and the width 135.31 micro meter were recorded. The absence of agglomeration and the coating of crystals clearly shows removal of impurities adhered to the gypsum crystal.

6. Calcination of Phosphogypsum

Studies on the calcination of beneficiated phosphogypsum and its assessment to produce quality grade gypsum plaster have been reported and discussed. The broad specifications of the Pilot Plant along with energy consumption used for the beneficiation of phosphogypsum. The results of beneficiation of using the new filter and the moisture content lost during the drying of wet gypsum in the drier as well as the dried product are also reported.

Calcination of Beneficiated Phosphogypsum

Gypsum is useful as an industrial material because it quickly loses its water of hydration when heated, producing partially or totally dehydrated calcined gypsum, and when water is added to this calcined gypsum, it reverts to the original dihydrate, the set and hardened gypsum product. Thus, this process, calcination and hydration, are the back bone of gypsum technology.

Calcination

$$CaSO_4.2H_2O \xrightarrow{\text{Heat}} CaSO_4.1/2H_2O + 3/2H_2O$$

$$CaSO_4.2H_2O \xrightarrow{\text{Heat}} CaSO_4 + 2H_2O$$

Hydration

$$CaSO_4.1/2\ H_2O + 3/2H_2O \longrightarrow CaSO_4.2H_2O + \text{heat}$$

$$CaSO_4 + 2H_2O \longrightarrow CaSO_4.2H_2O + \text{heat}$$

The industrial importance of gypsum is due to its calcined product *i.e.,* calcium sulphate hemihydrate or plaster of Paris. It is a metastable phase. It exists in two polymeric forms, *i.e.,* α- and β-varieties. The alpha form is well known as autoclaved plaster because it is produced in the autoclaves under steam pressure white β-form popularly known as stucco plaster or plaster of Paris is produced by the dry processes like open pan calcination in rotary kiln or kettles.

β-hemihydrate generally differs from α-hemihydrate in its application characteristics, heats of hydration and methods of preparation. The β-hemihydrate forms flaky, rugged small crystals where α-hemihydrate is composed of compact, well formed, transparent, large primary particles. However, β-hemihydrate *i.e.,* plaster of Paris (POP) is generally preferred over the α-form because of low density and medium strength properties which are desirable for the gypsum products like building boards, blocks, etc. In contrast, α-hemihydrate is dense, compact and brittle in nature and this is practically less preferred for mass use in building sector. Thus, the main emphasis was concentrated in the production and use of β-hemihydrate plaster from the phosphogypsum.

Production of Calcined Gypsum/β-Hemihydrate from Phosphogypsum
Phosphogypsum samples unbeneficiated and beneficiated by washing, wet sieving and treated in the pilot plant were calcined in the laboratory by spreading the phosphogypsum samples in trays and heating them in the oven at 145–150°C for a period of 4.0 to 4.5 hours with intermittent spatulation.

(*a*)

(*b*)

Fig. 3.12 Feeding of Beneficiated Phosphogypsum (*a*) and Production of Dried Phosphogypsum (*b*) in the Rotary Drier

The phosphogypsum samples were kept inside the ovens till the temperature reached 150°C. Initially the temperature drops to 90–100°C and then gradually rises to 145–150°C and maintained for a period of 4.0 to 4.5 hours. The spatulation of the gypsum is done every half-an-hour with the help of steel spatulas from top to bottom uniformly so that a thoroughly mixed plaster is produced. Precaution is taken to open the outlet in the oven for easy escape of water vapour released by the gypsum being calcined. Precaution is also taken not to open the oven for longer time so that the excessive loss of temperature may be avoided. The feeding of beneficiated phosphogypsopum and the production of calcined gypsum, *i.e* beta hemihydrate in the pilot plant are shown in Figs. 3.12(*a*), (*b*).

Here, it is necessary to mention that phosphogypsum even after calcination retains the similar particle shape as found in the uncalcined material. Therefore, instead of its already having fine particle size, it is imperative to further grind the material so as to prevent the recurrence of similar shape of particles. The additional grinding also helps in breaking up of the crystals in such a way to as to remove the thixotropic characteristics of the plaster.

After the calcination process is completed, the oven is closed, the material is allowed to cool to ambient temperature. The calcined material is ground in the ball mill to achieve a fineness passing 150 micron IS sieve, *i.e.,* equivalent to specific surface of 280–300 m^2/kg (Blaine's).

Table 3.8 Physical Properties of Gypsum Plaster (Unretarded) Produced from the Phosphogypsum (Tested as per IS : 2542–1978)

Sl. No.	Phospho Plaster	Residue on 1.18 mm (%)	Consistency (%)	Setting time (Minutes)	Transverse strength (N/mm^2)	Soundness	Mechanical Resistance (mm)	Exp. (%)
1.	—	Nil	60.0	4.0	4.0	No disintegration, Popping/Pitting	—	—
2.	Washed	Nil	60.0	12.0	6.72	No disintegration, Popping/Pitting	—	—
3.	Wet sieved	Nil	60.0	13.0	6.12	No disintegration, Popping/Pitting	—	—
4.	Pilot Plant Treated	Nil	60.0	10.0	5.40	No disintegration, Popping/Pitting	—	—
IS : 2547–1976 (Gypsum Building Plaster)		Max. 5.0	—	10.30	Min. 0.14	Set plaster shall not show any disintegration Popping/ Pitting	n.s.	n.s.

n.s: not specified

Table 3.9 Physical Properties of Gypsum Plaster (Retarded) produced from the Phosphogypsum (Tested as per IS : 2542–1978)

Sl. No.	Phospho Plaster	Residue on 1.18 mm (%)	Consis-tency (%)	Setting time (Minutes)	Transverse. strength (N/mm^2)	Soundness	Mechanical Resistance (mm)	Exp. (%)
1.	Un-processed	Nil	60.0	8.0	3.52	No disintegration, Popping/Pitting	—	—
2.	Washed	Nil	60.0	35.0	5.60	No disintegration, Popping/Pitting	—	—
3.	Wet sieved	Nil	60.0	30.0	5.06	No disintegration, Popping/Pitting	—	—
4.	Pilot Plant Treated	Nil	60.0	29.0	5.12	No disintegration, Popping/Pitting	—	—
IS : 2547–1976 (Gypsum Building Plaster)		Max. 5.0	—	10.30	Min. 0.14	Set plaster shall not show any disintegration Popping/Pitting	n.s.	n.s.

n.s: not specified

Testing and Evaluation of Hemihydrate Plaster

The calcined gypsum/plasters produced from the unprocessed phosphogypsum, washed, wet sieved as well as the pilot plant treated phosphogypsum were tested for the properties like consistency, setting time, soundness, mechanical resistance, compressive strength, bulk density and the flexural strength [transverse strength) as per IS : 2542 (Part-1) sec 1 to 12] - 1978, Methods of test for gypsum plaster, concrete and products, Part 1-Plaster & concrete and IS : 8272–1984, specification for gypsum plaster for use in the manufacture of fibrous plaster boards. The results of gypsum plaster tested as per IS : 2542 (Part 1)-1978 for the plaster to be used for Gypsum Building Plaster are listed in Tables 3.8 and 3.9.

The data show that the gypsum plasters comply with the requirements laid down in the standard except the setting of the unprocessed phospho plaster. The fast setting of the plaster is due to impurities of P_2O_5 and F which adversely affect the setting time. The transverse strength of the plaster is quite adequate and much above the minimum specified value of 0.14 N/mm^2. However, for commercial level and for mass application of the plaster, longer setting time is required. Hence, the plaster samples were retarded with the chemical retarder. A small dose of citric acid (0.05 to 0.15%) was added to the plaster to achieve the desired setting times.

Gypsum building plasters can be used for the manufacture of preformed gypsum building products which have the specific advantages of lightness and high fire resistance. The retarded hemi hydrate plaster is eminently suitable for undercoat (Browning plaster, Metal lathing plaster) and the Final Coat plaster (Finish plaster, Board finish plaster).

Pre design Cost Estimates for the Production of Beneficiated

Phosphogypsum

The predesign cost estimates for the beneficiation of phosphogypsum based on the pilot plant studies are presented here. The pilot plant has a processing load of 3 tonnes per day in 3 shifts. After processing 3 tonnes of phosphogypsum, about 2.85 tonnes (about 95%) of the beneficiated phosphogypsum are obtained, *i.e.*, 855 tonnes per year of 300 working days.

Pre design Cost Estimates

Product : Beneficiated Phosphogypsum
Capacity : 2.85 tonnes per day (3 shifts)
: 855 tonnes per year (300 working days)

Capital Investment:

		₹
A.	**Fixed Capital on Building**	
	(*a*) Industrial Land : 300 Sq.m @ ₹ 400/m^2	1,20,000
	(*b*) Building : 40 Sq.m @ ₹ 3000/m^2	1,20,000
	(*c*) Shed : 250 Sq.m @ 2500/m^2	6,25,000
	(*d*) Yard improvement (L.S.)	30,000
		8,95,000
B.	**Fixed Capital on Plant**	
	(*a*) Purchased Equipment (PE) cost:	6,40,000
	Slurry Mixers, 2 Nos.	40,000
	Slurry Pumps, 2 Nos.	50,000
	Ball Valves, 4 Nos.	7,000
	Vibratory Screen, 1 No.	60,000
	Rotary Vacuum filter/centrifugal Filter/ Filter Press, 1 No.	2,65,000
	Rotary Drier, 1 No	2,00,000
	Tube Well, 1 No.	18,000
	(b) Equipment erection @ 10% of PE	64,000
	(c) Electrical installation @ 15% of PE	96,000
	(*d*) Instruments and controls @ 5% of PE	32,000
	(*e*) Water services and drawing @5% of PE	32000
	(*f*) Water services and drawings @ 5% of PE	32,000
	(*g*) Laboratory & workshop @ 10% of PE	64,000
	(*h*) Engg. & supervision @ 10% of PE	64,000
	(*i*) Premium etc. (L.S.)	1,00,000
	(*j*) Miscellaneous and Contigencies (L.S.)	75,000
	Fixed Capital on Plant	**11,67,000**
	Total Fixed Capital (A + B)	**20,62,000**
	Working Capital @ 20% on the fixed capital	4,12,400
	Total Capital Investment (A + B + C)	24,74,400

Cost of Production:

			₹
1.	**Raw Material**		
	(*a*)	Unprocessed phosphogypsum 900 tonnes	90,000
		@ ₹100/- tonne (for site adjacent to the phosphaticfertilizer plant)	
2.	**Utilities**		
	(*a*)	Electric power ₹80/- tonne	68,400
	(*b*)	Water 3500 KL @ ₹5.0/KL	17,500
	(*c*)	L.P.G. @ ₹300/- tonne	2,56,500
			3,42,400
3.	**Labour & Supervision (L.S.)**		
	(*a*)	Mechanic-cum-operator 3 Nos. @ 3,000/- per month	1,08,000
	(*b*)	Mazdoor 500 mandays @ ₹50/- P.M.D.	25,000
			1,33,000
4.	**Maintenance & Repairs (M&R)**		
	(*a*)	Plant : 6% of B (fixed Capital on plant cost)	70,020
	(*b*)	Building : 2% of building & shed cost	14,100
			84,120
5.	**Depreciation**		
	Plant : 10% of B		1,16,700
	Building: 2.5% of building & shed		18,625
			1,35,325

Total Cost of production = ₹ 6,94,845
₹ 812.68 per tonne

3.3 Role of Phosphogypsum Impurities on Compressive Strength and Microstructure of β-Hemihydrate/Selenite Plaster

As described above phosphogypsum contains impurities of P_2O_5 as monocalcium phosphate, dicalcium phosphate and tricalcium phosphate, fluoride as sodium fluoride, sodium silico fluoride or calcium fluoride, organic matter and small quantity of soluble alkalies. These impurities impair workability, setting time and strength development of the plaster. The commercial importance of gypsum is due to its ability to form hemihydrate plaster. As phosphogypsum is a sole futuristic source of the gypsum industry world over, it is therefore, essential to beneficiate phosphogypsum to reduce harmful impurities present in it. Researches have shown that after beneficiation, the phosphogypsum is eminently suitable for making good quality plaster showing similar properties to natural gypsum plaster. The effect of phosphogypsum impurities has been reported on the properties of cement and plaster[29–30]. These impurities decrease strength and workability of the plaster which can be assigned to the negative effect of impurities associated with the phosphogypsum. Their role is crucial to decide industrial applications of gypsum plaster, hence, it is important to study the effect of these impurities on the physical properties of the gypsum plaster.

To ascertain whether these impurities really affect the physical properties of the gypsum plaster and its microstructure, investigations were undertaken to study the effect of impurities of phosphates, fluorides and organic matter and alkalies on the consistency, setting time, bulk density and compressive strength of the plaster. The results confirm that impurities greatly influence the physical properties and the morphoogy of the selenite plaster.

The selenite gypsum and phosphogypsum samples procured from Bikaner (Rajasthan) and Albright Morarji Pandit Ltd. Ambernath (Thane), India respectively were used for making calcined gypsum or gypsum plaster. Selenite gypsum of purity ($CaSO_4.2H_2O$) about 99.0 per cent was ground to pass 150 micron Indian Standard sieve and calcined at 150 – 160°C. Reagent grade chemicals similar to phosphatic, fluoride, organic matter and alkali impurities present in phosphogypsum were used for the hydration of gypsum plaster. The following reagents were used for the study.

Reagents: H_3PO_4, $Ca(H_2PO_4)_2.H_2O$, $CaHPO_4.2H_2O$, $Ca_3(PO_4)_2$, Na_2SiF_6, $Na_2O + K_2O$ and $C_{12}H_{22}O_{11}$

The reagents in the range 0.1–1.0 per cent as detected in phosphogypsum were uniformally mixed with selenite gypsum plaster.For making slides to be used for microscopic examination, the plaster was placed on the slide and blended with adequate quantity of solutions of phosphates, fluorides and organic matter. The plaster was allowed to set for 24 hours to form dihydrate gypsum. The dihydrate gypsum, thus formed in the chemical media was studied for its shape and size under the petrographic Leitz Pan Phot (Germany) microscop and physical properties.

The results of the effect of addition of different types of phosphatic chemicals similar to the impurities present in phosphogypsum on the morphology of selenite gypsum plaster are narrated below.

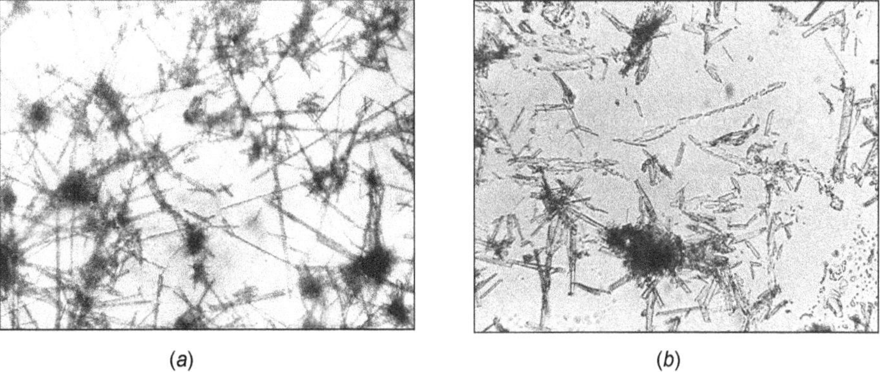

(*a*) (*b*)

Fig. 3.13 Microphotographs of (*a*) selenite plaster and (*b*) phosphogypsum plaster (x45).

Table 3.10 Physical Properties of Selenite and Phosphogypsum Plasters

Sl. No.	Plaster Sample	Consistency (%)	Setting Time (Minutes)	Bulk Density (kg/m^3)	Compressive Strength (MPa)
1.	Selenite	60.0	8.0	1300	14.2
2.	Phosphogypsum	68.0	4.0	1180	8.4

3.3.1 Morphology and Physical Characteristics of Selenite and Phosphogypsum Plasters

The morphology of set selenite gypsum plaster is shown in Fig. 3.13(*a*). The crystals are clusters of euhedral needles which are fairly elongated and well developed. The crystals have interlocking appearance and exhibit uniform distribution in all the directions. On the contrary, morphology of set phosphogypsum plaster [Fig. 3.13 (*b*)] show predominantly a mixture of badly developed prismatic, lath and needle shaped crystals forming agglomeration. The formation of such type of crystals confirm adverse effect of impurities present in phosphogypsum. The physical properties of selenite and phosphogypsum plasters are listed in Table 3.10.

It can be seen that selenite plaster has higher values for setting time, bulk density and compressive strength than the phosphogypsum plaster while the consistency is lower in the former than the latter. The fall in setting time and strength values in phosphogypsum plaster can be assigned to high consistency and also due to formation of clustered subhedral to anhedral deformed prismatic crystals instead of conventional needles as found in selenite plaster. This may be due to the negative effect of impurities of phosphates and fluorides which remain intact in phosphogypsum even during calcination.

3.3.2 Combined Effect of Phosphatic Compounds with and without Organic Matter on the Selenite Plaster

1. Calcium Phosphate [$Ca_3(PO_4)_2$] – Phosphoric Acid (H_3PO_4) – Monocalcium Phosphate Monohydrate [($Ca(H_2PO_4)_2.H_2O$] – Cane Sugar ($C_{12}H_{22}O_{11}$)

On the addition of 0.5% of $Ca_3(PO_4)_2$, H_3PO_4 and $Ca(H_2PO_4)_2.H_2O$ compounds in solution form to selenite plaster, mostly irregular, anhedral round bodies consisting of prismatic crystals were formed. Some needles were also detected On addition of 0.1% of organic matter in the form of cane sugar ($C_{12}H_{22}O_{11}$) solution in addition to above ternary mix, mostly prismatic crystals were formed. Further increasing concentration of each constituents of the ternary mix to 1.0% mixed with selenite plaster, mainly anhedral to

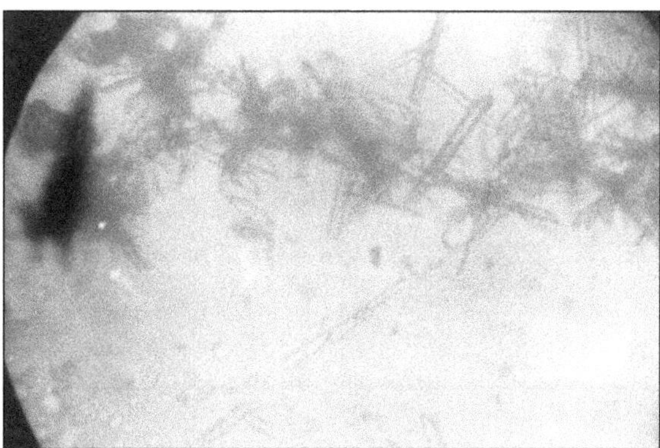

Fig. 3.14 Microphotograph of Selenite Plaster in presence of $Ca_3(PO_4)_2$(1%), H_3PO_4(1%) and $Ca(H_2PO_4)_2.H_2O$ (1.0%) (x45).

subhedral prismatic crystals of variable sizes and some lath shaped crystals were formed (Fig. 3.14).

The formation of prismatic crystals of anhedral to subhedral shape is considered to be due to suppression of crystal growth by the adsorption of phosphatic chemicals. The effect is further aggravated by the addition of $C_{12}H_{22}O_{11}$ solution. The physical properties of the hardened plaster containing above chemicals are listed in Table 3.11. It can be seen that the compressive strength and bulk density values were reduced considerably. The reduction in these two properties is due to the formation of ill developed prismatic crystals and progressive disappearance of needle shaped crystals. Moreover, the crystals were poorly stacked and have poor adhesion among themselves. The setting time of the plaster was found to be retarded due to formation of calcium phosphate coatings on the crystal surfaces formed by the interaction of PO_4^- ions and Ca^{++} ions respectively. These coatings suppress the setting time temporarily.

Table 3.11 Effect of Addition of $Ca_3(PO_4)_2 - H_3PO_4 - Ca(H_2PO_4)_2.H_2O - C_{12}H_{22}O_{11}$ on the Physical Properties of Selenite Plaster

Sl. No.	Concentration of compounds $Ca_3(PO_4)_2 - H_3PO_4 - Ca(H_2PO_4)_2.H_2O - C_{12}H_{22}O_{11}$	Consistency (%)	Setting time (Minutes)	Bulk density (kg/m^3)	Compressive strength (MPa)
1.	0.5 – 0.5 – 0.5 – 0	68.0	21.0	1170	9.1
2.	0.5 – 0.5 – 0.5 – 0.1	67.0	24.0	1110	8.0
3.	1.0 – 1.0 – 1.0 – 0	68.0	28.0	1140	8.3
4.	1.0 – 1.0 – 1.0 – 0.1	70.0	33.0	1070	7.6

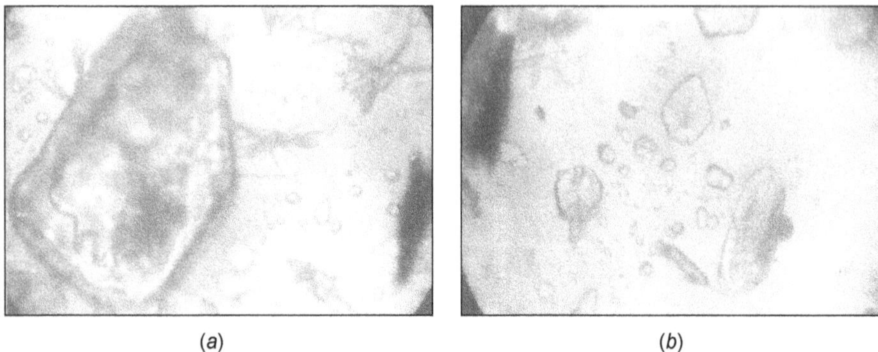

(*a*) (*b*)

Fig. 3.15 Microphotographs of Selenite Plaster in presence of (*a*) $Ca_3(PO_4)_2$(0.5%), H_3PO_4 (0.5%), and $CaHPO_4.2H_2O$ (0.5%) and (*b*) $Ca_3(PO_4)_2$ (1.0%), H_3PO_4 (1.0%), $CaHPO_4.2H_2O$ (1.0%) and $C_{12}H_{22}O_{11}$(0.1%) (x45)

2. Alcium Phosphate [$Ca_3(PO_4)_2$] – Phosphoric Acid (H_3PO_4) – Dicalcium Phosphate Dihydrate ($CaHPO_4.2H_2O$) – Cane Sugar ($C_{12}H_{22}O_{11}$)

On addition of 0.5% of $Ca_3(PO_4)_2 - H_3PO_4$ and $CaHPO_4$.$2H_2O$ without $C_{12}H_{22}O_{11}$ solution on the selenite plaster, mostly well defined prismatic crystals were formed [Fig.3.15 (a)]. On addition of 0.1% of $C_{12}H_{22}O_{11}$ to the selenite plaster, majority of prismatic crystals in association of some lath shaped crystals appeared. On increasing the concentration of each constituents to 1.0% without $C_{12}H_{22}O_{11}$ solution, anhedral to subhedral prismatic crystals were obtained. On addition of 0.1% of $C_{12}H_{22}O_{11}$ solution to the ternary mix, deformed anhedral prismatic crystals irregularly distributed were formed. Some rounded prismatic, radiating and lath shaped crystals were also identified [Fig. 3.15 (*b*)].

The crystal habit modification from needle-like to prismatic and lath shaped crystals can be assigned to differential take up of PO_4^- ions by gypsum dihydrate. The results corroborate the findings of Moriyama *et al.*[31] and Yamada *et al.*[32]. The physical properties of the hardened selenite plaster are shown

Table 3.12 Effect of Addition of $Ca_3(PO_4)_2 - H_3PO_4$–$CaHPO_4.2H_2O - C_{12}H_{22}O_{11}$ on the Physical Properties of Selenite Plaster

Sl. No.	Concentration of compounds $Ca_3(PO_4)_2 - H_3PO_4 - CaHPO_4.2 - H_2O - C_{12}H_{22}O_{11}$	Consistency (%)	Setting time (Minutes)	Bulk density (kg/m^3)	Compressive strength (MPa)
1.	0.5 – 0.5 – 0.5 – 0	66.0	19.0	1190	10.1
2.	0.5 – 0.5 – 0.5 – 0	67.0	22.0	1180	9.6
3.	1.0 – 1.0 – 1.0 – 0	66.5	26.0	1160	8.9
4.	1.0 – 1.0 – 1.0 – 0	70.0	29.5	1020	8.16

in Table 3.12. Data showed compressive strength and bulk density values were reduced. The fall in strength can be correlated with the formation of prismatic crystals of variable sizes of irregular boundaries and poor stacking of the crystals. The increase in consistency may be due to smaller size of the crystals which have presumably high surface area.

From the results reported above, it can be concluded that combinations of phosphatic compounds containing soluble phosphates in excess have more scatheus effect on the morphology and physical properties of hardened plaster than the combinations having lower content of soluble phosphates as shown below:

$$Ca_3(PO_4)_2 - H_3PO_4 - Ca(H_2PO_4)_2.H_2O - C_{12}H_{22}O_{11.} > Ca_3(PO_4)_2 - H_3PO_4 - CaHPO_4.2H_2O - C_{12}H_{22}O_{11}$$

3. Combined Effect of Phosphatic and Fluoride Compounds with and without Cane sugar Solution

Phosphoric Acid (H_3PO_4) – Sodium Silicofluoride (Na_2SiF_6) – Cane Sugar ($C_{12}H_{22}O_{11}$)

On addition of 0.5% H_3PO_4 and Na_2SiF_6 without $C_{12}H_{22}O_{11}$ solution to the selenite plaster, mostly ill developed needles of short length interspersed with prismatic crystals were obtained (Fig. 3.16). Further increasing to 1.0% Na_2SiF_6 without changing H_3PO_4 concentration, the hardened selenite plaster exhibited formation of thick radiating fibrous structures. On increasing H_3PO_4 to 1.0% and keeping Na_2SiF_6 at 0.5%, the quantity and size of the needles were reduced. On increasing concentration of H_3PO_4 and Na_2SiF_6 to 1.0% to the selenite plaster, the crystals habit was found to change to prismatic habit.

On addition of 0.1% $C_{12}H_{22}O_{11}$ solution to H_3PO_4–Na_2SiF_6 mix (0.5–0.5 per cent), subhedral to anhedral prismatic, lath and tabular crystals with cavities in them were formed. The formation of radiating fibrous crystals is due

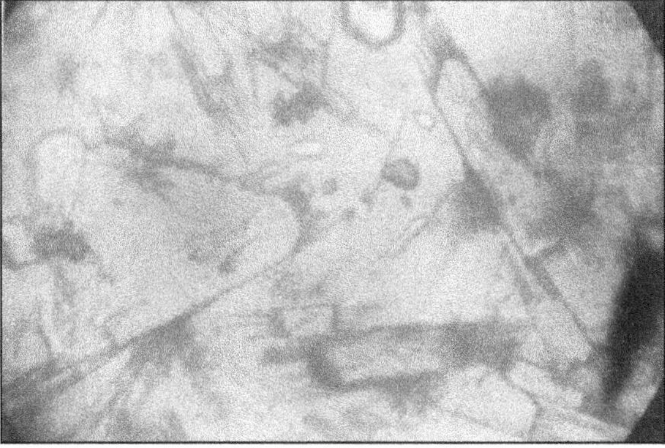

Fig. 3.16 Microphotograph of Hardened Selenite H_3PO_4(0.5%), Na_2SiF_6(0.5%) and $C_{12}H_{22}O_{11}$ (0.1%) (x45)

Table 3.13 Effect of Addition of $H_3PO_4 - Na_2SiF_6 - C_{12}H_{22}O_{11}$ on the Physical Properties of Seleniite Plaster

S. No.	Concentration of compounds $H_3PO_4 - Na_2SiF_6 - C_{12}H_{22}O_{11}$	Consistency (%)	Setting time (Minutes)	Bulk density (kg/m^3)	Compressive strength (MPa)
1.	0.5 – 0.5 – 0	63.0	6.0	1190	9.6
2.	0.5 – 1.0 – 0	64.0	4.0	1160	8.6
3.	1.0 – 0.5 – 0	64.0	5.5	1110	8.02
4.	1.0 – 1.0 – 0	65.0	4.5	1090	7.4
5	0.5 – 0.5 – 0.1	65.0	5.5	1140	8.2
6.	1.0 – 1.0 – 0.1	65.5	6.5	1100	6.64

to the predominant effect of Na_2SiF_6 on some of the growing crystal surfaces of dihydrate gypsum as explained earlier. Further reduction in needles like crystals and appearance of prismatic crystals along with tabular and laths may be attributed to the combined effect of PO_4^-, F^- and sugar solutions which act by reducing the growth of nuclei and modifying the crystal habit. The relevant physical properties are tabulated in Table 3.13.

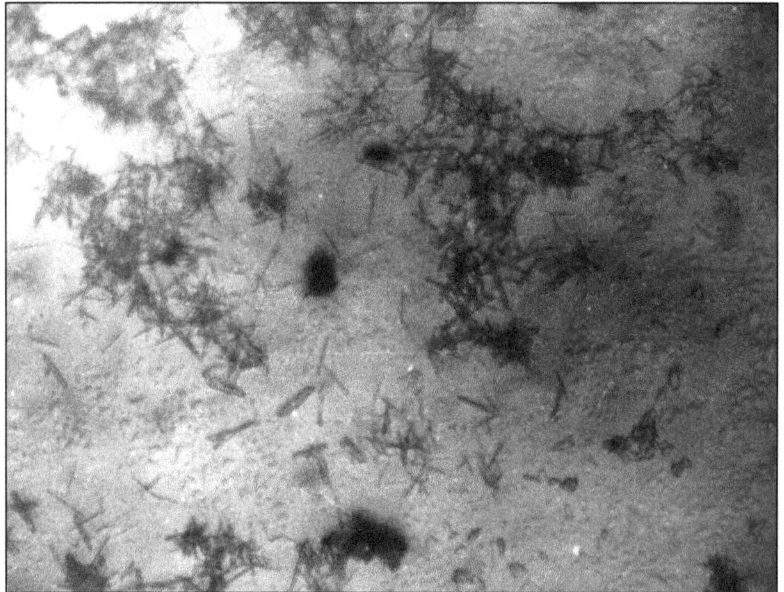

Fig. 3.17 Microphotograph of Hardened Selenite Plaster in presence of $Ca(H_2PO_4)_2.H_2O$ (0.5%), Na_2SiF_6(0.5%) and $C_{12}H_{22}O_{11}$ (0.1%) (x45)

Table 3.14 Effect of Addition of $Ca(H_2PO_4)_2.H_2O$– Na_2SiF_6 – $C_{12}H_{22}O_{11}$ on the Physical Properties of Selenite Plaster

Sl. No.	Concentration of compounds $Ca(H_2PO_4)_2.H_2O$ – Na_2SiF_6 – $C_{12}H_{22}O_{11}$	Consistency (%)	Setting time (Minutes)	Bulk density (kg/m^3)	Compressive strength (MPa)
1.	0.5 – 0.5 – 0	65.0	5.5	1170	9.3
2.	1.0 – 1.0 – 0	65.0	5.5	1160	8.5
3.	0.5 – 0.5 – 0.1	64.0	6.5	1140	8.68
4.	1.0 – 1.0 – 0.1	67.0	7.0	1130	7.7

Data show that reduction in compressive strength and bulk density occurred when 0.5 and 1.0% concentration of H_3PO_4 and Na_2SiF_6 were used. The strength values were further reduced when $C_{12}H_{22}O_{11}$ solution was added to above mixture of chemicals. The main reason for reduction in these properties can be ascribed to the formation of less defined prismatic and tabular crystals having deformed boundaries and presence of cavities and above all due to weak fusion among the adjacent crystals. The acceleration of setting of crystals is due to the presence of micro crystallites of high surface area as well as due to enhancement in the dissolution of hemihydrate plaster and precipitation of sodium sulphate ($CaSO_4.2H_2O + Na_2SiF_6 + xH_2O \longrightarrow CaSiF_6 + Na_2SO_4 + xH_2O$).

The Na_2SO_4 acts as activator of the hemihydrate plaster.

Monocalcium Phosphate Monohydrate [$Ca(H_2PO_4)_2.H_2O$] –Sodium Silico Fluoride (Na_2SiF_6) – Cane Sugar ($C_{12}H_{22}O_{11}$)

On adding 0.5% of $Ca(H_2PO_4)_2.H_2O$ and Na_2SiF_6 without $C_{12}H_{22}O_{11}$ solution to selenite plaster, mostly needle shaped crystals of small size were formed (Fig. 3.17). On increasing concentration of both $Ca(H_2PO_4)_2.H_2O$ and Na_2SiF_6 to 1.0%, the size and number of needles were further reduced. On addition of 0.1% of $C_{12}H_{22}O_{11}$ solution to the mixture comprising 1.0 % each of $Ca(H_2PO_4)_2.H_2O$ and Na_2SiF_6 added to selenite plaster, subhedral to anhedral prismatic crystals of irregular boundaries were obtained. On further increasing the $Ca(H_2PO_4)_2.H_2O$ and Na_2SiF_6 compounds to 1.0% each with 0.1% $C_{12}H_{22}O_{11}$, poorly distributed prismatic micro crystallites interspersed with lath crystals were formed. The physical properties of hardened selenite plaster in presence of above chemicals are listed in Table 3.14.

Data show that compressive strength and bulk density were reduced on increasing concentration of compounds. The addition of cane sugar solution further reduced strength and bulk density. The formation of poorly developed needles and micro prismatic crystals are responsible for fall in strength. The

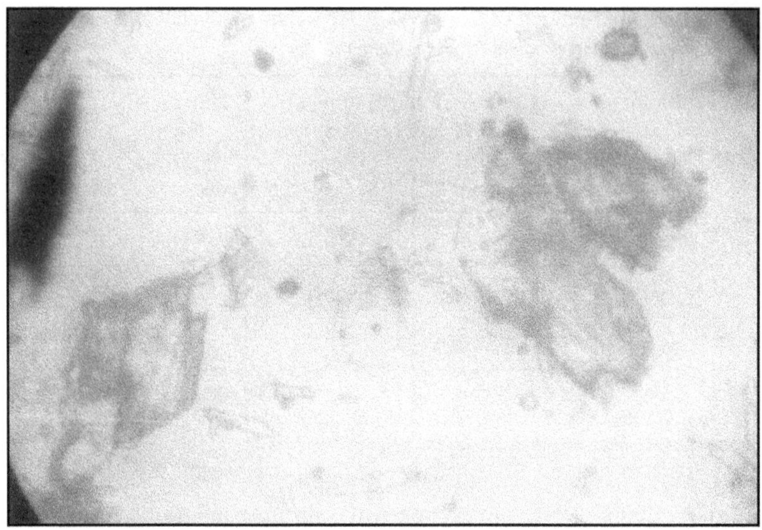

Fig. 3.18 Microphotograph of Hardened Selenite Plaster in presence of $CaHPO_4.2H_2O$ (1.0%) and Na_2SiF_6 (1.0%) (x45)

slight retardation of setting on addition of cane sugar solution can be ascribed to the formation of lath - like crystals.

Dicalcium Phosphate Dihydrate ($CaHPO_4.2H_2O$) – Sodium Silico Fluoride (Na_2SiF_6)

On addition of 0.5% $CaHPO_4.2H_2O$ and Na_2SiF_6 without $C_{12}H_{22}O_{11}$ solution to selenite plaster, needles like crystals having reduced size were formed in association with few tabular and prismatic crystals. On further increasing concentration to 1.0%, the size and number of needles were further reduced and few prismatic crystals with twinning were also formed (Fig. 3.18). On adding $C_{12}H_{22}O_{11}$ solution (0.15%) to mixture of $CaHPO_4.2H_2O$ – Na_2SiF_6 added to selenite plaster, the size and quantity of needles were reduced, rather tabular shaped crystals of improved size were obtained. The physical properties of set plaster in presence of above chemicals are reported in Table 3.15. The strength

Table 3.15 Effect of Addition of $CaHPO_4.2H_2O$ – Na_2SiF_6 – $C_{12}H_{22}O_{11}$ on the Physical Properties of Selenite Plaster

Sl. No.	Concentration of compounds $CaHPO_4.2H_2O$ – Na_2SiF_6 – $C_{12}H_{22}O_{11}$	Consistency (%)	Setting time (Minutes)	Bulk density (kg/m^3)	Compressive strength (MPa)
1.	0.5 – 0.5 – 0	66.0	6.5	1200	11.8
2.	1.0 – 1.0 – 0	66.0	6.5	1260	12.4
3.	1.0 – 1.0 – 0.15	68.0	8.5	1210	11.9

is due to needle like crystals. The slight retardation of the setting time is due to action of $C_{12}H_{22}O_{11}$ on the selenite plaster.

Phosphoric Acid (H_3PO_4) – Calcium Fluoride (CaF_2) – Cane Sugar ($C_{12}H_{22}O_{11}$)

On addition of 0.5% of H_3PO_4 and CaF_2 to the selenite plaster, prismatic crystals of uneven edges were formed along with radiating fibrous needles and rounded grains (Fig. 3.19). On addition of 0.1% $C_{12}H_{22}O_{11}$ to the plaster showed formation of lath shaped crystals of reduced sizes interspersed with prismatic and needles. Further increasing compound concentration to 1.0%, prismatic crystals of deformed edges were formed along with radiating laths and needle-like crystals. On addition of 0.1% $C_{12}H_{22}O_{11}$ solution along with 1.0% H_3PO_4 and CaF_2, deformed prismatic and lath shaped crystals were formed almost in equal proportion having poor distribution. The formation of prismatic crystals in particular and laths and needle in general are due to variable adsorption of the chemicals on the growing dihydrate crystal surfaces. The physical properties of the hardened plaster in presence of above chemicals are listed in Table 3.16.

It can be seen that the compressive strength and bulk density values were reduced. The fall in strength is attributed to the formation of deformed prismatic crystals as well as lath shaped crystals. These crystals owing to poor distribution and lack of proper adhesion among themselves thereby leading to fall in strength. The retardation of setting time may be assigned to the formation of coating of $Ca_3(PO_4)_2$ on the gypsum surface due to interaction of P_2O_5 and Ca^{++} ions which temporarily suppresses the hydration for sometime.

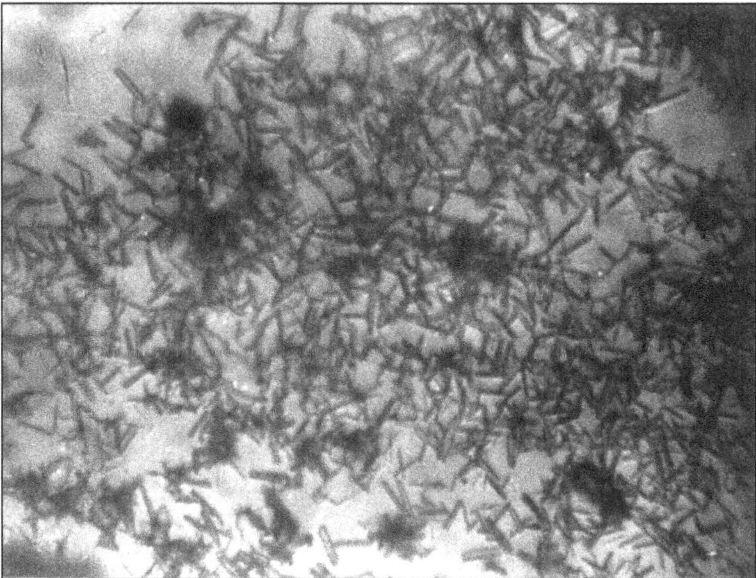

Fig. 3.19 Microphotographs of Hardened Selenite Plaster in presence of (*a*) H_3PO_4 (0.5%) and CaF_2 (0.5%) and (*b*) H_3PO_4(1.0%) and CaF_2(1.0%)

Table 3.16 Effect of Addition of H_3PO_4–CaF_2–$C_{12}H_{22}O_{11}$ on the Physical Properties of Selenite Plaster

Sl. No.	Concentration of compounds H_3PO_4–CaF_2–$C_{12}H_{22}O_{11}$	Consistency (%)	Setting time (Minutes)	Bulk density (kg/m^3)	Compressive strength (MPa)
1.	0.5 – 0.5 – 0	65.0	8.0	1180	9.2
2.	0.5 – 0.5 – 0.1	66.5	11.5	1160	8.8
3.	0.5 – 1.0 – 0	65.5	10.0	1180	9.0
4.	1.0 – 1.0 – 0	66.0	12.0	1150	8.9
5.	1.0 – 1.0 – 0.1	66.5	16.0	1120	8.1

The relative effect of phosphates and fluorides with and without cane sugar solution on the morphology and physical properties can be represented as follows:

$$H_3PO_4 – Na_2SiF_6 – C_{12}H_{22}O_{11} > Ca(H_2PO_4)_2.H_2O – Na_2SiF_6 – C_{12}H_{22}O_{11} > CaHPO_4.2H_2O – Na_2SiF_6 – C_{12}H_{22}O_{11} > H_3PO_4–CaF_2–C_{12}H_{22}O_{11}$$

The inference of study indicates when needle shaped crystals were formed, the compressive strength of the hardened selenite plaster was maximum. The other properties of plaster like setting time, bulk density and workability were normal. The combination of phosphatic compounds containing soluble phosphates in excess of 1.0% show deleterious effect on the morphology, setting time and strength development of the selenite plaster than the combinations having lower content of soluble phosphates. On addition of mixture of phosphatic, fluoride and organic compounds (cane sugar) to the plaster, mostly prismatic crystals of irregular boundaries having twinning and short length were obtained along with lath and tabular shaped crystals. The prismatic crystals of short length and irregular boundaries interspersed with lath and tabular shaped crystals having poor distribution and stacking give less compressive strength than the needle shaped crystals of selenite plaster[33].

References

1. Slack, A.V., Phosphoric Acid, Vol. 1, 1968, Marcel Dekker, Inc., New York.
2. Singh Manjit. Physico-chemico Studies on Phosphogypsum for Use in Building Materials, PhD Thesis, University of Roorkee, Roorkee, India, 1980.
3. Gutt, W, et.al., A Survey of Locations, Disposal and Prospective Uses of the Major Industrial By-products and Waste Materials, BRE Current Paper, February 1874, No. 19, pp. 36–51.
4. Dass, S.K., *et al.* Rapid Radio Tracer Technique of Phosphate Estimation, Fertilizer Technology, Vol. 10, 1973, pp. 182–185.
5. Token, J.M. and Black, B.M., Determinaton of Fluoride, Anal.Chem., Vol. 25, 1953, p. 1953.

6. Yamada, T, Kamata, K and Nagai, S., Gypsum by-product in phosphogric acid process (VI), Distribution of P_2O_5 in gypsum crystal, Sekko Sekkai, Vol. 53, 1961, pp. 164–68.
7. Murakami, K, Utilization of Chemical Gypsum for Portland Cement, Proceedings of the 5th International Symposium on Chemistry of Cement, Tokyo, Vol. IV, 1968.
8. Murakami, K, Tanaka, H, and Sato K, Chemical Analysis of Impurities of Phosphogypsum, Sekko Sekkai, Vol. 91, 1967, pp. 249–55.
9. Maki I, and Suzukawa U, Infrared Study on the By-product Gypsum from the Phosphoric Acid Process, J. Ceram. Ass. Japan, Vol. 71, 1963, pp. 80–85.
10. Badouska H, Characteristics of Phosphogypsum Waste from Gdansk and Police, Cement Wapno Gips, 30/42, Vol. 3, 1976, pp. 67–73.
11. Callusi I and Longe V, Thermal Decomposition of Calcium sulphate Ii, Cement, Vol. 2, 1974, pp. 75–98.
12. Mehedlopetrosyn, P. *et al.,* Thermokinetic Evaluation of quality of Gypsum, kattechnic, Vol. 25, No. 12, 1974, pp. 407–08.
13. Dahlgren S.E., Phosphate Substution in Calcium Sulphate from Phosphoric Acid Manufacture, Brit. Chem. Eng®, Vol. 10, No. 11, 1965, p. 776.
14. Barry. E.E. and Kuntz, R.A., Calcium sulphate Transition Temperature in the DTA of Lattice Substituted Gypsum, Chemistry and Industry, No. 38, Sept. 1972, p. 1073.
15. Singh Manjit., *et al.,* Possibilities of Utilizing Phosphogypsum as Building Materials, Cement & Building Materials from Industrial Wastes, July 24–25, 1992, Hyderabad, pp. 86–93.
16. Colling. R.K., Evaluation of Phosphogypsum for Gypsum Products, Canadian Institute of Mining and Metallurgy Bulletin, Sept.1972, Vol. 65, pp. 41–51.
17. Badowska, H., Osieeka, E and Majewski, B, Utilizaton of Phosphogypsum Waste in Plaster Production, Colloq. Int. Util. Sous-Prod. Dechets. Genie. Civ (C.R.), 1978, Vol. 1, pp. 187–192.
18. Kitchen, D.K. and Skinner, W.J., Chemistry of the By-product Gypsum and Plaster, Journ. Appld. Chem. and Biotech. 21 (1971) pp. 53–60.
19. Olmez, H., Erden, E., The Effect of Phosphogypsum on the Setting and Mechanical Properties of Portland and Trass Cement, Cem. Concr. Res. 1971, Vol. 1, pp. 663–678.
20. Singh, Manjit, Rehsi, S.S. Rehsi, Taneja, C.A., Beneficiation of Posphogypsum for Use in Building Mterials, National Seminar on Building Materials Their Science & Technology, 15–16 April, 1982, New Delhi, pp. II A 1–5.
21. Singh Manjit, Processing of phosphogypsum for the manufacture of gypsum plaster, Research and Industry, June 1982, Vol. 27, pp. 167–169.
22. Singh Manjit, Rehsi S.S. and Taneja, C.A., Beneficiating Phosphogypsum for the Manufacture of gypsum plaster and plaster products, Indian Ceramics, April, 1983 Vol. 26, No. 1, pp. 3–8.
23. Singh Manjit, A Chemical Process for Purifying Phosphogypsum, Environmental Health, Vol. 25, No. 4, 1983, pp. 300–306.
24. Singh, Manjit, Indian Patent, An Improved Process for Purification of Phosphogypsum, No. 1460/Del, Filed 1999.
25. Singh Manjit, Garg Mridul & Rehsi S.S., Purifying phosphogypsum for cement manufacture, Construction and Building Materials (U.K.) Vol. 7, No. 1, March 1993, pp. –7.
26. Singh Manjit, Rehsi, S.S. and Taneja, C.A., Rendering Phosphogypsum suitable for Plaster Manufacture, Indian Journal of Technology, 1984, Vol. 22, pp. 28–32.
27. Singh, M., Garg, M, Verma, C.L., Handa, S.K., Kumar, R, An Improved Process for the Purification of Phosphogypsum, Construction and Building Materials, 1900, Vol. 10, pp. 597–600.

28. Taneja, C.A., Singh, M., Evaluation of Phosphogypsum for Use in Preparation of Different Building Materials, 1976, Vol. 21, pp. 263–265.
29. Singh, M., Utilization of By-product Gypsum for Building Materials, Research No. 9, CBRI Publication, Roorkee, India, 1988.
30. Eipeltauer, E., The Applicability Phosphoric Acid Gypsum Sludge, Tonindustrie Zeitung, 1973, Vol. 97, pp. 4–8.
31. Moriyama I, Abe, T. Rate of Crystal Growth of Dihydrate in Phosphrous and Sulphuric Acid, Kogyo Kagaku Zasshi, Vol. 66, 1963, p. 4.
32. Yamada T, Kamata K, Nagai S. Gypsum By-produced in Phosphoric Acid Process (VI), Distribution of P_2O_5 in Gypsum Crystals, Sekko Sekkai, Vol. 53, 1961, pp. 164–168.
33. Singh Manjit, Effect of Phosphatic and Fluoride Impurities of Phosphogypsum on the Properties of Selenite Plaster, Cement and Concrete Research (USA), 2003, Vol. 33, pp. 1363–369.

Gypsum in Cement Industry

INTRODUCTION

Indian cement industry is second largest cement producer in the world (217 million tonnes/year) after China (2011) yet shares only 6% of the total world production. Out of installed capacity of 146.50 million tonnes in the country (year 2010–11), the total production of 127.50 million tonnes, of cement was from standard cement plants, the remaining 6.5 million tonnes comes from 365 mini-cement plants. It is tremendous growth since 1914 when first cement plant was set up in Porbandar with a capacity of 1000 tonnes and in 1947 total production was 2.22 million tonnes. Today the per capita consumption of cement in India is 85 kg as against world average of 260 kg. The industry's capital investment is around ₹ 18,000 crore and the manpower employment of 1,35,000 nos. The Current cement production in India stands at as 217 million tonnes (Cement Manufacturing Association). The major consumption of cement is in private housing sector (65%) and in government infrastructure is about 20%. Only about 1% of cement is sold in bulk form unlike US and Japan. India also exports cement (3.40 million tonnes) and cement clinker (1.27%) to many countries (2011–12 year), in the far east, middle east and Africa.

India produces mainly Ordinary Portland Cement (OPC), blast furnace slag cement Portland pozzolana cement, besides small quantities of high alumina cement, low heat OPC, high early strength OPC, oil well cement, super sulphated cement, sulphate resisting OPC, etc.

4.1 Manufacture of Cement

Large quantity of gypsum and anhydrite are used by the cement industry as a retarder for Portland cement. The percentage of sulphate contained in the cement is reckoned as SO_3. For Portland cement and blast furnace slag cement, the SO_3 is recommended up to 2.75 per cent and 4.0 per cent respectively[1].

Gypsum and its various calcined phases can play an important role in the production of clinker as mineralizer to lower the clinkering temperature or modifier in the recycling of alkalies both as binding components as used in the expansive cements and as controller of the hydration and the rheology of the cements[2–3]. It is generally accepted that the aluminate phase is responsible for the flash set and that gypsum influences the hydration of C_3A to prevent flash set. In the presence of gypsum, the kinetics of the reaction is slowed down thereby giving the desired setting properties. This is achieved through the formation of tri-calcium sulpho aluminate, *i.e.*, ettringite ($C_3A.3CS.H_{32}$) coating on the grains of anhydrous cement. The ferrite phase of the cement also undergoes a similar reaction and a compound similar to ettringite in structure is obtained. The mechanism is described below:

Aluminate phase reacts with water to give the following products:

$$2C_3A + xH \longrightarrow C_3AHx + C_2AHx \text{ (Hexagonal Hydrates)} \quad (1)$$

$$C_3A + xH \longrightarrow C_3AHx \text{ (Cubic Hydrate)} \quad (2)$$

The hydration in presence of gypsum takes place as follows:

$$C_3A + 3CSH_2 + 26\ H \longrightarrow C_3A.3CS.H_{32} \text{ (Ettringite Phase)} \quad (3)$$

$$C_2AnF_1 + CSH_2 + H \longrightarrow C_6(AF)S_3H_{32} \quad (4)$$

With the building up of crystallization pressure on the ettringite layers formed on the anhydrous grains – whose formation continuous— the ettringite film on the grains bursts but is formed again due to further formation of ettringite and sulphate bearing phases. The SO_3 in cement is practically consumed in 24 hours. The formation of sulphoaluminate (ettringite) hydrates in the early stages of cement hydration provides a loose structure in which the further precipitation of CSH gel takes place to fill the space which is then transformed into rigid structure of hardened cement paste. In addition to controlling setting time of cement, some amount of SO_4^{-3} ions enters into CSH gel lattice to contribute to strength development. In order to get desired setting and strength characteristics of cement paste, the presence of an optimum amount of $SO^{-3}{}_4$ ions is always necessary. A low $SO^{-3}{}_4$ ion content in the paste can lead to the transformation of ettringite into monosulphate-sulphoaluminate hydrate in following way :

$$3C_3A + CSH_2 + 3O\ H \longrightarrow 3C_3A.CS.H_{12} \text{ (Monosulphate Phase)} \quad (5)$$

An excess of SO_4^{-3} ions content can give rise to expansive forces and lower strength development in the cement.

While in the early stages, C_3A plays an important role C_3S content and its interaction with gypsum is equally important. Although C_3A and C_3S contents are important, there are other parameters in defining the optimum gypsum content, there are other parameters, such as specific surface, clinker composition and presence of alkalis which need consideration before choosing the right quality gypsum and its optimum quantity for controlling the setting and strength characteristics of cement paste.

Generally, the main source of SO_3 in Indian cement industry is natural gypsum or mineral gypsum. But now a days, phosphogypsum is used as alternative to natural gypsum either full or as partial replacement to natural gypsum in the manufacture of cement[4–5]. The phosphogypsum is used to lesser extent as the retarder to regulate setting of cement. The main reason for low consumption of phosphogypsum by the cement industry is the presence of small quantity of impurities of phosphates, fluorides and organic matter which interfere in unpredictable way with the setting of cement and adversely affect the strength development of cement mortar. In India, most of the phosphogypsum is produced by dihydrate process. With regard to level of impurities, the dihydrate process of phosphoric acid yields impure gypsum in the manufacture of cement.

The phosphogypsum, on an average, contains more than 90% $CaSO_4.2H_2O$ and so is richer in SO_3 content compared to cement grade gypsum which is expected to be of 80–85 per cent purity. This, in turn, will reduce the total requirement of gypsum for cement plants and can lead to saving both on the quantity and transportation costs of gypsum. The foregoing factors and the successful utilization of phosphogypsum as set controller in cement manufacture in some countries including Japan, *prima facie*, made a case for evaluation of phosphogypsum produced in India for its suitability as set controller. Table 4.1 lists the sources of phosphogypsum *vis-à-vis* the cement plants close to them.

The replacement of natural gypsum by the by-product gypsum has been the subject of investigations for several years. Studies have shown that impurities (P_2O_5, F, organic matter, etc.) if present, in relatively larger amounts, could delay setting times and decrease strength development of cements. Studies have been conducted to examine the influence of various types of impurities generally present in phosphogypsum on the properties, like setting time and strength development of cement. The data obtained confirm that the impurities adversely affect the setting time and strength of cement in varying degrees.

4.2 Effect of Phosphogypsum (Unprocessed) on the Properties of Portland Cement

For experimental studies, the cement clinker collected from M/s Shri Digvijay Co. Ltd., Gujarat, of chemical composition – SiO_2 24.17%, Al_2O_3 3.39%, Fe_2O_3 3.38%, CaO 62.42%, MgO 3.21%, SO_3 0.41%, Loss on Ignition 0.46%, two phosphogypsum (PG) samples of chemical composition *viz.*, (A) Albright Morarji Dharam Dassji Ltd. (AMDD), PG – P_2O_5 0.92%, F 1.50%, SiO_2 6.60%, R_2O_3 5.86%, CaO 27.76%, MgO 0.25%, Na_2O 0.46%, SO_3 35.65%, H_2O 17.85%, organic matter 0.11%, and (B) Southern Petrochemical Industries Ltd. (SPIC), Tuticorin PG – P_2O_5 0.42%, F 1.12%, SiO_2 0.96%, R_2O_3 0.06%, CaO 32.07%, MgO 0.01%, Na_2O 0.27%, SO_3 45.08%, H_2O 19.68%, organic matter 0.56%, were used for making cement.

Phosphogypsum was added to cement clinker to maintain 1.8% of SO_3 level and ground to a fineness of 320 m^2/kg (Blaine).

Table 4.1 Sources of Phosphogypsum and Cement Plants Nearest to Them

Sl. No.	Source	Nearest Cement Plant	Distance from Source of Gypsum (km)
1.	Coromondal Fertilizer Ltd., Visakhapatnam	Andhra Cements Ltd. Vijay Wada ACC, Kistna CCI, Mandhar (MP)	280 295 530
2.	Fertilizers and Chemicals Travancore Ltd., Udyogmondal	Travancore Cements Ltd.	45
3.	EID Parry Ltd., Ennor Tamilnadu	Dalmia Cement Ltd., Dalmiapuram India Cements Ltd., Sankarnagar	305 325
4.	Gujarat State Fertilizers Ltd., Vadodara	ACC, Sevalia	35
5.	Albright, Morarji & Dharam Dassji, Ambernath, Maharastara)	ACC, Sahabad (Karnataka) Bagalkot Udyog Ltd., Bagalkot	325 315
6.	RCF, Chembur,Mumbai	ACC, Sevalia ACC, Sahabad (Karnataka Bagalkot Udyog Ltd., Bagalkot	315 415 405
7.	Southern Petrochemical Industries Corpn., Tuticorin, T.N.	India Cements Ltd., Sankarnagar	23
8.	FCI, Haldia (West Bengal)	Durgapur Cements Work, Durgapur	190
9.	Hindustan Copper Ltd., Khetri (Raj.)	Dalmia Dadri Cements Ltd., Charkhi Dadri (Haryana)	90

The cement produced was tested for setting time and compressive strengths at 1, 3, 7 and 28 days as per IS : 4031–1968 using graded standard Ennore (Chennai) sand conforming to IS : 650–1966 in 1:3 cements and mortar.

The results showing effect of phosphogypsum on the properties of cement are reported in Table 4.2. Results showed that setting time is retarded and the strength is decreased from 1 day to 28 days. This behaviour of cement is due to presence of impurities.

4.2.1 Effect of Phosphates, Fluorides and Organic Matter

To know the effect of impurities on setting time and strength development, various type of chemical reagents like phosphates, fluorides, organic matter were mixed with the control sample of cement. The setting time and compressive strength of the cement containing various chemicals were examined as per relevant Indian standards.

Table 4.2 Properties of Cement Produced by Intergrinding Sikka Cement Clinker (Gujarat) and Phosphogypsum.

Gypsum	Purity SO_3 (%)	Properties of cement							
		Fineness m^2/kg	Setting time (minutes)		Compressive strength MPa			
			Initial	Final	1d	3d	7d	28d
1. Mineral	45.48	320	80	184	10.1	19.6	28.0	34.8
2. Phosphogypsum Sample 'A'	39.65	325	290	500	7.0	13.2	20.5	30.0
3. Phosphogypsum Sample 'B'	45.08	326	200	350	9.2	16.9	27.0	35.8

Chemical reagents:

Phosphates – H_3PO_4, $Ca(H_2PO_4)_2$, H_2O, $CaHPO_4.2H_2O$ and $Ca_3(PO_4)_2$
Fluorides – NaF and Na_2SiF_6
Organic Matter – Humic acid
Mixtures – H_3PO_4– Na_2SiF_6, H_3PO_4– Na_2SiF_6– Humic Acid

The effect of phosphates, fluorides and organic matter on the setting time of Portland cement is shown in Tables 4.3–4.5.

With the use of different phosphatic compounds, the effect on setting time is variable. H_3PO_4 gave maximum retardation of setting time. Similarly, fluorides and humic acid retard setting time to a great extent. However, retardation is increased remarkably when impurities are used in mixture form. These results corroborate the findings of Takemoto, Ito, and Suzuki[6], and Mori and Sudo[7].

4.2.2 Retarding Mechanism of Water-Soluble Phosphates and Fluorides

The liquid phase in cement paste is alkaline in nature with high pH due to high concentration of lime and alkali. It was found that insoluble calcium phosphate salts are produced near the cement particles soon after P_2O_5 enters into the liquid phase and, subsequently, they cover the surface of the cement grains. Which serve as protective layer of the particles. The formation of CaF_2 and $Ca_3(PO_4)_2$ can be shown by the following equations:

$$SiF_6^{-2} + 6OH^- \longrightarrow 6F^- + SiO_2^{-2} + 3H_2O$$

$$2F^- + Ca^- \longrightarrow CaF_2$$

and

$$HPO_4^{-2} + O \longrightarrow H_2O + PO_4^{-2}$$

$$2PO_4^{-2} + 3Ca \longrightarrow Ca_3(PO_4)_2$$

Table 4.3 Effect of Phosphates, Fluorides and Organic Matter on the Setting Time of Portland Cement

Chemical Agent	Percentage of addition to gypsum in cement, %			Setting time minutes	
	P_2O_5	F	Organic matter (Humic acid)	Initial	Final
Mineral gypsum	0.0	0.0	0.0	80	184
H_3PO_4	0.01	0.0	0.0	120	220
	0.05	0.0	0.0	124	225
	0.10	0.0	0.0	128	230
	0.50	0.0	0.0	148	240
	1.0	0.0	0.0	159	265
$Ca(H_2PO_4)_2.H_2O$	0.10	0.0	0.0	120	224
	0.50	0.0	0.0	140	240
	1.00	0.0	0.0	154	258
$CaHPO_4.2H_2O$	0.50	0.0	0.0	100	230
	1.00	0.0	0.0	116	240
$Ca_3(PO_4)_2$	0.50	0.0	0.0	92	190
	1.00	0.0	0.0	109	220

Table 4.4 Effect of Phosphates Fluorides and Humic Acid on the Setting Time of Portland Cement

Chemical Agent	Percentage of addition to gypsum in cement, %			Setting time minutes	
	P_2O_5	F	Organic matter (Humic acid)	Initial	Final
NaF	0.0	0.05	0.0	110	250
	0.0	0.10	0.0	130	259
	0.0	0.50	0.0	155	264
	0.0	1.0	0.0	170	276
	0.0	1.5	0.0	184	289
Na_2SiF_6	0.0	0.05	0.0	170	260
	0.0	0.10	0.0	182	277
	0.0	0.50	0.0	206	290
	0.0	1.0	0.0	240	312
	0.0	1.5	0.0	257	333

(Contd...)

Chemical Agent	Percentage of addition to gypsum in cement, %			Setting time minutes	
	P_2O_5	F	Organic matter (Humic acid)	Initial	Final
Humic acid	0.0	0.0	0.1	125	240
	0.0	0.0	0.2	135	269
	0.0	0.0	0.4	151	280
	0.0	0.0	0.5	171	289
	0.0	0.0	0.6	186	297

Table 4.5 Effect of Mixtures of Phosphates, Fluorides and Organic Matter on the Setting Time of Portland Cement

Chemical Agent	Percentage of addition to gypsum in cement, %			Setting time minutes	
	P_2O_5	F	Organic matter (Humic acid)	Initial	Final
Mixtures H_3PO_4–Na_2SiF_6	0.5	0.5	0.0	250	330
	1.0	1.0	0.0	260	370
	1.0	1.5	0.0	279	410
H_3PO_4–Na_2SiF_6 Humic acid	0.5	0.5	0.1	286	375
	1.0	1.0	0.1	330	394
	1.0	1.5	0.5	359	440

The other retardation concept was reported by Tabikh and Miller[8]. It can be attributed to the ability of the complex phosphates and fluorides to serve as "bridges" or "cross-linking agent" between the hydrating cement particles or quasi "bidentate ligands" on cement particles through Si–O–Si or Si–O–P–O–Si linkages. These structures which involve chemical bonding, forming a more rigid and ordered protective coatings than a simple depositional covering over the cement particles.

4.2.3 Effect of Phosphates, Fluorides, and Organic Matter on the Compressive Strength of Cement

Data (Figs. 4.1 and 4.2) showed that compressive strength decreases from 1 day to 28 days on adding H_3PO_4 and $Ca(H_2PO_4)_2.H_2O$ to the Portland cement. The level of strength reduced as the concentration of H_3PO_4 is increased. At 1.0% addition of H_3PO_4, the fall in strength was maximum. On addition of sparingly soluble phosphate *i.e.*, $Ca(H_2PO_4)_2.H_2O$ the strength is too reduced.

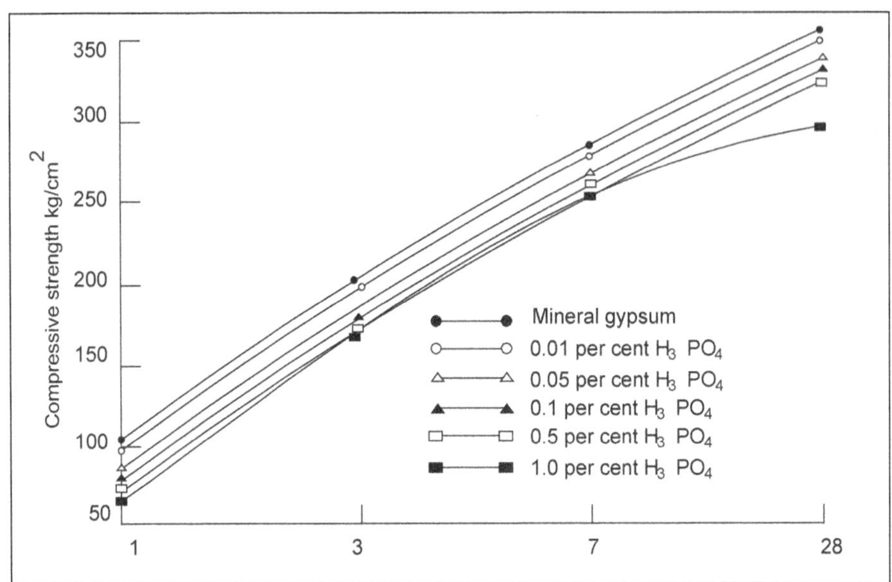

Fig. 4.1 Effect of addition of H_3PO_4 on the Compressive strength of Portland Cement

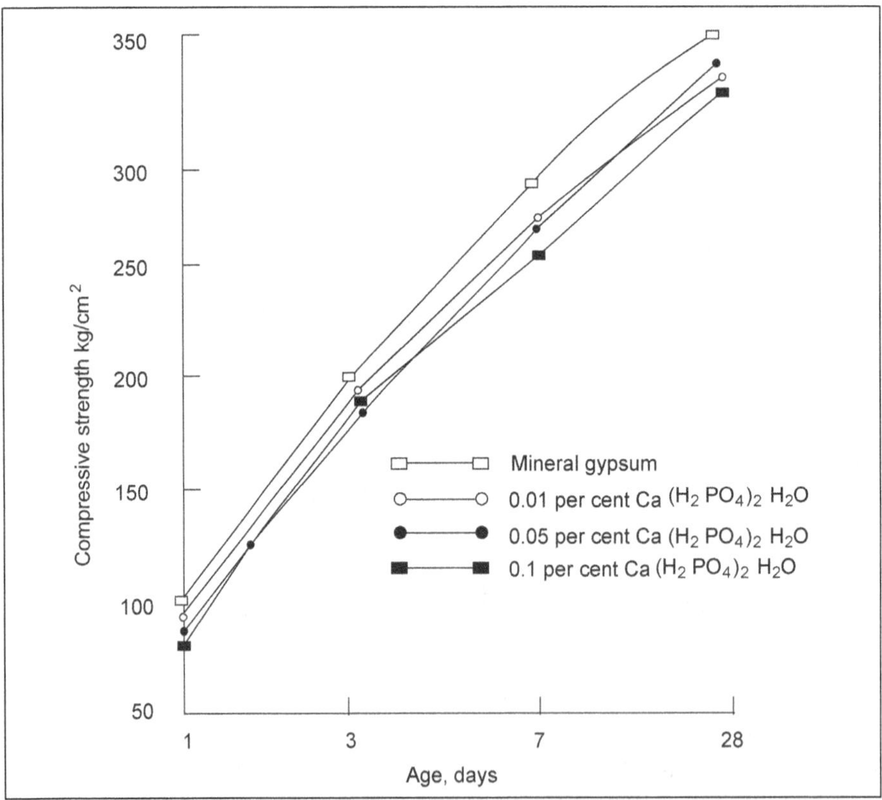

Fig. 4.2 Effect of addition of $Ca(H_2PO_4)_2$ on the Compressive strength of Portland cement

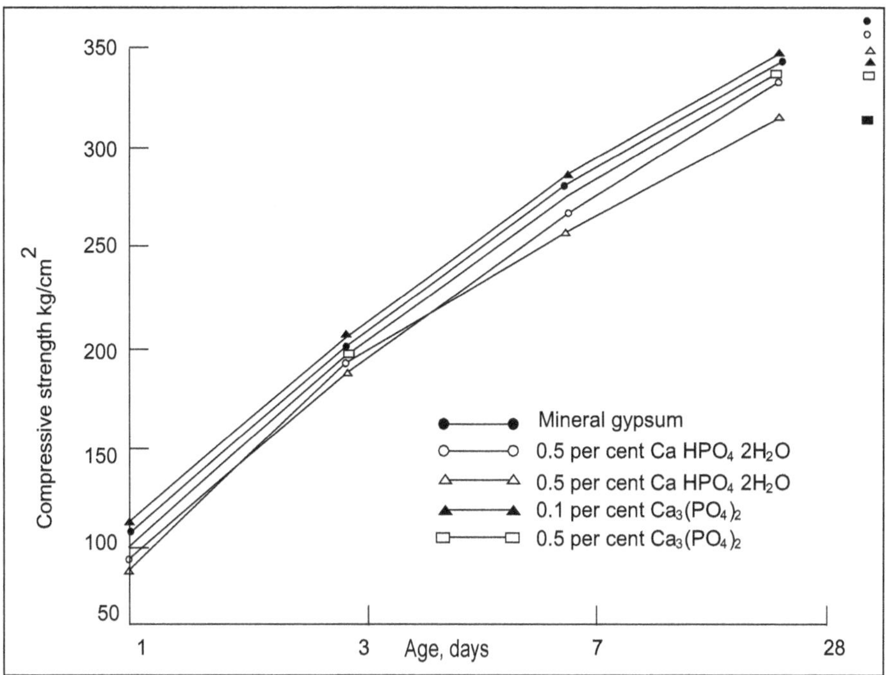

Fig. 4.3 Effect of addition of $CaHPO_4.2H_2O$ and $Ca_3(PO_4)_2$ on the compressive strength of Portland cement

However, the level of strength reduction is lower than the addition of H_3PO_4 and $Ca(H_2PO_4)_2$ compounds to the cement. Figure 4.3 showed addition of less soluble $Ca_3(PO_4)_2$ does not affect the development of strength, rather it is improved. The results clearly shows that soluble phosphates have a more scatheous effect as regards the development of strength than the sparingly or less-soluble phosphates. At 1.0% addition of H_3PO_4, the fall in strength was maximum. On addition of sparingly soluble phosphate *i.e.,* $CaHPO_4.2H_2O$, the strength is too reduced. However, the level of strength reduction is lower than the addition of H_3PO_4 and $Ca(H_3PO_4)_2.H_2O$ compounds to the cement. Data showed addition of less soluble $Ca_3(PO_4)_{2,}$ does not affect the development of strength, rather it is improved. The results clearly shows that soluble phosphates have a more scathing effect as regards the development of strength than the sparingly or less-soluble phosphates.

The effect of NaF and Na_2SiF_6 compounds on the strength of cement is plotted in Figs. 4.4 and 4.5 respectively. The trend of results is similar to the behaviour of phosphates. However, the fall in strength can be considered lower than the phosphatic compounds. The addition of organic matter *i.e.,* humic acid, also decreases strength at all ages of hydration (Fig. 4.6). The humic acid was added to cement as a substitute of organic matter, as the humic acid is very close to the organic matter generally found in phosphatic/fluorophosphatic rocks.

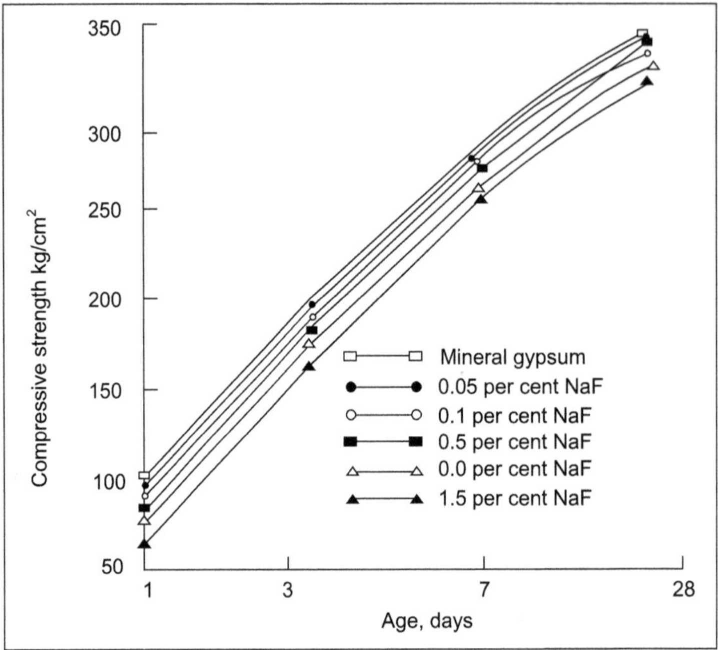

Fig. 4.4 Effect of addition of NaF on the compressive strength of Portland cement

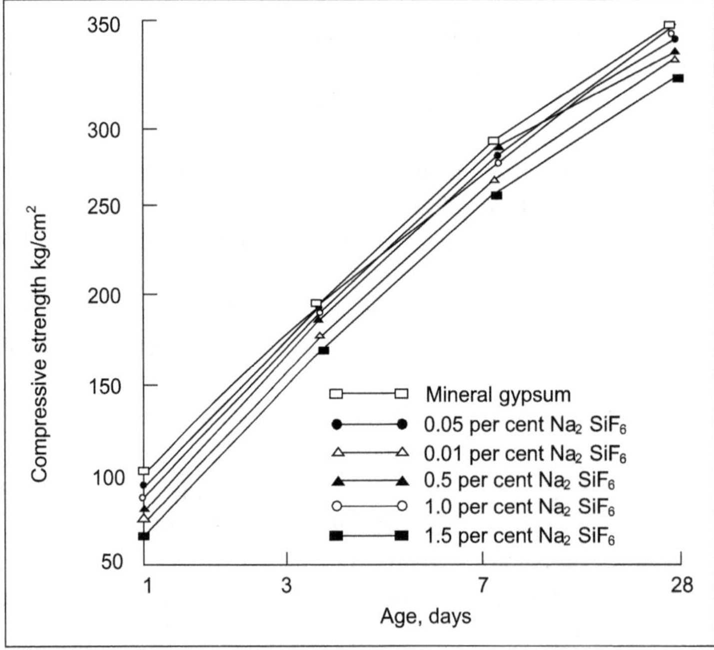

Fig. 4.5 Effect of addition of Na_2SiF_6 on the compressive strength of Portland cement

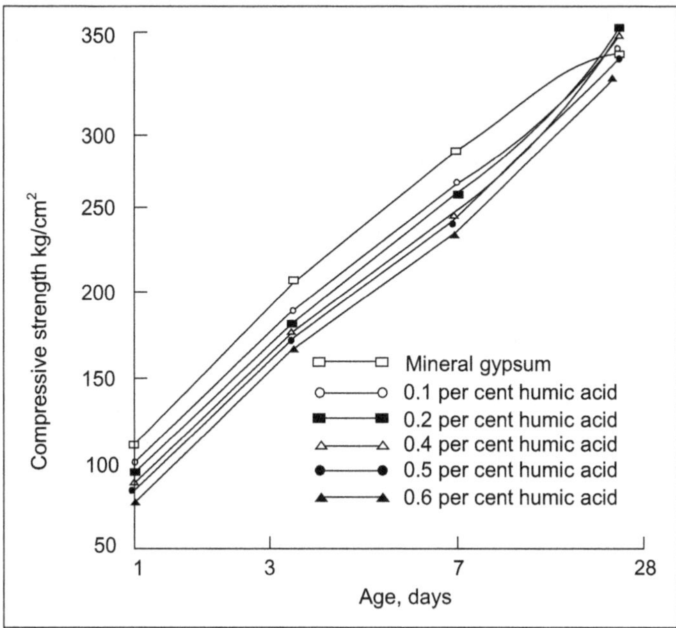

Fig. 4.6 Effect of addition of Humic acid on the compressive strength of Portland cement

Further data showed that strength was reduced considerably when the mixture of phosphates, fluorides and organic matter was employed (Fig. 4.7). The decrease in strength was, however, more than the strength values attained by the phosphates, fluorides and organic matter separately. The fall in strength is due to impurities of phosphogypsum which form protective coatings on the surface of cement particles, thus temporarily suppress the hydration reaction of tricalcium or dicalcium silicates (C_3S and C_2S). Murakami[9] also supported this concept of suppression of hydration by the impurities of phosphogypsum.

4.2.4 Effect of Beneficiated Phosphogypsum on Properties of Portland Cement

The effect of CFL and SPIC phosphogypsum samples unprocessed and beneficiated/processed by different methods, on the physical properties of cement clinker (Rajgangpur, Orissa) of chemical composition SiO_2 23.89%, Al_2O_3 5.63%, Fe_2O_3 1.90%, CaO 63.56%, MgO 4.30%, Loss of ignition 1.10%, is shown in Tables 4.6 and 4.7.

Data showed that on using unprocessed PG, setting time is retarded and the compressive strength is reduced. On using the beneficiated PG, the setting time is accelerated and the strength is increased .The retardation of setting time is due to the water-soluble impurities of PG which gradually enter into the aqueous phase of cement and affect the hydration of cement. The mechanism

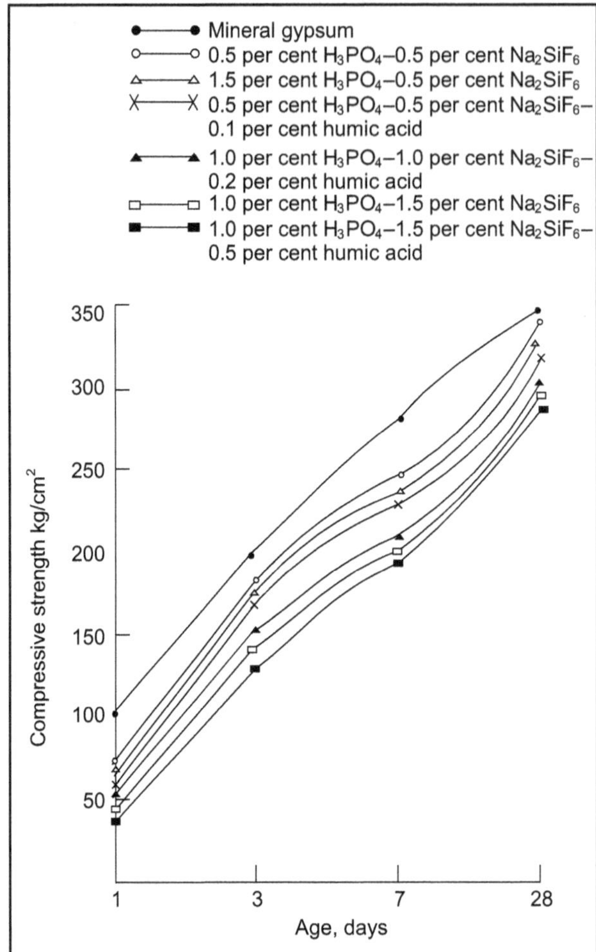

Fig. 4.7 Effect of the mixture of phosphates, fluorides, and humic acid on the compressive strength of Portland cement

of effect of impurities on setting and strength of cement has been explained above under 4.2.2.

It can be seen that there is improvement in compressive strength of cement from 1 to 3 days with the use of beneficiated PG. Best results were obtained when PG calcined at 850°C was used and SO_3 content was increased to 2.0%. The anhydrite not effected the setting time adversely of the cement since the impurity of $CaHPO_4.2H_2O$ on heating is converted into the insoluble and un reactive pyrophosphate as per following equation :

$CaHPO_4$ on heating → Meta phosphate → Pyrophosphate ($Ca_2P_2O_7$) "Insoluble, Unreactive"

Table 4.6 Effect of Beneficiated PG on Properties of Portland Cement produced by intergrinding Rajgangpur Cement Clinker (Orissa) and Phosphogypsum Sample CFL

Sl. No.	Gypsum Used	Treatment	SO_3 Content (%)	Fineness m^2/kg	Setting Time (minutes)		Compressive Strength				Autoclave Expn. (%)
					*I	*F	1d*	*3d	7d	28d	
1.	Mineral	As such	1.8	317	180	212	19.4	36.8	42	53.9	0.20
2.	*PG	*UP	1.8	320	310	395	13.5	29.2	42.4	52.5	0.38
3.	PG	Wet sieved through 150 micron IS sieve	1.8	318	185	240	19.2	36.2	46.6	54	0.28
4.	PG	Calcined at 160°C and washed	1.8	318	275	342	16.9	34.7	42.2	51	0.32
5.	PG	Calcined at 85°C	2.0	320	215	305	19.6	34.9	43.2	55.8	0.21
6.	PG	Treated with 55% H_2SO_4–5% Reactive SiO_2 Mixture and Washed	1.8	317	140	170	17.1	33.9	41.6	53	0.28
7	IS : 269–2003 limits	—	Max. 2.75	Min. 225	Min. 30	Max. 600	Note less than				Max.0.8
							—	16	22	33	

*UP : Unprocessed, *PG : Phosphogypsum, I : Initial, F : Final, d : Day(s)

In case of phosphoanhydrite, the amount of SO_3 has been increased to 2.0% in stead of 1.8% SO_3 as used in experimental cements produced by using other processed PG samples. There was apprehension that the presence of enhanced SO_3 content would have an adverse effect on soundness of cement. In order to access the extent of likely adverse effect of the uncombined gypsum in cements hydrated for different periods was determined. The results are reported in Table 4.8. Data showed that SO_4^{2} was almost consumed in both mineral and phosphogypsum within 24 hours forming ettringite (3CaO. Al_2O_3.3$CaSO_4$.32H_2O). The phosphogypsum anhydrite acted equally well as natural gypsum and was not likely to cause unsoundness. These results, confirmed the findings of Copland, *et al.*[10].

4.2.5 Studies on Heat of Hydration

Since the hydration of cement is an exothermic reaction, the measurement of the liberated heat can help in understanding the influence of impurities in

Table 4.7 Effect of Beneficiated PG on Properties of Portland Cement produced by intergrinding Rajgangpur Cement Clinker (Orissa) and Phosphogypsum Sample SPIC

Sl. No.	Gypsum Used	Treatment	SO_3 Content (%)	Fineness m^2/kg	Setting Time (minutes)		Compressive Strength				Autoclave Expn. (%)
					*I	*F	1d*	*3d	7d	28d	
1.	Mineral	As such	1.8	317	180	212	19.4	36.8	42	53.9	0.20
2.	*PG	*UP	1.8	318	240	320	15.6	33.2	39.2	54.5	0.35
3.	PG	Wet sieved through 150 micron IS sieve	1.8	322	200	285	18.8	36.0	41.0	55	0.26
4.	PG	Calcined at 160°C and washed	1.8	330	220	290	18.4	35.0	39.9	533	0.28
5.	PG	Calcined at 850°C	2.0	320	215	305	20.9	38.0	44.2	56.8	0.21
6.	PG	Treated with 55% H_2SO_4–55% Reactive SiO_2 Mixture and Washed	1.8	317	114	160	18.0	34.0	40.0	53	0.20
7.	IS.269–2003 limits	—	Max. 2.75	Min. 225	Min. 30	Max. 600	Not less than				Max. 0.8
							—	16	22	33	

*UP : Unprocessed, *PG : Phosphogypsum, I : Initial, F : Final, d : Day(s)

Table 4.8 Solubility of SO_3 in Lime Water

Sl. No.	Cements Containing	Acid Soluble SO_3(%)	Period of Curing (Hrs.)	Lime Soluble SO_3 (%)	SO_3 % as Calcium Sulphoaluminate
1.	Mineralgypsum	1.8	6.0	0.556	1.224
		—	12.0	0.110	1.690
		—	24.0	0.005	1.795
		—	30.0	Nil	1.800
2.	Phosphogypsum	2.0	6.0	0.147	1.863
		—	12.0	0.081	1.919
		—	24.0	0.009	1.991
		—	30.0	Nil	2.000

Table 4.9 Heat of Hydration of Trial Cements with Various Gypsum

Gypsum	Heat, K Cal/g	
	Unwashed gypsum	Washed gypsum
Mineral gypsum	51	38
Phosphogypsum	—	—
AMDD	40	37
SPIC	41	35

phosphogypsum on the setting characteristics of trial cements. Murakami[9] has analysed this aspect and observed that the protective coatings of trisulphate hydrate formed on the surface of tricalcium aluminate grains suppress the subsequent hydration of C_3A. When C_3A is mixed with phosphogypsum containing impurities, the suppression of heat liberation is more than that observed with pure gypsum. This tendency is, however, maximum when Portland cement just comes in contact with water. C_3S which also is responsible for heat liberation in early stages does not behave likewise. The heat liberation in Portland cement is rather complex due to the presence of various clinker minerals and alkalies, and cannot be explained on the basis of direct measurements. The results of heat of hydration (Table 4.9) showed that the heat liberated in cements with phosphogypsum was lower than that in cements with natural gypsum. Also the heat of hydration was relatively less for trial cements made with phosphogypsum, among which, in turn it was minimum for the trial cement made with phosphogypsum SPIC which was more acidic. It is interesting to note that heat of hydration of trial cements made with washed gypsum samples including natural gypsum was more or less the same indicating that the earlier difference was due to soluble impurities.

4.3 Effect of By-product Gypsum on Hydration and Strength of Blended Cements

Studies on effect of use of by-product gypsum on properties of Portland cement have been reported by various researchers above. As we know waste gypsum such as phosphogypsum, fluorogypsum and marine gypsum is available in India and Indian sub continent in the close vicinity of the cement plants, their effect on blended cements has not been studied in detail. To examine this aspect, the effect of phosphogypsum, fluorogypsum and marine gypsum has been reported on the properties such as hydration and strength development of blended cements like Portland Slag Cement (PSC) and Portland Pozzolana Cement (PPC) as an additive to cement clinker in place of natural gypsum. The results showed retardation of setting time and fall in strength of cements at initial stage of hydration. However, later age strength is less effected irrespective of type of cement used. The increase in strength and reduction of setting time of cements noticed with the use of treated gypsum. No direct correlation was established between ettringite formation and strength development or setting time of the blended cements at early stage of hydration.

Studies on effect of tartaro-,titanogypsum, boro, citro and desulphogypsum have been reported on early hydration of Portland cement at 0.5 water-cement ratio. It was found that cement containing tartarogypsum produced much more ettringite than those with titanogypsum and natural gypsum[11]. It was further observed that more ettingite was formed on early hydration in cement with commercial samples of borogypsum, citrogypsum and to a much lesser extent when desulphogypsum was ground with Portland cement. These by-product gypsum gave higher water consistency and setting time values as well as satisfactory strength development at 3, 7 and 28 days. No correlation was found between degree of ettringite formed and either water consistency or setting time. Comparisons were also made with previous investigations of Portland cement containing Boro-, Citro-, and Desulphogypsum[12–13].

4.4 Influence of Blended Gypsum on the Properties of Portland Cement and Portland Slag cement

Researches have been made by several workers to use processed phospho and fluorogypsum in the manufacture of cement. As already has been discussed that the impurities of P_2O_5, F, organic matter, akalies present in the gypsum do adversely affect the setting and hardening of cement. Attempts were, therefore, made to study the effect of unprocessed phospho and fluorogypsum on the properties of Portland cement and Portland slag cement with and without natural gypsum. The natural gypsum was blended with by-product gypsum to reduce the bad effect of obnoxious impurities of P_2O_5 and F present in the waste material The role of blended gypsum *vis-a-vis* by-product gypsum on the grind ability properties of cements was also examined. The results of these findings are discussed below.

The phosphogypsum, fluorogypsum, natural gypsum, cement clinker and granulated blast furnace slag of chemical composition (Table 4.10) were used to prepare the Ordinary Portland Cement (OPC) and Portland Slag Cement (PSC) respectively. Data showed phosphogypsum and fluorogypsum contain impurities of P_2O_5, F, organic matter and alkalies.

The Ordinary Portland Cement (OPC) was produced by intergrinding cement clinker with unprocessed phosphogypsum (UPPG) and fluorogypsum (F.G.) with and without natural gypsum (N.G.) at 2.0% SO_3 content to a fineness of 317 to 330 m^2/kg (Blaine).

The Portland slag cement was produced by intergrading the granulated blast furnace slag and cement clinker with 5 per cent blended gypsum to a fineness of 416 to 420 m^2/kg (Blaine).The cements produced were tested for consistency,

Table 4.10 Chemical Composition of Raw Materials

Constituents	Per cent				
	Phospho-gypsum	Fluoro-gypsum	Natural gypsum	Cement clinker	Granulated slag
P_2O_5	0.47	—	—	—	—
F	0.86	1.2	—	—	—
Organic matter	0.59	—	—	—	—
Na_2O	0.27	—	—	—	—
SiO_2+insoluble in HCl	0.98	0.55	8.80	24.19	35.85
Al_2O_3+Fe_2O_3	1.29	1.50	0.66	7.9	19.12
CaO	32.04	40.70	30.25	64.40	39.62
MgO	0.54	0.50	0.10	3.34	4.09
SO_3	43.21	55.60	39.60	0.31	0.13
MnO	—	—	—	—	0.63
LOI	19.40	—	20.00	0.70	0.54

setting time and compressive strength as per IS : 4031. The grind ability of the cement samples was studied as per Bond's grind ability test [14].

The effect of unprocessed phospho- and fluorogypsum as well as blended gypsum on the physical properties of OPC and PSC are reported in Tables 4.11 and 4.12 respectively.

It can be seen that setting time of cements is prolonged while the compressive strength reduced on the addition of phospho- and fluorogypsum alone. However, the retardation of setting time has been found lower with the fluorogypsum than the phosphogypsum. The retardation of setting time can be attributed to the protective coatings of $Ca_3(PO_4)_2$ and CaF_2 compounds as inert and inactive substances formed by the impurities of phosphates and fluorides on the hydrating cement particles,[15–16] thereby causing temporary suppression of the hydration of cement grains. On addition of blended gypsum (UPPG/FG + natural gypsum) to the cement clinker, the setting time gets accelerated and the compressive strength improved considerably (Figs. 4.8 and 4.9). This may be ascribed to the dilution of impurities with the addition of natural gypsum. It is interesting to note that strength development of cement mortars were not affected by the impurities of fluoride of hydrofluoric acid, hence fluorogypsum does not so markedly retard the hydration of cement as the water soluble fluoride of phosphogypsum[17]. The retardation is due to formation of silica gel or CSH ($Na_2SiF_6 + Ca(OH)_2 \rightarrow CaF_2$ + C–S–H + NaOH) on the surface of cement grains. The autoclave expansion of experimental cements was within the maximum specified value of 0.8 laid down in IS : 269–1989.

Table 4.11 Physical Properties of Ordinary Portland Cement in Presence of Blended Gypsum

Sl. No.	Type of Gypsum		Fineness (m^2/kg) (Blaine)	Consistency (%)	Setting time (minutes)	
					Initial	Final
1. Natural gypsum (N.G.)			317	26.1	186	262
2(*a*) Unprocessed phosphogypsum (UPPG)			320	26.0	325	409
2(*b*) UPPG + N.G.						
	40	60	323	26.3	188	270
	50	50	320	26.1	190	280
	60	40	330	26.4	191	286
3(*a*) Fluorogypsum (F.G.)			320	24.5	155	226
3(*b*) F.G. + N.G.						
	40	60	320	26.9	132	198
	50	50	318	27.2	140	202
	60	40	326	26.8	150	220

Table 4.12 Physical Properties of Portland Slag Cement in Presence of Blended Gypsum

Sl. No.	Type of Gypsum		Fineness (m^2/kg) (Blaine)	Consistency (%)	Setting time (minutes)	
					Initial	Final
1. Natural gypsum			416	22.8	238	305
2(*a*) Unprocessed phosphogypsum (UPPG)			420	23.7	270	380
2(*b*) UPPG + N.G.						
	40	60	400	23.0	201	275
	50	50	418	22.8	208	290
	60	40	418	23.0	222	299
3(*a*) Fluorogypsum (F.G.)			417	22.2	190	226
3(*b*) F.G. + N.G.						
	40	60	420	22.6	188	230
	50	50	420	23.0	200	240
	60	40	416	23.0	202	250

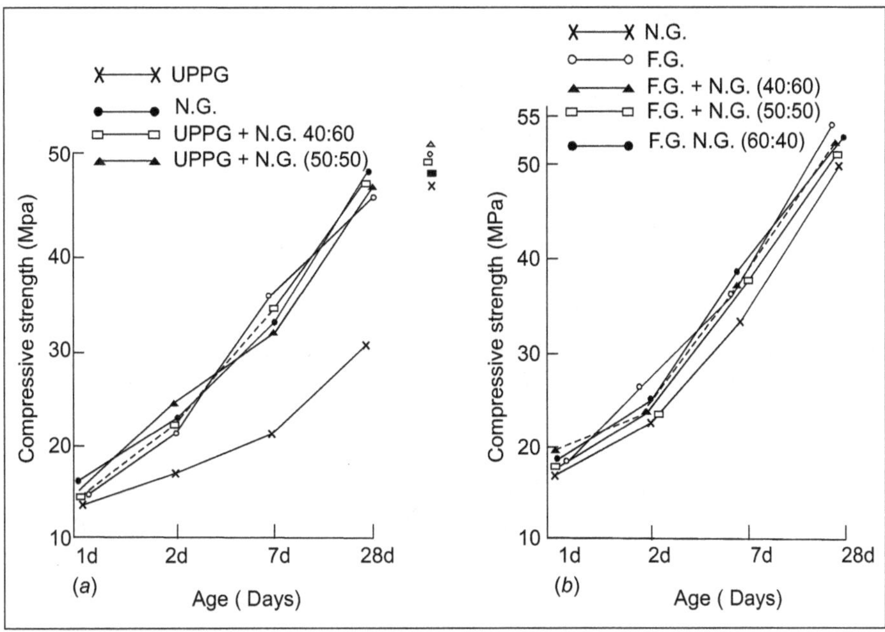

Fig. 4.8 Effect of Blended Gypsum (*a*) Natural + Phosphogypsum and (*b*) Natural + Fluorogypsum on the Compressive Strength of Portland Cement

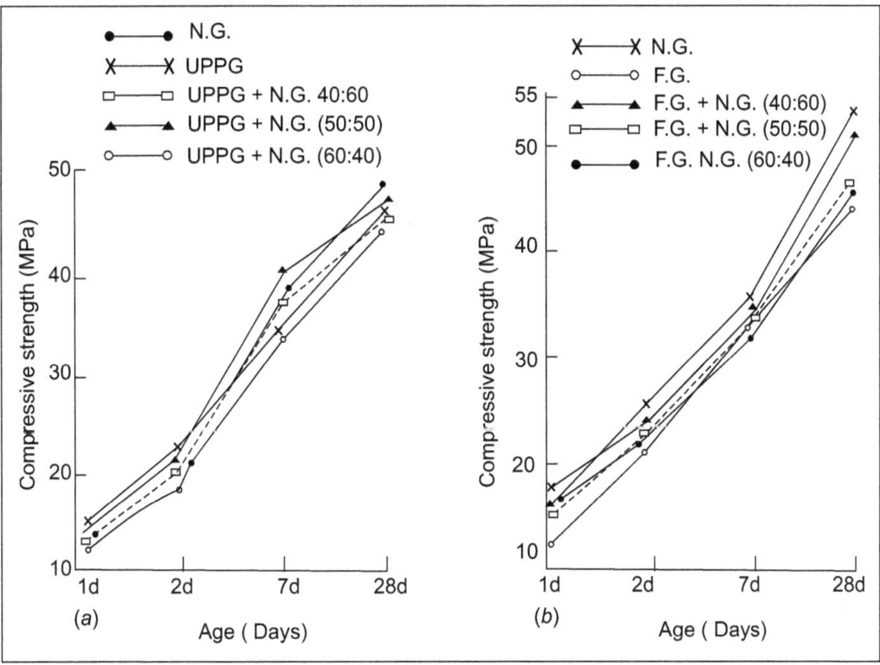

Fig. 4.9 Effect of Blended Gypsum (*a*) Natural + Phosphogypsum and (*b*) Natural + Fluorogypsum on the Compressive Strength of Portland Slag Cement

4.4.1 Energy Saving

The data on grind ability of Ordinary Portland Cement (OPC) and Portland slag cement (PSC) obtained in presence of blended gypsum is reported in Table 4.13. The power consumption for grinding cements with fluorogypsum was mitigated on blending it with natural gypsum and found comparable to the values attained with natural gypsum. The presence of argillaceous matter in natural gypsum apparently reduces the time of grinding thus, affecting power consumption.

The above study concluded that decrease in setting time and increase in the compressive strength of cements was observed on addition of blended gypsum. Maximum attainment of compressive strength achieved on using blended gypsum having proportion of 40:60, phospho-/fluorogypsum-natural gypsum by weight. The power consumption for grinding cement mitigated with the use of blended gypsum (fluorogypsum plus natural gypsum) to a level of grinding cement with natural gypsum. The use of blended gypsum (phospho -/fluoro+ natural gypsum) is recommended in the manufacture of cements to eliminate costly processing of the by-product gypsum.

Table 4.13 Grind Ability of Ordinary Portland Cement and Portland Slag Cement in Presence of Blended Gypsum

S. No.	Type of gypsum		Grind ability (g/r)	Work index (KWH/t)	Fineness (m^2/g) (Blaine)	Power (KWH)
OPC 1. Natural gypsum (N.G.)			0.86	20.50	320	30.1
2(*a*) Fluorogypsum			0.90	21.50	320	32.6
2(*b*) F.G. + N.G.						
	40	60	0.86	20.66	320	30.2
	50	50	0.84	20.40	320	30.2
	60	40	0.83	20.42	320	30.1
	PSC		—	—	—	—
1. N.G.			0.80	23.80	400	58.0
2(*a*) UPPG			0.88	24.90	400	59.1
2(*b*) UPPG + N.G.						
	40	60	0.82	23.70	400	58.04
	50	50	0.84	24.20	400	58.10
	60	40	0.82	23.90	400	58.22

References

1. Bensted, J., Early Hydration Behaviour of Portland cement containing Chemical by-product Gypsum, World Techn., 1979, Vol. 10, pp. 404–410.
2. Frigione, G, Gypsum in cement, Advances in Cement Technology, IInd Edition, edited by Ghosh, S.N. Tech Books, International, New Delhi, 2002, pp. 87–169.
3. Singh, Manjit, Treating Waste Phosphogypsum for Cement & Plaster Manufacture, Cem. Conr. Res., (USA) 2002, Vol. 32 pp. 033–1038.
4. Murakami, K., and Tanaka, H, Properties of By product gypsum, Sekko Sekkai, Vol. 61, 1962, p. 73.
5. Murakami, K., Tanaka, M., Saito, K., Hasimoto, K., Ichimma, K., Yogyo Kyokaishi, 1968, Vol. 76, pp. 253–255.
6. Takemoto, K., Ito, I, and Suzuli, S, Influence of Chemical By-product Gypsum on the Quality of Portland cement, Semento Gijutsu Nenpo, Vol. 11, 1957, pp. 55–61.
7. Mori, H, and Sudo, G, Influence of Phosphates, on the Hardening of Portland cement, Semento Gijutsu Nenpo, 1960, Vol. 14, p. 7.
8. Tabhikh, A.A, Miller, F. M., The Nature of Phosphogypsum Impurities and their Influence of Cement Hydration, Cement and Concrete Research, Vol. 1, 1971, pp. 663–678.
9. Murakami, K., Utilization of Chemical Gypsum for Portland cement, Proceedings of the 5th International Symposium on the Chemistry of Cement, Part IVth, Admixtures & Special Cements, Vol. 4, 1968, pp. 457–510.
10. Copeland, L.E., and Kantro, D.L., Hydration of Portland Cement, Porceedings of the 5th International Symposium on Chemistry of Cement, Part II, 1968, 1968, pp. 4–16.
11. Bensted John, Early Hydration Behaviour of Portland cement containing Tartaro- and Titanogypsum, Cement & Concrete. Research., Vol. 11, 1981, pp. 219–226.
12. Bensted, John and Bye, G.C., Silicates Industries, 1978, Vol. 43, No. 6, pp. 117–122.
13. Bensted, John, Early Hydration Behaviour of Portland cement containing Boro-, Citro- and Desulphogypsum, Cem. Con. Res., Vol. 10, No. 2, 1980, pp. 165–171.
14. S. Kanneurf, Rock Products, Vol. 65, 1957, p. 86.
15. Singh Manjit, Influence of Phosphogypsum Impurities on Two Properties of Portland Cement, Indian Concrete Journal, Vol. 61, 1987, pp. 186–190.
16. Teoreanu, I and Hvymh, T, The Influence of Fluoride Compounds on the Crystallization and Composition of the Mineralogical Consituents of Portland Cement Clinker, BRE Library Communication, 1666, 1972.
17. Kobayashi, K., Influence of Impurities of Phosphogypsum on the Development of Strength and the setting time of Portland Cement, Semento Gijutsu Nenpo, Vol. 18, 1968, pp. 79–85.

5

Use of By-product Gypsum in the Production of Super Sulphated Cement

INTRODUCTION

Super-sulphated cement (SSC) is produced by inter-grinding or intimately blending a mixture of ground granulated blast furnace slag and gypsum anhydrite with a small amount of cement clinker or Portland cement which acts as activator.

In Belgium, it is used to large extent. It is famous as 'Ciment Metallurgic Sur Sulphate Cement' (NBN 132). In France and Germany, this cement is called as 'Cement Sulphutam Zement' and is easily available in the market. In England, this cement is used to great extent and covered under BS : 4298.

In India, the raw materials for making SSC are available in plenty, but not in nearby areas. For instance, granulated blast furnace slag is produced at steel plants which are located in Eastern and Southern Eastern India. The resources of mineral gypsum are, on the other hand, available in Rajasthan which is in the west. The transportation of mineral gypsum to a site near steel plants, thus involves heavy expenditure. It is seen that some of the superphosphate plants producing waste phosphogypsum are located near slag granulation plants. Since phosphogypsum, on calcination at higher temperature, produces anhydrite, studies were, therefore, carried out to find out the suitability of anhydrite obtained from phosphogypsum for the production of SSC.

5.1 Raw Materials

5.1.1 Phosphoanhydrite

Phosphoanhydrite can be obtained by calcining phosphogypsum samples AMD, CFL and SPIC at 650 to 850°C for 4 hours. The chemical composition of phosphogypsum samples and the contents of impurities in phosphoanhydrite have already been discussed (under Chapter III).

5.1.2 Granulated Blast Furnace Slag

Two granulated slag samples were procured from Indian Iron and Steel Co., Burnpur, West Bengal (A) and Bokaro Steel Plant, Bokaro (B), Jharkhand, respectively. These slag samples were dried at 105°C + 2 in an oven to constant weight, cooled, ground to pass 100 mesh (150 micron – IS sieve) and analysed for SiO_2, Al_2O, Fe_2O_3, MnO, CaO, MgO, SO_3 and sulphur contents, and for their bulk density, glass content and refractive index as per relevant Standard specifications and test procedure[1].

5.1.3 Cement Clinker

A sample of cement clinker was procured from M/s ACC Cement Works, Kymore, Madhya Pradesh. Its chemical composition shows SiO_2 24.19%, Al_2O_3 5.29%, Fe_2O_3 2.62%, CaO 64.04%, MgO 3.34% and L.O.I. 0.70%.

5.2 Preparation of Super-sulphated Cement, Concrete and its Assessment

Each of the two slags and Kymore clinker were ground separately in a ball mill to fineness of 400 m^2/kg (Blaine's). Gypsum anhydrite was separately ground to 300 m^2/kg. The ingredients were then mixed in the proportions of 75:20:5, 70:20:10, and 65:20:15 (by weight %) respectively in a powder mixer for 4 hours. Each mix was transferred to a ball mill and ground further to more than 400 m^2/kg. The nine mixes thus prepared (3 for each phosphyoanhydrite calcined at 650°, 750° and 850°C) were tested for physical properties as per IS : 4031–1968 using Ennore standard sand as per IS : 650–1966 for casting cubes for compressive strength.

5.2.1 Concrete Making Properties of SSC

Concrete cubes (size, 10 cm × 10 cm × 10 cm) were cast according to IS : 516–1969 using SSC of optimum mix proportions, gravel (passing 19 mm and retained on 5 mm sieve) and Badarpur sand (passing 5 mm sieve) in the proportions 1:1.5:3. The water-cement ratio was fixed at 0.5. Concrete cubes were cured in water and tested for compressive strength after different periods.

5.2.2 Effect of Temperature on Hydration of SSC

The Indian Standard on super-sulphated cement IS : 6909–1990, recommends a maximum temperature of 40°C for curing of SSC mortar and concrete. In view of high temperature prevailing in the tropics during a part of the year, the effect of temperature of curing was considered worth investigating.

Cubes (sizes 5 cm) were cast 1:3 cement-sand mortar at a flow of 105–110% as per ASTM specification C 109. These cubes were cured separately in water at 27, 35, 45 and 50°C in the oven and tested at 3, 7 and 28 days. The cement was also tested for setting time as per IS : 4031–1968 methods of physical tests for hydraulic cement.

5.2.3 Grind-ability of Slag and Cement Clinker

Grind-ability of Burnpur and Bokaro slags and of Kymore clinker was studied as per Bond's grind-ability test[2]. Work index was calculated from the grind-ability by the empirical equation:

$$W_i = \left(\frac{16}{G:82}\right)\frac{\sqrt{\gamma}}{100}$$

Where

W_i = The work index

G = Ball mill grind-ability

γ = Micron size of the mesh of grind.

The work index is defined as the kWH required to grind one tonne of material to 80 per cent passing 100 γ.

Approximate power required to grind the material to fineness of 400 m^2/kg (Blaine's) was calculated from the equation.

$$W = \frac{10\ W_i}{\sqrt{P}} - \frac{10\ W_i}{F}$$

Where

W = The work in kWH/tonne.

W_i = The work index.

P and F are the 80 per cent passing size in micron of the product and feed material.

5.2.4 Heat of Hydration

Heat of hydration of SSC was determined as per IS : 4031–1968 at 3,7 and 28 days of hydration.

5.3 Phosphoanhydrite

It has been discussed in detail earlier that phosphogypsum contains impurities of phosphates, fluorides and organic matter which adversely affect the development of strength, if used as an additive to cement clinker. Soluble impurities adhering to the surface of crystals or in inter-granular spaces can be washed out with water. However, HPO_4^{2-} and AIF_5^{2-} ions are found to be present in gypsum-crystal lattice substituting SO_4^{2-} ions. These impurities dissolve out slowly on hydration and adversely affect the hydration characteristics and strength development of cement. On calcining phosphogypsum from 650° to 850°C, lattice bound impurity of $CaHPO_4.2H_2O$ gets converted to inert beta Ca-pyrophosphate. Suitability of the phosphonahydrite thus, produced was investigated for making SSC.

It may be pointed out in this context that on heating phosphogypsum to phosphoanhydrite, some of the fluoride is volatilized which may attack the refractory lining of calcining kiln. Therefore, due precaution should be taken to protect the lining of the kiln or calcining unit.

5.4 Granulated Blast-Furnace Slag

The composition of blast-furnace slag resembles that of Portland cement in their main constituents like lime, silica and alumina but contains less lime than does Portland cement. The chemical composition of slag depends on the composition of iron ore as well as on the impurities present in the limestone added to the furnace burden. Granulated blast-furnace slag which is produced by cooling rapidly molten slag consists mostly of glass, which is reactive and suitable for the manufacture of SSC.

Chemical composition of the Burnpur (Slag A) and Bokaro (Slag B) granulated slags are reported in Table 5.1. Data showed that these slag samples were characterized by low CaO/SiO_2 and SiO_2/Al_2O_3 ratios. Normally, slag which contain more than 12% alumina are considered suitable for making super sulphated cement. Recent studies of Keil and Locher[3] and of Tanaka and Yamne[4] have clearly defined the field of maximum activity of slags in the system CaO–SiO_2–Al_2O_3, with and without MgO for sulphate activation. According to Tanaka and Yamne, synthetic slags of the composition in the range of 24–34% SiO_2, 16–20% Al_2O_3 and 46–50 % of CaO are best for SSC. However, Chopra and Lal[5] showed that, by carefully working out the optimum conditions of activation

Table 5.1 Chemical Composition

Constituents	Per cent composition	
	Slag A	Slag B
SiO_2	31.92	35.85
Al_2O_3	24.49	18.07
FeO	0.85	1.05
MnO	1.13	0.63
CaO	36.36	39.62
MgO	5.23	4.09
SO_3	Tr	0.13
S	0.62	0.54

regarding the quantity and solubility of anhydrite in relation to time and soluble alumina, the strength could be raised substantially so as to make Indian slags suitable for commercial exploitation.

5.4.1 Bulk Density, Glass Content and Refractive Index of Slags

Result of bulk density, glass content and refractive index of slag are tabulated in Table 5.2. It can be seen from the data that slag B has much lower bulk density than slag A. This is due to the fact that the granulation of slag B was carried out close to furnace at a relatively higher temperature. As per recommendation glass content in the slags of good hydraulicity should not be less than 90 per cent. Both these slags passed this requirement. The recommended value of refractive index is 1.65. Both slag samples showed somewhat less values of refractive index. However, refractive index requirements are not so critical as to debar the use of a particular slag.

Table 5.2 Bulk Density, Glass Content and Refractive Index Values

Slag designation	Bulk density passing 6 mm sieve (kg/m^3)	Glass content per cent	Refractive index
A	1028	95	1.636
B	303	98	1.644

5.4.2 Cement Clinker

The addition of cement clinker to a mixture of slag and anhydrite provides the desired alkalinity for the necessary chemical reactions to take place and for the development of strength through the formation of calcium sulphoaluminate hydrates and calcium silicate hydrate.

5.5 Evaluation of Super-sulphated Cement

5.5.1 Properties of Super-sulphated Cement

The optimum portion of SSC as determined by Tanaka *et al.* (1958) are slag 80–85%, anhydrite 10–15% and Clinker 2.5%. However, while investigating the suitability of TISCO slag for making SSC, Chopra and Lal[6] showed that the optimum proportions using Indian slags are slag 70–75% anhydrite 20–25% and clinker 5%. Higher percentage of anhydrite used for activation is because of the high alumina content of Indian slags.

Physical properties of SSC mixes prepared from Burnpur and Bokaro slags and Kymore clinker with AMD, CFL and SPIC phosphoandhydrite prepared at 650°, 750° and 850°C are shown in Tables 5.3–5.8. The data (Tables 5.3–5.4) showed that SSC mixes having mix proportions 65:20:15 (650°C); 70:20:10

Table 5.3 Physical Properties of SSC prepared by using Burnpur Slag, Gypsum Anhydrite Sample AMD and Cement Clinker

Phospho-anhydrite Temp. °C	Mix. Proportion BS:GA:C*	Setting Time (Minute)		Compressive Strength (MPa)			Cold expansion (mm)
		Initial	Final	3d	7d	28d	
650°C	75:20:5 70:20:10 65:20:15	360 320 325	480 495 460	10.55 13.0 18.5	15.0 20.0 22.0	32.5 30.5 35.0	2.00 2.00 3.00
750°C	75:20:5 70:20:10 65:20:15	280 265 220	495 360 310	12.0 15.5 20.0	17.0 20.8 25.0	32.0 34.6 38.6	2.5 2.5 2.8
850°C	75:20:5 70:20:10 65:20:15	300 298 310	390 420 400	12.5 14.0 17.0	15.0 20.9 24.0	30.6 32.0 35.0	2.1 1.8 2.0
IS : 6909–1990 Limits	—	Min. 30	Max. 600	Min. 15.0	Min. 22.0	Min. 30.0	Max. 5 mm

BS : Burnpur slag
GA : Gypsum anhydrite
C : Cement Clinker

Table 5.4 Physical Properties of SSC Prepared by Using Bokaro Slag, Gypsum Anhydrite Sample AMD and Cement Clinker

Phospho-anhydrite Temp. °C	Mix. Proportion BS:GA:C*	Setting Time (Minute)		Compressive Strength (MPa)			Cold expansion (mm)
		Initial	Final	3d	7d	28d	
650°C	75:20:5 70:20:10 65:20:15	320 340 290	475 500 425	120. 13.5 17.5	17.0 22.0 24.0	28.0 30.0 34.6	2.5 2.4 2.0
750°C	75:20:5 70:20:10 65:20:15	239 270 220	435 420 305	14.0 15.2 19.6	20.0 22.5 26.5	37.0 38.0 39.5	3.0 3.0 2.6
850°C	75:20:5 70:20:10 65:20:15	305 295 300	399 440 430	13.0 15.6 15.8	17.0 18.5 22.9	27.0 29.8 31.5	2.2 2.0 2.8

BS : Burnpur slag
GA : Gypsum anhydrite
C : Cement Clinker

and 65:20:15 (750°C) and 70:20:10 and 65:20:15 (850°C) complied with the requirements of IS : 6909–1990 for their physical properties. Optimum temperature of calcination of AMD phosphogypsum was fixed at 750°C, since at this temperature, maximum strength was obtained and setting time was also reduced considerably as compared to the other mix proportions. The amount of cement clinker required for activation was 10–15%. It can be seen from Table 5.5

Table 5.5 Physical Properties of SSC Prepared by Using Burnpur Slag, Gypsum Anhydrite Sample CFL and Cement Clinker

Phospho-anhydrite Temp. °C	Mix. Proportion *BS:GA:C	Setting Time (Minute)		Compressive Strength (MPa)			Cold expansion (mm)
		Initial	Final	3d	7d	28d	
650°C	75:20:5	315	405	13.0	18.0	30.6	2.0
	70:20:10	295	425	14.0	22.4	31.0	2.5
	65:20:15	289	390	21.0	29.1	40.0	2.5
750°C	75:20:5	220	395	13.3	19.7	30.9	2.8
	70:20:10	225	255	17.8		37.6	2.9
	65:20:15	125	210	22.0	30.7	40.6	3.0
850°C	75:20:5	275	395	12.5	17.9	27.2	3.0
	70:20:10	270	402	12.9	20.6	29.8	3.0
	65:20:15	280	390	15.1	22.9	34.7	2.8

BS : Burnpur slag

Table 5.6 Physical Properties of SSC Prepared by Using Bokaro Slag, Gypsum Anhydrite Sample CFL and Cement Clinker

Phospho-anhydrite Temp.°C	Mix Proportion *BS:GA:C	Setting Time (Minute)		Compressive Strength (MPa)			Cold expansion (mm)
		Initial	Final	3d	7d	28d	
650°C	75:20:5	280	425	14.0.	19.5	31.0	2.2
	70:20:10	280	450	15.0	21.0	33.0	2.6
	65:20:15	250	370	19.0	28.0	29.5	2.8
750°C	75:20:5	229	405	15.0	20.0	40.0	2.5
	70:20:10	240	388	16.5	24.6	38.0	2.5
	65:20:15	180	245	22.5	29.0	41.0	2.8
850°C	75:20:5	245	370	13.0	18.0	26.0	3.0
	70:20:10	280	428	14.0	21.0	30.6	2.2
	65:20:15	270	398	15.6	24.0	33.5	2.4

BS : Bokaro slag

that SSC mixes corresponding to mix proportions 65:20:15 (650°C); 70:20:10 and 65:20:15 (750°C) and 65:20:15 (850°C) passed the specified requirements Table 5.6 showed that the SSC mixes having in mix proportions 70:20:10 and 65:20:10 (650°C); 70:20:10 and 65:20:15 (750°C) and 65:20:15 (850°C) also complied with the requirements of IS specification. The optimum temperature of calcination of CFL phosphogypsum was fixed at 750°C and the quality of cement clinker required was 10–15 per cent.

The data (Tables 5.7–5.8) showed that SSC mixes of all mix proportions (75:20:5, 70:20:10 and 65:20:15) complied with the requirements of IS : 6909–1990 for physical properties. Highest strength was obtained with 70:20:10 mix prepared using phosphogypsum samples SPIC fired at 750°C.

Table 5.7 Physical Properties of SSC Prepared by Using Burnpur Slag, Gypsum Anhydrite Sample SPIC and Cement Clinker

Phospho-anhydrite Temp. °C	Mix. Proportion *BS:GA:C	Setting Time (Minute)		Compressive Strength (MPa)			Cold expansion (mm)
		Initial	Final	3d	7d	28d	
650°C	75:20:5	260	410	16.7	22.0	34.0	2.3
	70:20:10	236	325	20.9	29.5	40.6	2.7
	65:20:15	260	406	18.7	26.0	38.9	2.6
750°C	75:20:5	240	365	16.8	22.7	41.0	2.4
	70:20:10	175	236	24.0	33.0	43.0	2.3
	65:20:15	230	370	21.9	30.0	41.7	2.4
850°C	75:20:5	260	400	16.0	21.6	31.6	2.6
	70:20:10	240	350	17.0	23.0	34.0	2.6
	65:20:15	240	381	16.1	22.0	32.6	2.6

BS : Burnpur slag

Table 5.8 Physical Properties of SSC Prepared by Using Bokaro Slag, Gypsum Anhydrite Sample SPIC and Cement Clinker

Phospho-anhydrite Temp. °C	Mix. Proportion *BS:GA:C	Setting Time (Minute)		Compressive Strength (MPa)			Cold expansion (mm)
		Initial	Final	3d	7d	28d	
650°C	75:20:5	280	390	15.0	23.1	34.0	2.1
	70:20:10	260	340	23.0	31.2	42.0	2.2
	65:20:15	271	398	19.0	30.0	39.5	2.4
750°C	75:20:5	200	280	16.9	26.7	38.0	2.4
	70:20:10	160	205	23.0	31.6	42.7	2.1
	65:20:15	180	260	19.0	30.1	40.0	2.3
850°C	75:20:5	265	380	15.8	21.0	32.0	2.8
	70:20:10	265	360	16.7	30.6	39.0	2.6
	65:20:15	260	366	15.0	29.5	37.0	2.6

BS : Bokaro slag

This mix proportion was found to give more strength with Burnpur Slag (Table 5.7) than with Bokaro slag (Table 5.8). The optimum quantity of cement clinker required for activation was 10 per cent.

5.5.2 Concrete Making Properties

Concrete with SSC is designed, prepared, compacted and cured in the same way as with Portland cement. It has been observed that for the same water/ cement ratio super-sulphated cement gives a more workable mix than Portland cement.

Results of compressive strength of concrete mixes are reported in Table 5.9. The data showed that strength was comparable to those of Portland cement

Table 5.9 Compressive Strength of Concrete Mixes

Phosphoanhydrite Temperature,°C	Optimum mix proportion S*: P.A.: CC	Period of Curing (days)	Compressive Strength (MPa)	
			Burnpur Slag	Bokaro Slag
AMD (750°C)	65:20:15	3 7 28	13.20 20.20 28.40	12.40 22.00 30.00
CFL (850°C)	65 :20:15	3 7 28	13.00 20.20 30.8	12.84 19.91 32.00
SPIC (850°C)	70:20:10	3 7 28	14.0 25.00 35.00	13.8 24.4 30.0

*S = Super-sulphated Cement, P.A. : Phosphoanhydrite, CC : Cement clinker

concrete of M-20 grade. Super-sulphated cement made with SPIC phosphoanhydrite (850°C) has given the highest compressive strength.

SSC concrete is highly resistant to aggressive environment *viz.*, seawater, certain trade effluents and sulphur oxidizing bacteria etc. SSC prepared from Indian slags has been shown to be resistant to weak acids[7], which ordinarily attack normal Portland cement concrete.

5.5.3 Effect of Temperature on Hydration of SSC

Apprehensions have been expressed[8] about the possible deterioration in the strength of super sulphated cement mortar and concrete on curing under tropical conditions.

Data showing effect of temperature on compressive strength and setting of super-sulphated cement-sand mortar is given in Table 5.10. It can be seen that compressive strength increased with the increase in temperature of curing up

Table 5.10 Effect of Temperature on Compressive Strength and Setting time of Super-sulphated Cement-sand Mortar (CFL 750°C, 65:20:15)

Curing Temperature	Compressive strength (kg/cm^2)			Setting time (Minutes)	
	3d	7d	28d	Initial	Final
27	9.00	11.0	30.5	180	245
35	9.62	16.0	31.2	170	220
45	12.0	19.0	31.5	156	160
50	13.0	22.0	31.8	140	154
55	12.8	20.5	30.9	160	170

to 50°C. It was believed[9] that cementing action and development of strength in supersulphated cement at early ages up to 28 days is mainly due to the formation of tricalcium sulphoaluminate hydrate ($3CaO.Al_2O_3.3CaSO_4.32H_2O$). As can be seen from the continued rise in strength the stability of tricalcium sulpho-aluminate hydrate was not affected by curing at temperature up to 50°C. Later age strength in SSC is attributed to formation of calcium silicate hydrate [C–S–H(1)][10]. The data showed that the setting time was not appreciably affected by rise in temperature of curing. Taneja, *et al.*[11] and Singh[12–13] also found similar results while working with different type of Indian slags and waste phosphogypsum samples. Hence, this type of cement can be safely recommended for use in summer months also.

5.5.4 Grind-ability of Slag and Clinker

Grind-ability has been defined broadly as the response of a material to grindability effect, which can be usefully employed to study grinding installations. It is determined on the basis of Bond's formula of grind-ability which can be defined as "the total work useful in breakage which has been applied to a stated weight of homogeneous broken material is inversely proportional to the square root of the diameter of the product particles".

The results of the grind-ability studies on Burnpur, Bokaro slags and Kymore Cement Clinker are shown in Table 5.11.

The data (Table 5.11) showed that power consumption for grinding slag was considerably higher than that for cement clinker. Additional power is also required in various operations such as winning, crushing, grinding of raw materials and running of rotary kiln etc. in the manufacture of cement clinker.

5.5.5 Heat of Hydration

The heat of hydration is the quantity of heat in calories per gram of unhydrated cement, evolved upon complete hydration at a given temperature. It is largely influenced by the mineralogical composition of the cement. The fineness of the cement also influences the rate of development as an increase in fineness increases hydration of cement. The total amount of heat liberated, is however, not affected by the fineness of cement.

The results of heat hydration determined for mix proportions 65:20:15 (750°C) are tabulated in Table 5.12. IS:6909–1990 and BS:4248–1968 for super-sulphated

Table 5.11 Grind-ability of Slag and Clinker

Materials	Grind-ability Gm/Revolution	Work Index KWH/Tonne	Fineness m^2/kg (Blaine's)	Power required KWH
Burnpur Slag	0.74	22.84	400	56.0
Bokaro Slag	1.62	12.09	400	29.6
Kymore Clinker	0.85	20.41	320	50.0

Table 5.12 Heat of Hydration of SSC

Phospho-anhydrite Temp. °C	Mix. Proportion S:PA:C	Heat of dissolution (Cals/g)				Heat of Hydration		
		Unhydraed	Hydrated			3d	7d	28d
			3d	7d	28d			
AMD 650°C	65:20:15* 65:20:15**	526 519	501.70 492.91	485.60 479.30	465.09 459.24	24.30 26.09	40.40 39.70	60.91 59.76
CFL 750°C	65:20:15* 65:20:15**	515 520	494.2 497.4	476.7 496	459.3 462	20.9 22.6	38.4 34.0	55.7 58.0
SPIC 750°C	70:20:10* 70:20:10**	516 506	495.4 489.4	438.7 476	465.0 463.6	20.60 16.60	32.30 30.0	51.0 42.40
IS : 6909–1990 Limits	—	—	—	—	—	—	Max. 60	Max. 70

*Burnpur slag, **Bokaro slag

cement specify maximum limits of 60 and 70 cals/g, heat of hydration at 7 and 28 days respectively. The test cements complied with these specifications. Due to its low heat of hydration, SSC is eminently suitable for mass concrete construction. Its relatively higher sulphate-resistance *vis-à-vis* Portland cement makes it more suitable for use in marine structures.

Since cost of transport of mineral gypsum from Rajasthan to the different steel plants producing by-product granulated slag is very high and the waste phosphogypsum is available within a short distance, the latter can prove to be an economical substitute of natural mineral gypsum for making anhydrite for the production of SSC. Its utilization will bring in the market a new type of cement having special properties of low heat of hydration and sulphate resistance.

Investment for manufacture of SSC will be much lower than that of normal Portland cement, because besides grinding and blending equipment, no special machinery or kiln is required. In the present day shortage of power, its manufacture is an economic advantage *vis-a-vis* Portland cement. Moreover, the major raw material required for its production is slag which is waste product of the steel industry and is available in plenty. Prospects for manufacture of this cement in West Bengal appear to be bright because of the availability of large quantities of slag from Indian Iron and Steel Co., Burnpur and Bokaro Steel Plant, Bokaro and the requirements of the cement in the state for general concrete construction and marine structures.

5.6 Use of Fluorogypsum in Making Super-sulphated Cement

Scarcity of gypsum of high purity required to produce anhydrite for super-sulphated cement, and its requirement is much larger quantities (20–25%) compared to 4–5% in case of ordinary Portland cement have been the stumbling blocks to the production of supersuphated cement in India. In this context, an endeavour has been made to utilize anhydrite available as a waste product from the hydrofluoric acid industry. Anhydrite is obtained during the interaction of fluorspar with concentrated sulphuric acid as follows.

$$CaF_2 + H_2SO_4 = CaSO_4 + 2HF$$

In this chapter, an attempt has been made to produce SSC by blending the ground slag with fluoroanhydrite and cement clinker. The SSC was tested and evaluated for concrete making properties and the effect of different temperatures on the properties and performance of the cement. The properties of SSC and the hydration characteristics as evaluated by differential thermal analysis (DTA), XRD and scanning electron microscopy (SEM) have been discussed.

5.6.1 Materials and Methods

The samples of granulated slag, fluorogypsum and Portland cement clinker procured from M/s Indian Iron & steel Co. (IISCO), Burnpur, M/s Everest Refrigerant, Mumbai and M/s Kymore Cement Plant respectively were analysed for various chemical constituents as per IS : 1727–1967, Methods of tests for pozzolanic materials, IS : 1288–1983, Methods of test for mineral gypsum and as per standard test procedures. The results of chemical analysis are shown in Table 5.13. Data indicate that the fluorogypsum contain impurities of F and the free acidity.

5.6.2 Preparation of Super-Sulphated Cement, Concrete and its Assessment

The samples of slag and cement clinker were ground separately in a ball mill to fineness of 400 m^2/kg (Blaine's). Gypsum anhydrite was separately ground

Table 5.13 Chemical Composition of Raw Materials Used

Constituent %	Granulated slag	Anhydrite	Portland cement
SiO_2	33.70	0.71	21.40
Al_2O_3	20.75	0.58	6.20
Fe_2O_3	0.58	—	3.50
MnO	1.04	0.03	—
CaO	36.40	41.80	63.20
MgO	8.80	0.10	Tr.
S	0.70	0.04	—
SO_3	—	56.20	—
F	—	1.10	—
$H_2O + CO_2$	—	0.81	—

to 300 m^2/kg (Blaine's). The ingredients were then mixed in the different proportions (by weight)-Mix 1-slag : Clinker : anhydrite :: 70:20:10, Mix 2 slag: Clinker:anhydrite :: 65 : 20 : 15 and Mix 3 slag : Clinker:anhydrite :: 75 :15 : 10 in a powder mixer for 5 hours. Each mix was transferred to a ball mill and ground further to more than 400 $m^{2/}/kg$ (Blaine's). The experimental mixes thus prepared were tested for physical properties including heat of hydration at 3, 7 and 28 days of hydration. Ennore standard sand as per IS : 650–1970, specification for coarse and fine aggregates from natural resources, for concrete, was used for casting cubes (7.06 cm × 7.06 cm × 7.06 cm) for determining compressive strength.

5.6.3 Concrete Making Properties of SSC

Concrete cubes (10 cm × 10 cm × 10 cm) were cast using SSC of optimum mix proportions, gravel (passing 19 mm and retained on 5 mm sieve) and Badarpur sand (passing 5 mm sieve) in the proportion 1 : 1.5 : 3. The water-cement ratio was fixed at 0.5. Concrete cubes were cured in water and tested for compressive strength after different periods.

5.6.4 Effect of Temperature on Hydration of SSC

The Indian Standard on super-sulphated cement IS : 6909–1990, recommends a maximum temperature of 40°C for curing of SSC mortar and concrete. In view of high temperature prevailing in the tropics (April to September) during a part of the year, the effect of temperature of curing was considered worth investigating.

Cubes (5 cm × 5 cm × 5 cm) were cast using 1:3 cement-sand mortar at a flow of 105–110%. These cubes were cured separately in water at 27, 35, 45 and 50°C in the oven and tested at 3, 7 and 28 days. The cement was also tested for setting time as per IS : 4031–1988, methods of physical tests for hydraulic cement material.

5.6.5 Properties of Super-Sulphated Cement

The optimum proportion of SSC as determined by Tanaka *et al.* is slag 80–85%, anhydrite 10–15% and Clinker 2.5%. However, while investigating the suitability of TISCO slag for making SSC, Chopra and Lal showed that the optimum proportion using Indian slags are slag 70–75%, gypsum anhydrite 20–25% and cement clinker 5%. Higher percentage of gypsum anhydrite used for activation is because of the high alumina content of Indian slags. SSC concrete is highly resistant to aggressive environment *viz.*, seawater and sulphur oxidizing bacteria, etc. SSC prepared from Indian slags has been shown to be resistant to weak acid, which ordinarily attack Portland cement concrete.

Physical properties of SSC mixes are depicted in Table 5.14. The three cements conformed to IS : 6909 (1990), specification for super-sulphated cement. The

Table 5.14 Physical Properties of Supersulphated Cements

Sl. No.	Property	SSC 1	SSC 2	SSC 3	IS : 6909–1990 Limits
1.	Fineness, m^2/kg	400	400	400	Min. 300
2.	Setting time, min				
	Initial	180	190	190	Min. 30
	Final	320	310	320	Min. 600
3.	Compressive strength, MPa				
	3	28.3	26.1	21.9	Min. 15.0
	7	32.1	30.4	30.8	Min. 22.0
	28	42.2	41.2	43.1	Min. 30.0
	90	54.6	49.0	52.20	—
	180	61.0	52.4	52.4	—
	360	67.1	63.0	58.40	—
4.	Soundness				
	Le Chatelier's expansion, mm	0.15	0.50	0.40	Max. 5.0
	Autoclave expansion, %	0.50	0.62	0.60	N.S.

strength of cement samples is much higher than the specified limit. The SSC sample 1 showed higher strength development than the SSC samples 2 and 3.

Strength development in cement samples has been supplemented by the differential thermal analysis (DTA, Perkin Elmer Diamond TG/TGA)) and the X-ray diffractometry (XRD, CuKα). The DTA of SSC 1 is shown in Fig. 5.1. The thermogram shows appearance of endotherms at 160°, 165°, 170°C at all ages (3, 7, 28 and 90 days) of hydration. The endotherms are due to ettringite

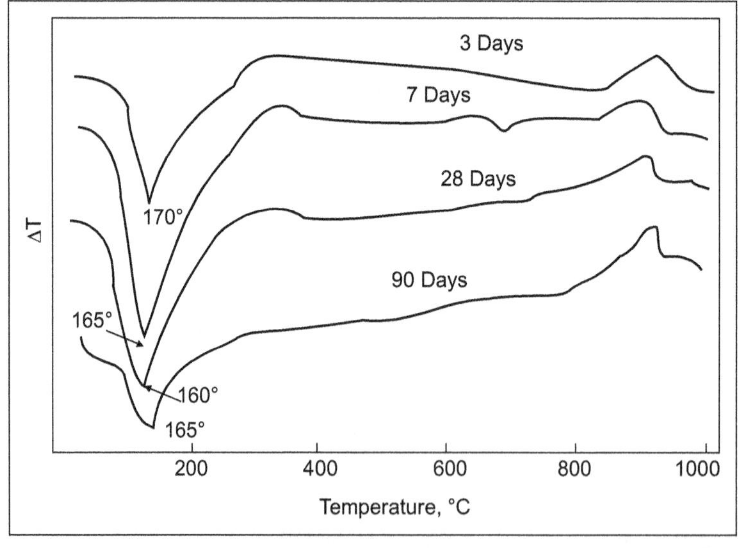

Fig. 5.1 D.T.A. of SSC sample 1 The XRD of the SSC 1 (28 days cured) is plotted in Fig. 5.2.

phase in the SSC. A small inflection at 260–270°C may be ascribed to the decomposition of hydroxyl bound to aluminium. According to Kalousek, Taneja and Singh *et al.*[10–13], the high sulphoaluminate form appears at 160°C but Turriziani and Schippa[14] in their studies on quaternary solids in the system $CaO–Al_2O_3–CaSO_4–H_2O$ found larger endothermic peaks of high sulphate sulphoaluminate at 160–170°C.

However, the intensity of ettringite endotherm increased to maximum level at 28 days and then it is declined. D.T.A. study does not show direct formation of CSH. However, an unexplained endotherm at 700°C at 7 days of hydration may be considered for the CSH peak or may be probably it has gone into solid solution with the ettringite phase.

XRD data reported in Fig. 5.2 is well in agreement with the literature reported in Ref.[15] which indicates that the composition of the cementitious phase is similar to ettringite of composition $3CaO.Al_2O_3.3CaSO_4.31H_2O$. At all other ages the ettringite probably contains lower SO_3 and Al_2O_3 ratios. And as per Midgley and Rosa man[16] the phase might have $SO_3–Al_2O_3$ ration more than 2.50. The ettringite phase may also contain structure of $3CaO.Al_2O_3.3CaSO_4$ aq in solid solution with $3CaO.Al_2O_3.3Ca(OH)_2$ aq as reported by Nurse[17] as the cementing compound formed in the hydration of SSC

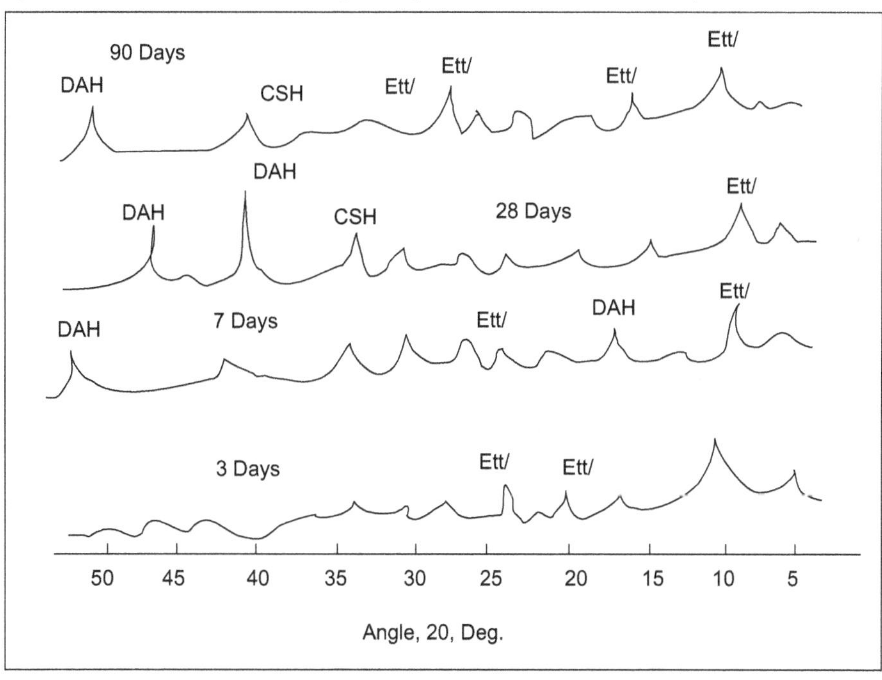

Legends
DAH : Dicalcium aluminate hydrate, Ett. : Ettringite,
CSH : Calcium silicate hydrate

Fig. 5.2 X-Ray Spectrophotometeric analysis SSC Sample 1

produced out of slags of low CaO-SiO_2 ratios. The lines at 5,02, 2.54 and 1.90Å in the 7 days cured SSC sample shows presence of dicalcium aluminate hydrate (DAH).The appearance of CSH of Stratling's compond or Gehlnite hydrate is also indicated by line at 4.20, 2.65, 2.52 and 2.42 Å in the 7 days cured SSC sample.

On the basis of DTA and XRD studies the hardening mechanism of SSC is not fully clear but as per results it can be concluded that ettringite is responsible for strength development in SSC up to 28 days intermixed with formation of CSH type compounds and other phases like gehlnite or gibbsite which are present in amorphous or poorly crystalline state.

The hydration studies are further supplemented by the Scanning Electron Microscope (SEM, LEO 438 VP) of the SSC Sample 1 (Fig. 5.3) cured at 7 and 28 days. It shows anhedral to subhedral fibrous and prismatic crystals of variable sizes and stacking indicating formation of ettringite and platy agglomerates of CSH bodies. These findings corroborate the results of DTA and XRD and results reported by Midgley and Pettifer[18].

5.6.6 Concrete Making Properties

Concrete with SSC is designed, prepared, compacted and cured in the same way as with Portland cement. It has been observed that for the same water/ cement ratio super-sulphated cement gives more workable mix than the Portland cement. Results of compressive strength of concrete mixes are reported in Table 5.15.

5.6.7 Effect of Temperature on Hydration of SSC

Apprehensions have been expressed about the possible deterioration in the strength of super-sulphated cement mortar and concrete on curing under tropical conditions. Data showing effect of temperature on compressive strength and

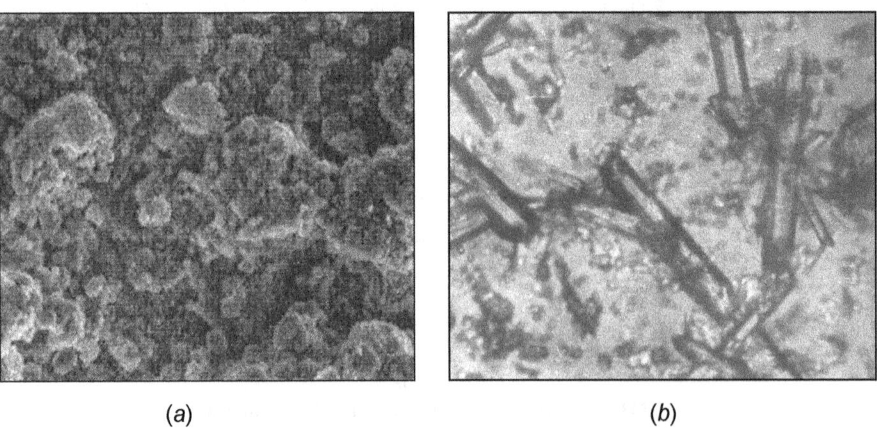

(*a*) (*b*)

Fig. 5.3 SEM of hydrated SSC Sample 1 at 7 days (a) and 28 days (b) *(Scale bar (a) 20 (b)10 μm)*

Table 5.15 Compressive strength of concrete mixes

Period of curing, Days	Compressive strength, MPa		
	Cement 1	Cement 2	Cement 3
3	12.1	13.6	12.2
7	26.1	17.8	27.1
28	35.2	21.4	32.1
90	50.4	34.6	50.1

Table 5.16 Effect of temperature on compressive Strength and Setting Time of Super-sulphated Cement and Mortar (1:3) (Cement 1)

Curing temp. °C	Compressive strength, MPa			Setting time, Minutes	
	3 days	7 days	28 days	Initial	Final
27	12.1	14.1	36.1	175	320
35	12.5	19.4	37.0	160	315
45	15.2	22.2	38.2	152	250
50	16.4	26.1	37.4	146	230

setting of super-sulphated cement-sand mortar is given in Table 5.16. It can be seen that compressive strength increased with the increase in temperature of curing up to 50°C.

It can be seen with continued rise in strength the stability of tricalcium sulpho-aluminate hydrate was not affected by curing at temperature up to 50°C. Later age strength in SSC is attributed to formation of calcium silicate hydrate (C–S–H) as shown by XRD and SEM studies. The data showed that the setting time was not appreciably affected by rise in temperature of curing. Hence, this type of cement can be safely recommended for use in summer months also.

5.6.8 Heat of Hydration

The heat of hydration is the quantity of heat in calories per gram of unhydrated cement, evolved upon complete hydration at a given temperature. It is largely influenced by the mineralogical composition of cement. The fineness of the cement also influences the rate of development as an increase in fineness increases hydration of cement. The total amount of heat liberated, is however, not affected by the fineness of cement.

The results of heat of hydration determined for two SSC is tabulated in Table 5.17. IS : 6909–1990 and BS : 4248–1968 for supersuphated cement specify maximum limits of 60 and 70 cals/g, heat of hydration at 7 and 28 days

Table 5.17 Heat of Hydration of SSC

Super sulphated cement	Heat of dissolution (cals/g)				Heat of hydration (cals/g)		
	Unhydraed	Hydrated			3d	7d	28d
		3d	7d	28d			
Cement 1	510	484.5	470.0	458.3	25.5	40.0	51.7
Cement 2	515	491.0	478.0	460.0	24.0	37.0	55.0
IS : 6909–1990 Limits	—	—	—	—	—	Max. 60	Max. 70

respectively. The test cements complied with these specifications. Due to its low heat of hydration, SSC is eminently suitable for mass concrete construction. The SSC was evaluated for sulphate resistance test as per IS : 12330–1988, Specification for sulphate resistant Portland cement and found to comply the requirements of the standard. Its relatively higher sulphate-resistance *vis-à-vis* Portland cement makes it more suitable for use in marine structures.

Since cost of transport of mineral gypsum from Rajasthan or Jammu to the different steel plants producing by-product granulated slag is very high and the waste fluorogypsum is now available at many places in the country, the latter can prove to be an economical substitute of natural mineral gypsum for making SSC.

Investment for manufacture of SSC will be much lower than that of normal Portland cement, because besides grinding and blending equipment, no special plant and machinery or kiln is required. In the present day crises of power, its manufacture is an economic advantage compared to Portland cement. Moreover, the major raw materials is from HF industry. Prospects for manufacture of this cement in West Bengal and other states appear to be bright because of the availability of large quantities of slag from Indian Iron and Steel Co., Burnpur and Bokaro Steel Plant, Bokaro and the necessity of the cement for general concrete construction and offshore marine structures. Use of fluorogypsum (an anhydrite), a byproduct of HF acid industry now easily available in the country may be used for making SSC and may add economy to the country.

5.7 Super-sulphated Cement through Activation of β-Hemi-hydrate Plaster ($CaSO_4.½H_2O$)

Generally the slags are blended with 10 to 15 per cent gypsum anhydrite in presence of 2 to 5 per cent of cement. The Indian slags are characterized by low lime and high alumina content and as such are less reactive than the foreign slags. To overcome this problem, researches have been carried out which show that by increasing the anhydrite content to 20–25 per cent, the strength of cement can be increased considerably. However, for the production of anhydrite, high thermal energy (0.5–0.55 × 10^6 Kcal/tonne) is required.

Investigations were therefore, undertaken to activate the slag with calcium sulphate hemihydrates (β.$CaSO_4$.½ H_2O) made at much lower temperature (0.30–0.35 × 10^6 Kcal/tonne). Experimental cements were produced by blending the ground granulated slag with hemihydrate plaster and Portland cement clinker in suitable proportions and their characteristics were compared with the cement made using anhydrite as well as Portland cement. The SSC were also produced by partial replacement of granulated slag with the phosphatic slag. The hydration of these cements was checked by DTA and X-ray diffraction. The heat of hydration and sulphate resistance of these cements was determined and discussed.

5.7.1 Raw Materials

The raw materials such as granulated blast furnace slag, phosphatic slag, phosphogypsum and Portland cement clinker collected from various sources in the country were used to formulate SSC. The chemical composition of these materials is depicted in Table 5.18.

5.7.2 Preparation of Super-sulphated Cement (SSC)

The SSC cements were produced by intimately blending the finely ground granulated blast furnace slag (410 m^2/kg, Blaine) with the calcium sulphate

Table 5.18 Chemical Composition of Raw Materials

Constituents	Per cent			
	GBFS	Phosphatic Slag	Phospho-Gypsum	Portland Cement Clinker
P_2O_5	—	0.56	0.47	—
F	—	—	0.86	—
Organic matter	—	—	0.59	—
SiO_2 + insoluble in HCl	33.83	40.14	0.29	22.50
Al_2O_3 + Fe_2O_3	22.93	5.66	0.54	9.80
CaO	34.93	49.02	31.09	61.70
MgO	7.46	3.94	1.31	2.80
SO_3	0.84	—	43.21	0.06
Na_2O	—	—	0.29	Tr.
LOI	0.20	—	18.38	2.1
Cl	—	—	—	—
Mn_2O_3	0.10	—	—	—

GBFS : Granulated blast furnace slag

Table 5.19 Composition of SSC Cements

Cement Design	Mix Composition (% by mass)			
	GBFS	$CaSO_4 \frac{1}{2} H_2O$	Cement Clinker	Retarder
A	75	5	10	—
B	75	15	10	0.2 (Borax)
C	75	15	10	0.1 (Tartaric acid)
D	75	15 (Anhydrite)	10	—
E	—	—	96	4.0 (Gypsum)

hemihydrates (320 m^2/kg, Blaine) in different proportions (Table 5.19) While the calcium sulphate hemihydrate (β $CaSO_4.1/2\ H_2O$) was produced by heating phosphogypsum at 150°C, the anhydrite [$CaSO_4$(II)] was made at 850°C. The experimental cements were made by partial replacement of granulated slag with phosphatic slag (420 m^2/kg Blaine's). A Portland cement was also produced for comparative study. The SSC cements were tested for their various properties as per methods given in IS : 4031–1988. The strength development in SSC cements was evaluated with DTA (Stanton Red Croft, U.K.) and X-ray diffraction (Philips, Holland). The sulphate resistance of SSC cement was checked as per procedure laid down in IS : 12330–1988[19].

5.7.3 Properties of SSC

The physical properties of optimum mixes of SSC cements produced by blending granulated blast furnace slag (GBFS) duly activated with calcium sulphate hemihydrates and anhydrite in presence of Portland cement clinker are reported in Table 5.20. It can be seen that cements set fast on addition of water. The

Table 5.20 Properties of SSC Cement

Cement Design	Setting Time		Compressive Strength (MPa)			Soundness (Cold Expan.) (mm)
	Initial	Final	3d	7d	28d	
A	10	65	16.1	23.7	35.4	1.0
B	60	120	15.9	22.3	43.1	1.1
C	55	106	16.9	38.4	56.1	1.2
D	62	170	15.8	35.2	53.3	0.9
E	110	205	17.5	28.0	46.0	1.6
IS : 6909 Limits	Min. 30	Max. 600	Min. 15	Min. 22	Min. 30	Max. 5.0

fast setting of SSC cement may be ascribed to the quick setting of calcium sulphate hemhihydrate on account of much higher concentration of SO_4^{3-} ions in aqueous media containing high alumina. To overcome this problem, large number of chemical retarders such as borax, tartaric acid were found effective to give satisfactory setting times desired by the standard. The retarder concentration beyond 0.2 per cent was avoided to cause any detrimental effect on the strength of SSC.

It is interesting to note that compressive strength of SSC cements increased appreciably with the progress in hydration of cements. It can be seen that on retardation, the cements do not show any fall in the strength, rather strength values are enhanced. The SSC cements have been found sound as their cold expansion was much below the maximum specified value of 5.0 mm.

The SSC cements compared fairly well with the traditional super-sulphated cement produced by blending the slag with anhydrite and Portland cement. The results confirm that SSC cement can be produced by activating the slag with $CaSO_4.1/2.H_2O$ instead of $CaSO_4$ (anhydrite)[20].

The properties of SSC cement produced by substituting GBFS with 10 per cent (by mass) phosphatic slag are reported in Table 5.21. Data show that SSC cement complied with the requirements laid down in IS : 6909–1990. No detrimental effect was noticed on the strength development of cement when GBFS was replaced with 10.0% of the phosphatic slag. The findings are quite encouraging and may be applied to other metallurgical slags such as Fe, Mn, Si, Mn, Cu, Ni, etc.

The hydraulic products formed during the hardening of SSC cements have been identified with the help of DTA and XRD. DTA of SSC cements namely 'C' and 'D' are shown in Figs. 5.4 and 5.5 respectively. In case of cement 'C' (Fig. 5.4), the endotherms formed at 100°, 125°, 150°–230°, 640°–830°C and the exotherms at 890°C and 920°C may be attributed to the formation of gel water, ettringite, dehydration of calcium sulphate hemihydrate, C_4AH_{13}, formation of tobermorite and devitrification of slag respectively. The intensity of ettringite endotherms increased with depletion of calcium sulphate hemihydrates peaks. Similarly tobermorite endortherms are enhanced with the increase in curing period. The appearance of endotherms at 490°C at 3 days may be due to $Ca(OH)_2$.

Table 5.21 Properties of SSC Cement Produced by Partial Replacement of GBFS with Phosphatic Slag

Properties					
Setting Time (Minutes)		Compressive Strength (MPa)			Soundness (Cold Expan.) (mm)
Initial	Final	3d	7d	28d	—
70	142	15.7	23.0	44.7	0.8

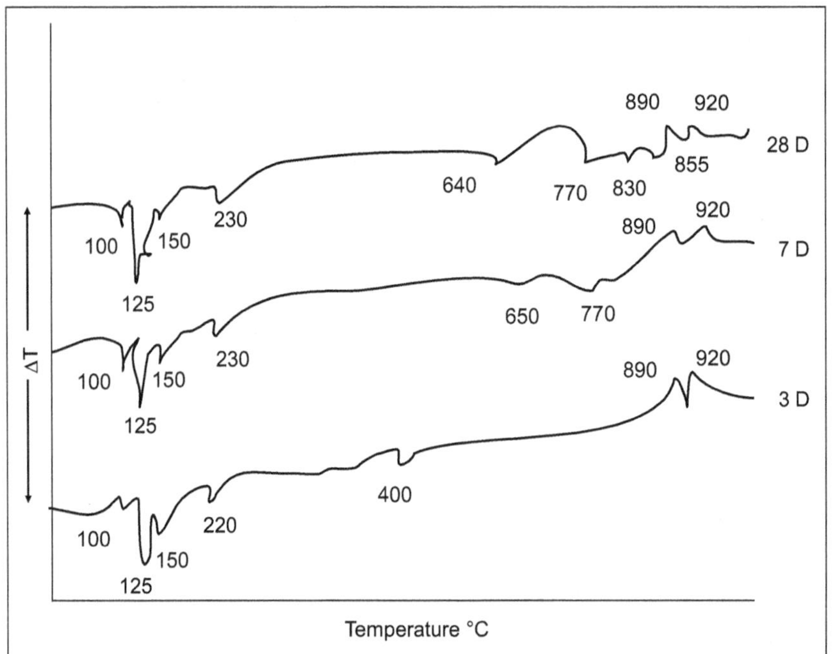

Fig. 5.4 Differential Thermograms of SSC (Mix C) Hydrated for Different Periods

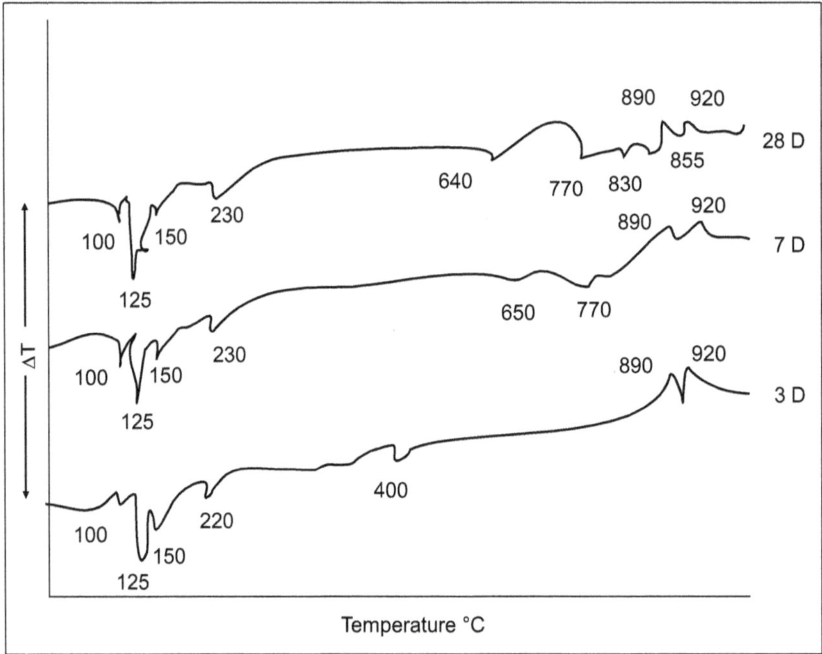

Fig. 5.5 Differential Thermograms of SSC (Mix D) Hydrated for Different Periods

DTA of cement 'D' (Fig. 5.5) shows development of endotherms at 150°, 180°, 230°, 500°, 700–800°C and exotherms at 845° – 920°C which may be assigned to the formation of ettringite, dehydration of calcium sulphte hemihydrates, C_4AH_{13}, formation of $Ca(OH)_2$, appearance of tobermorite and devitrification of slag respectively. An increase in intensity of ettringite and tobermorite endotherms with the enhancement in hydration was significant. No $Ca(OH)_2$ peak was detected at 28 days.

Thus, data confirms that formation of ettringite, tobermorite and C_4AH_{13} are essentially responsible for the strength of SSC cements.

The XRD patterns of SSC cements 'C' and 'D' are plotted in Figs. 5.6 and 5.7 respectively. It can be seen that major peaks are ettringite, tobermorite and C_4AH_{13}. The intensity of etrringite peaks increased with the increase in curing period. However, the intensity of gypsum and $Ca(OH)_2$ reflections are reduced. No peak for $Ca(OH)_2$ was recorded. In essence, the XRD data corroborate the findings of DTA.

The heat of hydration of SSC cements is reported in Table 5.22. It can be seen that the heat of hydration of cements is within the requirements of IS : 6909–1990 and hence can be recommended for use in mass concrete.

The sulphate resistant test conducted for SSC cements C and D showed the sulphate expansion values in the range 0.01 – 0.03 per cent against the maximum specified value of 0.045%. Hence, the cement is sulphate resistant.

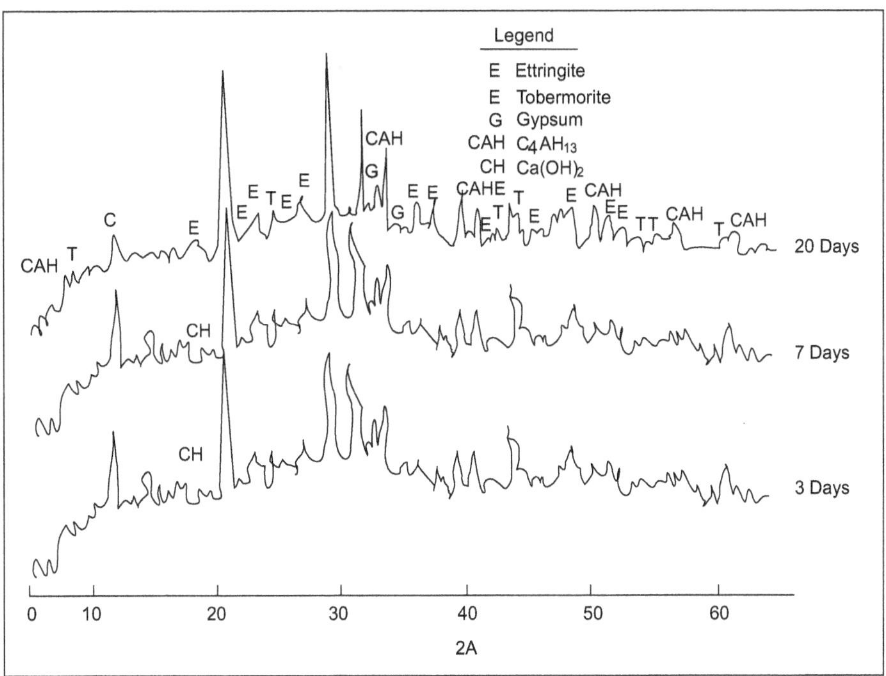

Fig. 5.6 XRD Patterns of Hydrated Cement (Mix C)

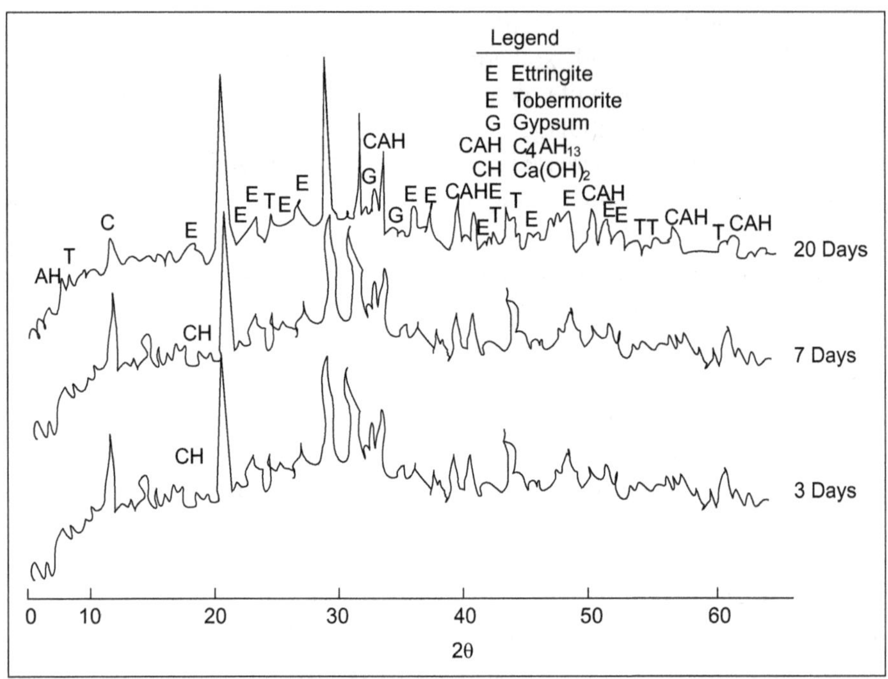

Fig. 5.7 XRD Patterns of Hydrated Cement (Mix D)

Table 5.22 Heat of Hydration of SSC Cements.

Cement Design	Heat of Dissolution (Cal/g)				Heat of Hydration (Cal/g)		
	Unhydrated	Hydrated			3d	7d	28d
		3d	7d	28 d			
C	505.6	529.8	539.2	555.8	24.2	33.6	50.5
D	522.2	549.3	559.2	576.8	27.1	37.0	54.6
IS : 6909–1990 Limits					—	Max. 60	Max. 70

The SSC can be produced by the activation of slag with calcium sulphate hemihydrates formed at 150°C against the traditional use of anhydrite *i.e.,* hard burnt gypsum produced at 850°C. Thus, net saving of energy can be achieved. The major phases responsible for strength development in SSC cement have been identified as etrringite, tobermorite and C_4AH_{13}. The rapid setting of these cements can be overcome by the small addition of borax or tartaric acid retarders. The use of calcium sulphate hemihydrate is recommended for making SSC cement.

References

1. Scott, W.W. and Furman, H.W., 'Standard Methods of Chemical Analysis', 1952, Vol. 2, pp. 2436–2437, London.
2. Kanneurrf, A.S. 'Research Pushes Grind ability Guesses into the Background', Rock Products, 1957, Vol. 60, No. 5, pp. 86–91.
3. Keil, F. and Locher, F.W., Zement-Kalk-Gips, 1958, Vol. 11, p. 245.
4. Tanaka, T., Sakai, T. and Yamane, J. Zement-Kalk-Gips, 1958, Vol. 11, p. 50.
5. Chopra, S.K. and Lal K., J. Scient. Ind. Res., 1961, 20D, p. 218.
6. Chopra, S.K. and Lal K., Indian Conc. Joun, 1961, Vol. 38, p. 114.
7. Building Research Station (U.K.), Digest, 1960, Vol. 130, p. 2.
8. Lea F.M., 'The Chemistry of Cement and Concrete', (Edward Arnold Ltd., London), p. 483, 1970.
9. Nurse R.W., 'The Chemistry of Cements', Vol. 2, H.F.W. Taylor (Academic Press), 1964, p. 53, London.
10. Kalousek G.L. and Adams M., J. Am. Concr. Inst., 1951, Vol. 48, p. 80.
11. Taneja C.A. Singh Manjit, Tehri, S.P. & Raj Tilak, Super-sulphated cement from Waste Phosphogypsum, 12th International Conference on Silicate Industry & Silicate Science, Budapest (Hungary), June 1977, pp. 621–627.
12. Singh Manjit, Chapter on 'Production of Super-sulphated Cement', Ph.D. Thesis, entitled 'Physico-Chemical Studies on Phosphogypsum for Use in Building Materials, 1979, pp. 165–187, University of Roorkee, Roorkee, India.
13. Singh Manjit, Super sulphated Cement, J. Inst.Eng. (H), 1980, Vol. 61, pp. 55–59.
14. Thuriziani R and Chippa GS, (1954) Re. Sci. 1954; 24, p. 2356–63.
15. Midgley HG (1957) Mg. Concr. Res., 9:24.
16. Midgley HG & Rosaman D (1960) The composition of ettringite in set Portland cement, Nat. Bur. Stnd. (U.S.), Monograph, Washington, 2:43.
17. Nurse RW (1960) Slag composition & its effect on the properties of super-sulphated cement, paper presented at the Convention on Production and Application of Slag Cement, Naples.
18. Midgley HG & Pettifer K (1971) The microstructure of hydrated super sulphated cement,Cem.Concr. Res., 1:101–104.
19. IS : 12330–1988, Specification for Sulphate Resistant Portland Cement, 1988, Bureau of Indian Standard, p. 3.
20. Singh Manjit and Garg Mridul, Investigation into Activation of Slags by Calcium Sulphate Hemihydrate, 5th International Conf. On Concrete Technology for Developing Countries, New Delhi, 17–19 No. 1999, Vol. 1, pp. 1–45–1–53.

6

Gypsum Building Products

INTRODUCTION

Gypsum has a wide variety of uses, but the most important by far in the manufacture of gypsum building plaster and plaster board and as a retarder in Portland and the blended cements which account for the greater chunk of the gypsum produced. For use in plaster industry, the gypsum is calcined to produce hemihydrate or anhydrite plaster forms. Ground gypsum is used as a soil conditioner, particularly on grassland to provide good calcium and sulphur deficiencies in soil, to mitigate soil alkalinity and to flocculate soil particles thereby improving drainage. Gypsum has important application in dressing land which has been flooded by the sea, by displacing sodium ions from clay minerals and thus, allowing these ions to be removed by percolation of ground water. Gypsum is also used as the filler in many industries and several relevant standards have been laid for their proper applications. It is an agent for 'burtonising' or artificially hardening water used for brewing beer. The variety of gypsum and their availability and the detailed characteristics have been explained in the foregoing chapters.

The importance of industrial application of gypsum is due to its conversion into variety of calcined products, *i.e.*, hemihydrate plasters (alpha and beta) and different type of anhydrite plasters ranging from soluble to insoluble anhydrite or dead burnt gypsum. All type of these plasters when mixed with water, set and hardened to the dihydrate gypsum of variable properties and form basis of thc gypsum products. Off course different type of admixtures like retarders, activators, polymers, plasticizers, super-plascitizers, flocculating agents, fibres, etc. are added to the gypsum plaster to formulate *in situ* or preformed building products. Extensive R&D efforts have been made across the world to synthesize variety of plasters, binders, boards, blocks and many other useful building products. Based on these researches valuable cost effective gypsum building products covering vast technological aspects have been launched in the market to fulfill needs of the people for ecstasy, aesthetic, comforts and safety. Some of the valuable building materials based on gypsum (natural/by-products) are described and discussed in this useful chapter.

6.1 Water-Resistant Gypsum Binder

The use of gypsum products is not permitted in the external situations due to solubility of gypsum (2 g/litre) in the exposed damp situations. Hitherto, gypsum products are safe to use in those exposed situations where annual rain fall is less than 10 inches. If the water-resistant quality is improved in the gypsum matrix, the use of natural as well as waste gypsum may be enhanced considerably. Extensive R&D efforts have been made at CBRI to produce water-resistant gypsum binder[1–2]. Both natural/mineral gypsum can be made water-resistant for their wide applications in building sector.

6.1.1 Production and Properties of Gypsum Binder

Phosphogypsum, beneficiated in the pilot plant, was calcined at 150–160°C for a period of 4–5 hours to form b-hemihydrate plaster. It was ground in a ball mill to a fineness of 300–320 m^2/kg (Blaine's). The plaster was then blended with the dried fly ash/granulated blast furnace slag, ordinary Portland cement (43 grade) and tartaric acid retarder in different proportions in a powder mixer for a period of one hour followed by grinding the whole mix in a ball mill for a period of half-an-hour to form uniform water-resistant binder[3–4]. The chemical composition and physical properties of the fly ash and the OPC are shown in Tables 6.1 and 6.2 respectively. The final fineness of the binder was kept at 300–330 m^2/kg (Blaine's). Several mixes of the binder (Table 6.3) were prepared to arrive at an optimum proportion. The binder was tested for various chemical and physical properties as per IS : 4031–1968 and IS : 4032–1968 respectively. The soundness of the binder was checked by cold expansion test as specified in IS : 6909–1990. The properties of the binders are given in Tables 6.4 and 6.5.

It can be seen from the Table 6.4 that binders prepared from the unprocessed phosphogypsum showed less retardation of the setting time on the addition of

Table 6.1 Chemical Composition of Fly Ash, Granulated Blast Furnace Slag and Ordinary Portland Cement

Constituents	Fly ash (%)	Ordinary Portland Cement (%)	Granulated Blast Furnace Slag (%)
SiO_2 + insoluble in HCl	62.80	22.41	33.83
$Al_2O_3 + Fe_2O_3$	28.40	8.60	22.93
CaO	1.52	61.50	34.93
MgO	0.78	4.21	7.46
SO_3	0.26	0.42	0.84
$Na_2O + K_2O$	0.80	0.80	—
LOI	1.50	1.41	—
Mn_2O_3	—	—	0.099

Table 6.2 Physical Properties of Fly Ash and Ordinary Portland Cement

Sl. No.	Property	Results	
		Fly ash	OPC
1	Fineness, specific surface Area (m^2/kg) (Blaine's)	300 (Min. 250)	330 (Min. 225)
2	Setting time (Minutes)		
	Initial	n. s.	80 (Min. 30)
	Final	—	280 (Max. 600)
3	Lime Reactivity, N/mm^2	4.5 (Min. 4.5)	n. s.
4	Compressive strength (N/mm^2)	n.s.	
	3 day	—	44.4 (Min. 23)
	7 day	—	52.4 (Min. 33)
	28 day	—	61.6 (Min. 43)
5	Soundness		
(i)	Lechatlier's Expansion, mm	n.s.	1.0 (Max. 10)
(ii)	Autoclave Expansion	0.15 (Max. 0.8)	0.25 (Max. 0.8)

Table 6.3 Proportioning of the Water-Resistant Binder from the Unprocessed and the Beneficiated Phosphogypsum Plaster and other Ingredients

Binder Designation	Mix Proportion (% by wt.)			
(A-A5=Unprocessed plaster & B-B5=Beneficiated plaster)	Gypsum	Fly ash	OPC	Retarder
A	75	15	10	—
A1	75	15	10	0.1
A2	70	15	15	—
A3	70	15	15	0.1
A4	70	20	10	—
A5	70	20	10	0.1
B	75	15	10	—
B1	75	15	10	0.1
B2	70	15	15	—
B3	70	15	15	0.1
B4	70	20	10	—
B5	70	20	10	0.1

Table 6.4 Properties of Gypsum Binders Prepared from Beneficiated Phosphogypsum Plaster (Made in Pilot Plant), Fly Ash and Ordinary Portland Cement.

Binder Designation	Consistency	Setting time (Minutes)		Compressive strength (MPa)			Soundness
	(%)	Initial	Final	3-day	7-day	28-day	Cold Expansion mm
A	42.0	3.0	20.0	9.0	15.4	19.2	1.2
A1	42.4	20.0	90.0	7.2	13.5	18.6	1.2
A2	43.1	2.5	22.0	8.2	14.1	18.2	1.1
A3	43.5	13.5	100.0	7.0	12.2	17.1	1.2
A4	43.1	2.0	21.0	6.8	13.9	17.5	1.3
A5	43.6	15.0	110.0	6.0	13.2	16.1	1.0

Table 6.5 Properties of Gypsum Binders Prepared from Beneficiated Phosphogypsum Plaster (Made in Pilot Plant), Fly Ash and Ordinary Portland Cement.

Binder Designation	Consistency	Setting time (Minutes)		Compressive strength (MPa)			Soundness (Cold Expansion mm)
	(%)	Initial	Final	3-day	7-day	28-day	
B	40.0	5.0	6.0	10.8	17.9	22.6	1.0
B1	40.2	26.0	135.0	9.4	15.4	19.6	1.2
B2	42.1	3.5	25.0	9.4	17.5	19.4	0.8
B3	43.2	20.0	120.0	8.2	16.4	18.2	1.1
B4	42.6	3.0	25.0	8.5	16.1	17.4	1.1
B5	43.4	18.0	140.0	7.5	15.8	16.9	1.2

tartaric acid retarder and the strength development was also low. On the other hand, the binders prepared from the beneficiated phosphogypsum plaster (Table 6.5) showed sufficient retardation of the setting time as well much improvement in the strength than the binder based on unprocessed phosphogypsum plaster. The improvement in the strength of the plaster can be attributed to the removal of impurities from the beneficiated phosphogypsum as well as decrease in consistency for proper workability.

Similar mixers were attempted with granulated slag also in place of fly ash to know the water-resistance in the gypsum matrix. The physical and chemical properties of slag based gypsum binders are listed in Tables 6.6 and 6.7 respectively.

Data show that gypsum binder based on granulated slag (Table 6.6) developed much high strength, longer setting time, and higher water resistance than the gypsum binder based on fly ash (Tables 6.4–6.5).

Table 6.6 Properties of Gypsum Binders Prepared from Beneficiated Phosphogypsum Plaster (Made in Pilot Plant), Blast Furnace Slag and Ordinary Portland Cement.

Properties	Values
Fineness, m^2/g Setting time, min	310–320
Intial	70–75
Final	145–160
Bulk density, g/cm^3	1.2–1.28
Compressive strength, MPa (28 days)	30–35
Soundness, mm	1.0–1.6
Water absorption, %	5.0–6.0
pH	11.0–11.5

Table 6.7 Chemical Composition of Gypsum Binder

Constituents	Percent by Wt.
SiO_2 + insoluble in HCl	8.20
$Al_2O_3 + Fe_2O_3$	9.00
CaO	37.30
MgO	1.80
SO_3	39.65
P_2O_5	0.15
F	0.058
Organic matter	0.090
$Na_2O + K_2O$	0.089
LOI	4.10

6.1.2 Hydration Mechanism of Water-Resistant Gypsum Binder

The strength development of gypsum binders (B/B1) with the hydration/curing period was investigated by DTA (Fig. 6.1). Data show appearance of double dehydration of gypsum at 140–170°C and 190–210°C due to inversion of $CaSO_4 \rightarrow CaSO_4.1/2\,H_2O$ and $CaSO_4.1/2\,H_2O \rightarrow CaSO_4$ (*b*) respectively. An endotherm at 750–780°C shows formation of CSH phase. The intensity of gypsum dehydration peaks increased with the increase in curing period confirming conversion of the gypsum plaster into the gypsum dihydrate. Whereas increased intensity of CSH peaks clearly manifest the enhancement in the compressive strength of the gypsum binder.

In case of gypsum binder containing granulated blast furnace slag, DTA shows dehydration endotherms obtained at 140–150°C and at 190–210°C are increased with curing period, confirming conversion of hemihydrate into gypsum. A small endotherm at 145°C and 760–775°C are due to decomposition of ettringite ($C_3A.3CaSO_4.32H_2O$) and CSH. Therefore, gypsum, ettringite and CSH are the main strength gaining products formed in gypsum binder hydration[5]. XRD shows increase in intensity of reflections with curing at 7.68880°, 4.2872°, 3.4055° and 1.4336° due to tobermorite phase including gypsum as predominant phase. After 3 and 7 days of hydration, the binder shows ettringite at 5.6775°, 3.8635°, 3.354°, 2.7859°, 2.2046°, 2.1204°, 2.0786°, 1.8937° and 1.8468° in addition to gypsum and tbermorite. At 28 days of curing, binder indicates appearance of C_4AH_{13} (1.8503°, 1.7477° and 1.6488°) with gypsum as the major phase[6].

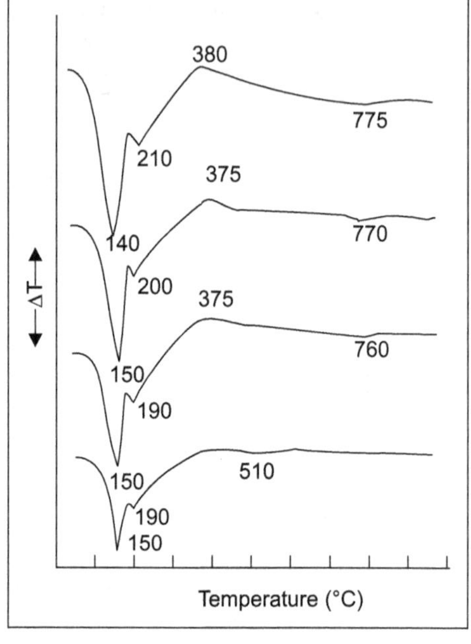

Fig. 6.1 Differential Thermograms of Gypsum Binder Hydrated at 27°C for Different Period

The Scanning electron microscopy (SEM) of the gypsum binder/cementitious binders are reported in Fig. 6.2. The SEM shows formation of radiating needles

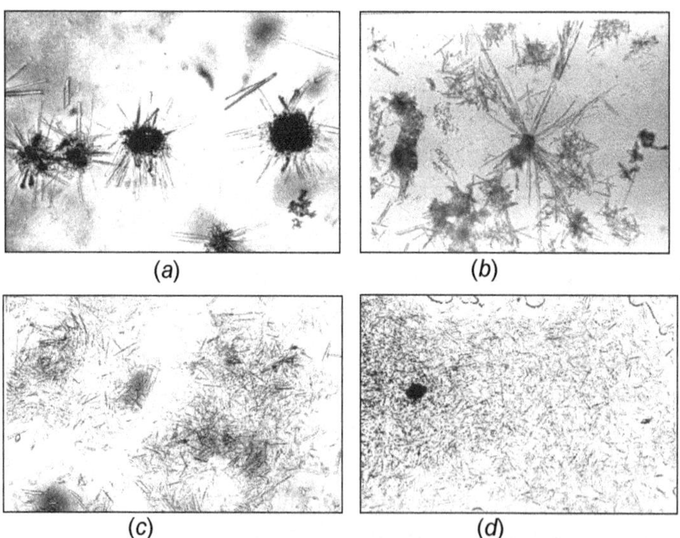

Fig. 6.2 SEM of Gypsum Binder (Slag Based) at 1 (*a*), 3 (*b*), 7 (*c*) and 28 days (*d*) of Hydration (x45)

of ettringite and hydrated plates of C_3S, C_3S and C_4AF at 1 & 3 days of hydration (Fig.6.2 (*a* and *b*)and 28 days cured binder (*c* and *d*) develops euhedral clusters of needles, lath and prismatic shaped crystals of gypsum,ettringite and CSH [7].

6.2 Durability of Gypsum Binder

The durability of the water-resistant gypsum binder was evaluated on the basis of its performance on exposure to (*i*) natural weathering (outside exposure) and (*ii*) high temperature and high humidity and by (*iii*) measuring its water absorption and its tendency to efflorescence. The results obtained are described below:

6.2.1 Effect of Outside Exposure on the Gypsum Binder

2.5 cm × 2.5 cm × 2.5 cm cubes were cast at normal consistency followed by exposure to outside atmosphere for a period of 1 day to 28 days in the months of January to March 2003. The atmospheric temperature varied between 15° – 27°C (maximum) and 3° – 12°C (minimum) the humidity variation was 70–75%. The cubes after hardening were tested for compressive strength. The results are reported in Table 6.8. It can be seen that with increase in exposure period the strength of binder fell below the strength attained on hardening the gypsum binder at 27°C for the same period. It can be recorded here that the strength decreased, yet there was progressive increase in the strength development of the binder cubes with the increase in curing period. No variation in the weight of the binder cubes was observed during the exposure test.

The progressive increase in the compressive strength was studied by the DTA of the hardened binder cubes. The results of the binder hardened for 1, 3, 7 and 28 days showed that the double dehydration peak of gypsum was smaller in size from 1 to 3 days of curing, indicating, thereby, low conversion of hemihydrate into the gypsum dihydrate. However, with the increase in hydration of the binder from 7 days to 28 days, the size of dehydration endotherms increased, confirming progressive increase in the conversion of hemihydrate into gypsum matrix. The small endotherm at 760–775°C was due to calcium silicate hydrate (CSH) compound. The increase in the intensity of double dehydration endotherms was attributed to the increase in the strength of the binders from 1 day to 28 days, respectively.

Table 6.8 Compressive Strength of Gypsum Binder in Exposed Atmosphere

Curing Period (Days)	Compressive Strength (MPa)
1	6.53
3	8.70
7	9.40
28	16.52

6.2.2 Performance of the Gypsum Binder in High Atmospheric Temperature and High Humidity

In order to find out the effect of high atmospheric temperature and high humidity on the performance of the binders, 2.5 cm × 2.5 cm × 2.5 cm cubes were cast out of the binder at normal consistency. The cubes were removed from the moulds and stored at the temperatures 40°C, 50°C and 60°C in the desiccators over the water. The compressive strength of the binders was determined for a period of 1 day to 28 days. The results are reported in Table 6.9.

The data show that with increase in curing temperature from 40°C to 60°C, the compressive strength was increased from 1 day to 28 days. At 50°C to 60°C, the gain in strength was found to be higher than the strength obtained at 40°C at 1, 3 and 7 days. However, at 28 days, the increase in strength was, not significant. The increase in the strength at 50°C to 60°C may be ascribed to the pozzolanic action of fly ash added to the plaster[8]. In fact, the amorphous SiO_2, Al_2O_3 and Fe_2O_3 constituents of the fly ash react with the $Ca(OH)_2$ released from the cement and forms tobermorite (CSH) in the presence of moisture. The CSH helps in increasing the strength of the binder[9].

6.2.3 Water Absorption of Gypsum Binder

To find out the water absorption, 2.5 cm × 2.5 cm × 2.5 cm cubes of gypsum binder, plain gypsum plaster and ordinary Portland cement were cast at normal consistency. While the gypsum binder and cement cubes were cured for different periods from 1 day to 28 days at 27 ± 2°C in more than 90 per cent relative humidity, the plain plaster cubes were cured at room temperature for 24 hours only. The cubes after respective curing were dried at 42°C to constant weight, cooled and then immersed in water for 2, 8 and 24 hours to find out the level of water absorption or leaching in the water. The results are reported in Table 6.10.

It can be seen form the Table 6.10 that with the increase in the hardening period from 1 day to 28 days, the water absorption of the binder and cement cubes was reduced. However, the water absorption increased with the length of immersion in water

Table 6.9 Effect of Rising Temperature on the Compressive Strength of the Gypsum Binder

Curing Period (Days)	Compressive Strength (MPa)		
	40°C	50°C	60°C
1	8.2	9.6	10.9
3	10.5	14.2	15.7
7	12.2	17.5	19.5
28	20.6	21.8	22.4

Table 6.10 Performance of Gypsum Binder in Water

Curing Period (Days)	Immersion Period (Hours)	Water Absorption (%)		
		Gypsum Binder	Ordinary Portland Cement	Phospho Plaster
1	2.0	13.92	19.66	26.92
	8.0	15.60	20.50	30.62
	24.0	15.91	20.50	32.10
3	2.0	10.60	17.57	33.70
	8.0	11.30	18.41	—
	24.0	11.37	18.82	—
7	2.0	9.90	16.59	Leaching
	8.0	10.80	16.59	—
	24.0	10.77	16.60	—
28	2.0	9.26	14.60	Leaching
	8.0	9.71	15.60	—
	24.0	10.08	16.00	—

(2 hours to 24 hours) for all the hardened gypsum binder as well as cement cubes. These findings clearly indicated that there was no leaching of the binder. Whereas the plain phosphogypsum plaster showed an increase in the water absorption up to 3 days of immersion and thereafter, leaching of plaster took place[10].

6.2.4 Efflorescence Studies on Gypsum Binder

To measure the level of efflorescence in the gypsum binder, the binder was mixed with 0.1% iron oxide pigment to impart red colour in order to distinguish white patches of efflorescence caused by the addition of various types of materials to the gypsum plaster. Test specimens in the form of discs of 10 cm diameter and 0.5 cm thickness, were cast at normal consistency using distilled water. After curing in 90% humidity for a period of 28 days, the discs were dried at 42°C to constant weight, and these were then kept in a tray filled with the distilled water up to 2.5 cm height. The water was allowed to rise in the specimens followed by drying at ambient temperature. The test cycles were repeated. The specimens were examined for efflorescence after a period of 1, 3, 7 and 28 days. No efflorescence was observed up to a period of 3 days. However, after a period of 7 and 28 days, slight efflorescence was observed on the edges of the specimens.

6.3 Suitability of Gypsum Binder for Masonry Mortars

The gypsum binder was examined for its suitability for mortar making properties[11]. The properties such as compressive strength, bond strength and water retention were studied.

6.3.1 Compressive Strength

To determine compressive strength of the mortars, 5 cm × 5 cm × 5 cm cubes were cast at 110 ± 5% flow using binder and sands (FM 2.0 and FM 1.25) in the mix proportions 1:2, 1:3, 1:4 and 1:5 by weight, respectively. The Portland cement: sand mortars cubes of the mix proportions 1:4, 1:5, 1:6 by weight were also prepared and cured in water. The results are reported in Table 6.11. It can be seen that mortar prepared with 1 part of binder and 2 or 3 parts of sand gave maximum compressive strengths. The strengths of these mortars was higher than the 1:4, 1:5 and 1:6 cement-sand mortars. It can be further seen that with the increase in sand content, the strength of mortar cubes prepared, was decreased. However, the mortar prepared with 1 part of binder and 3 or 4 parts of sand by weight developed strength which is higher than the 1:5 or 1:6 cement-sand mortars.

6.3.2 Bond Strength

Bond strength was measured by the force necessary to separate the masonry units bonded with the mortar. Factors affecting the development of bond between the mortar and the brick are water absorption of the brick, consistency, workability and water retaining capacity of the mortar against suction. The bond strength was determined using cross-brick couplets as per the method suggested by Pearson[12]. The results are tabulated in Table 6.12. These results indicate that the binder mortars, develop better bond strength than the conventional cement-sand (1:6) mortar.

6.3.3 Water Retaining Capacity

The water retaining capacity against suction of the mortars prepared with the binder was evaluated using sands of fineness modulus 2.0 and 1.25 as per IS : 4031–1968. The results are given in Table 6.12 The mortar prepared from the gypsum binder showed high water retaining capacity as compared to 1:6 cement-sand mortar. Higher water retentivity of the mortar indicates its high workability when used with the porous materials possessing high suction property.

6.3.4 Predesign, Cost Estimates for the Production of Gypsum Binder

Production	Water Resistant Gypsum Binder
Capacity	50 tonnes/day (Three 8-hour shifts) 15000 tonnes/year of 300 working days Capital Investment

Table 6.11 Compressive Strength of Mortars

Sl. No	Mix Proportion (By Weight)		Compressive Strength (MPa)			
	Gypsum Binder	Sand	1-day	3-day	7-day	28-day
1	1	2	8.61	10.82	11.60	12.84
	(F.M. 2.0)					
	1	3	6.21	6.26	6.63	6.86
	(F.M. 2.0)					
	1	4	4.61	3.92	4.12	47.20
	(F.M. 2.0)					
	1	5	2.21	2.32	2.61	3.24
	(F.M. 2.0)					
2	1	2	6.91	8.21	9.91	10.11
	(F.M. 1.25)					
	1	3	4.61	4.41	4.72	6.60
	(F.M. 1.25)					
	1	4	2.50	2.44	2.51	3.20
	(F.M. 1.25)					
	1	5	2.16	1.84	1.99	2.38
	(F.M. 1.25)					
	Cement	Sand				
3	1	4	1.54	1.95	4.18	7.60
	(F.M. 1.91)					
	1	4	1.22	1.60	2.50	4.65
	(F.M. 1.25)					
4	1	5	1.05	1.42	2.70	4.20
	(F.M. 1.91)					
	1	5	0.99	1.18	1.82	3.60
	(F.M. 1.25)					
5	1	6	0.85	1.02	2.50	3.82
	(F.M. 1.91)					
	1	6	0.75	0.96	1.45	1.96
	(F.M. 1.25)					

Table 6.12 Bond Strength and Water Retention of Mortars

Mix Proportion (By Weight)		Bond Strength N/mm^2		Water Retention (%)
Binder	Sand	7-day	28-day	
1	4	0.149	0.240	65.20
(F.M. 1.91)				
1	4	0.130	0.182	62.50
(F.M. 1.25)				
1	5	0.121	0.169	62.00
(F.M. 1.91)				
1	5	0.120	0.156	60.50
(F.M. 1.25)				
1	6	0.090	0.138	20.00
(F.M. 1.91)				
1	6	0.085	0.128	10.50
(F.M. 1.25)				

Capital Investment

	₹
A. Fixed Capital on Building	
Land 400 sq.m @ ₹500/m^2	2,00,000.00
Building 50 sq.m @ ₹4000/m^2	2,00,000.00
Shed 300 sq.m @ ₹4000/m^2	12,00,000.00
Yard improvement (lump sum)	50,000.00
	16,50,000.00

	₹
B. Fixed Capital on Plant	
(*a*) Purchased Equipment (PE) Cost	
Gypsum Calcinator 8 units @ ₹2,00,000/ unit	16,00,000.00
Roller/Ball mill 1 No.	1,00,000.00
Blender 1 No.	50,000.00
	17,50,000.00
(*b*) Equipment erection @ 10% of PE	1,75,000.00
(*c*) Electrical Installation @ 15% of PE	2,62,500.00
(*d*) Instruments & Controls @ 5% of PE	87,500.00
(*e*) Water services & Drainage @ 5% of PE	87,500.00
(*f*) Laboratory & Workshop @ 10% of PE	1,75,000.00
(*g*) Engg. and Supervision @ 10% of PE	1,75,000.00
(*h*) Premium etc. (L.M.)	50,000.00
(*i*) Contingencies (L.S.)	75,000.00
Fixed Capital on Plant	28,37,500.00
Total Fixed Capital (A+B)	**44,87,500.00**
C. Working Capital @ 20% on the Fixed Capital	**8,97,500.00**
Total Capital Investment (A+B+C)	**53,85,000.00**

Cost of Production

			₹
1.	**Raw Material**		
	(*a*)	Processed Phosphogypsum 13,392/tonnes @ ₹810 per tonne	1,08,47520.00
	(*b*)	Fly ash 2,250 tonnes @ ₹50/tonne inclusive of transportation charges	1,12,500.00
	(*c*)	Ordinary Portland cement 1500 tonne @ ₹2500/tonne	37,50,000.00
	(*d*)	Tartaric acid retarder 15 tonnes @ ₹50,500/tonne (Commercial grade)	7,57,500.00
			1,54,67,520.00
2.	**Utilities**		
	(*a*)	Electrical power 2,68,560 KWH @ ₹2/kWH	5,37,120.00
	(*b*)	Coal 603 tonnes @ ₹2000/tonne	12,06,000.00
	(*c*)	Packing of bags (L.S.)	55,000.00
			17,98,120.00
3.	**Labour and Supervision (L&S)**		
	(*a*)	Supervisor-cum-Manager 3 Nos. @ ₹5000/- P.M.	1,80,000.00
	(*b*)	Mechanic-cum-Operator 3 Nos. @ ₹3000/- P.M.	1,08,000.00
	(*c*)	Skilled Mazdoor 10 Nos. @ ₹2000/- P.M.	2,40,000.00
	(*d*)	Mazdoor 700 mandays @ ₹75/-P.M.D.	52,500.00
			5,80,500.00
4.	**Maintenance and Repair (M&R)**		
	(*a*)	Plant: 6% of B for fixed plant cost	1,70,250.00
	(*b*)	Building: 2% of building and shed cost	33,000.00
			2,03,250.00
5.	**Operating Supplies** 10% of maintenance and repair		20,325.00
6.	**Taxes and Insurance 2% of fixed capital A+B**		**89,750.00**
7.	**Plant and overheads 10% of L&S, M&R**		**78,375.00**
8.	**Depreciation**		
	Plant: 10% of B		2,83,750.00
	Building: 2.5% of Building and Shed cost		41,250.00
			5,13,450.00
9.	**Interest on total capital investment**		
	(*a*)	On fixed capital & 15%/annum	6,73,125.00
	(*b*)	On working capital @ 18%/annum	1,61,550.00
			8,34,675.00
		Annual Cost of Production	1,93,97,515.00
		Production per tonne	1293.16

D. Profitability Analysis ₹

(*a*) Gross annual income from sales @1430/tonne 2,14,50,000.00

(*b*) Total annual cost of production 1,93,97,515.00

(*c*) Gross annual return (*a* – *b*)

(*a* – *b*) × 100% Profitability :

$$\text{Total capital investment } \frac{2052485 \times 100}{53,85,000} = 38.10\%$$

6.4 Masonry Cement from Phosphogypsum

Suitability of unprocessed phosphogypsum has been examined for making masonry cement for use in construction in place of ordinary Portland cement mortars[13]. Thus, three masonry cement mixes were prepared by grinding together Portland Cement and phosphogypsum in the proportions 60:40 (A), 70:30 (B) and 80:20 (C) by weight. One batch of 70:30 mix was prepared with natural gypsum to observe the effect of impurities of phosphogypsum. These three batches of masonry cements were tested as per methods laid down in IS : 4031–1968. Results obtained, along with the Indian Standard requirements are given in Table 6.13.

PC = Portland Cement, PG = Phosphogypsum, A, B, C = Masonry Cements with Phosphogypsum, N = Masonry cement with Natural gypsum, *d* = days

It can be seen that masonry cement batches B and C pass the requirement for setting time and compressive strength. Water retention is above minimum specified value of 60%. There is no adverse effect of phosphogypsum on compressive strength which is at par with that of masonry cement produced with natural gypsum (mix N). Suitability of masonry cement batches B and C was examined for mortar making properties. The masonry mortars in 1:4 or 1:5 cement sand proportions (as produced abroad) were prepared with sands of three different fineness modules 1.25, 1.75 and 2.0 for the masonry cements B

Table 6.13 Properties of Masonry Cement
(Portland cement and phosphogypsum mixtures)

Masonry Cement	Composition		Setting time (Minutes)		Water Retention	Compressive Strength (N/mm²)	
(Batches)	**PC**	**PG**	**Initial**	**Final**		**7-d**	**28-d**
A	60	40	290	480	70	21	33
B	70	30	265	460	67	30	52
C	80	20	255	442	65	50	92
N	70	30	244	430	68	32	52
Requirement as per IS : 3466–1967							
		< 90 >1440 < 60 >25 < 50					

and C separately, water was added to each of the six batches of mortars to give the flow 110 ± 5°C. Cubes of 5 cm × 5 cm × 5 cm were cast from each batch of mortar and cured under 90% R.H. for 7 days, 28 days and 90 days respectively and tested for compressive strength. The results are given in Table 6.14.

Loose bulk density of masonry cement – 1000 kg/m^3, loose bulk density of sand = 1400 kg/cm^3.

The data show that the strength of 1:4 and 1:5 masonry cement-sand mortars is slightly less than 1:6 Portland Cement mortar but possess more water-retentivity (50–60%) than the cement sand mortar (18–20%). Most of the mortars fall in the grade MM 1.5 of IS : 2250–1981, code of practice for preparation and use of masonry mortar. From these studies, it is quite clear that phosphogypsum can be used without beneficiation or calcination for making masonry cement for use in masonry mortars. The use of phosphogypsum will be economical and may replace lime in composite mortars.

6.5 Gypsum Plaster Boards

The internal surfaces of walls and ceilings of most of the buildings are finished internally by applying plaster in one or more coats. In order to reduce the demand of site labour, the use of building board such as gypsum plaster board, fibre hard

Table 6.14 Compressive Strength of Masonry Mortars (N/mm^2)

Sand used (Fineness modulli)	Curing (days)	Portland Cement Sand Mortar	Masonry Cement B		Masonry Cement C	
		(1:6 by Vol)	1:4 by Vol. or 1:5.6 by wt.	1.5 by Vol. or 1:7 by wt.	1:4 by Vol. or 1:5.6 by wt.	1:5 by Vol. or 1:7 by wt.
2.0	7	—	1.01	0.75	1.21	0.92
	28	—	1.45	0.82	1.41	1.42
	90	—	2.01	1.15	2.21	1.51
1.75	7	1.54	1.18	0.90	1.35	1.22
	28	2.30	1.80	1.15	2.21	1.45
	90	2.56	2.80	1.60	2.91	1.61
1.25	7	—	1.10	0.92	1.21	1.30
	28	—	1.68	1.10	1.88	1.75
	90	—	2.60	1.45	2.70	1.80

board, cement coir boards and asbestos cement building board as covering for walls and ceiling has been increasing steadily. Gypsum boards have the specific advantage of being lighter than the boards of similar nature, such as fibre hard boards and asbestos cement building boards. Gypsum boards also possess better fire-resisting, thermal and sound insulating properties.

Gypsum boards may be manufactured as plain, laminated and reinforced boards. Reinforcing materials generally used are glass, vegetable fibres, etc. The boards may be used to provide dry lining finishes to masonry walls, to steel or timber frame partitions, or as ceilings to structural steel columns and beams, or in the manufacture of prefabricated partition panels. Laminated gypsum boards are used for laying of concrete ceiling. It combines firmly with concrete and represents readymade interior plastering. Glass Reinforced Gypsum (GRG) boards are pseudo ductile materials having reasonably high flexural and impact strengths. GRG can be sawn, drilled, screwed or nailed like timber. It is resistant to white ant and termite and completely non-combustible. Being isotropic in character, thin GRG panels may be used compared to timber panels, hence cost effective. GRG composites can be used as substitute of timber for panel door, wall paneling, partitions, false ceiling, etc. and also as furniture components. The gypsum boards may be fixed by nailing, screwing, or sticking with gypsum based or other adhesives.

6.5.1 Production of Fibrous Gypsum Plaster Boards

Fibrous gypsum plaster board consists of 12 mm thick boards of set gypsum plaster reinforced with organic fibre. The board is known for its high thermal insulation and fire resistance. It is a commercially lining material used abroad. The main use for such plaster board is as low cost substitute for plywood and fibre boards of the insulating cellular type used for light-weight partitions and false ceilings. The board may be used as wall surfacing in place of plaster for dry and faster methods of construction[14–17].

6.5.2 Raw Materials

Following materials were used in the manufacture of plaster boards:

(*i*) **Gypsum Plaster:** Plaster produced from the beneficiated phosphogypsum.

(*ii*) **Reinforcing Material:** Teased sisal fibre and the glass fibre (E-Type, 12 mm long).

(*iii*) **Stripping Agent:** A small quantity of oil is used as a stripping agent for releasing the board from the substrate. Stearine dissolved in kerosene was used for the purpose. Alternatively, a mixture of soap and mustard oil or mobile oil can be used.

1. Equipment Required

(*i*) Concrete Casting Table – 10 meter long, 2 meter wide and 0.8 meter high. The surface of the table is trowelled to a high gloss so that is perfectly smooth and truly level or alternatively Perspex (Poly methyl methiacrylate) sheets may be used as substrate.

6.5.3 Properties of Fibrous Plaster Boards

To arrive at optimum quantity of fibre and the strength/breaking load, the plaster boards of size 30 cm × 30 cm × 12 mm were cast at 70% consistency with the reinforcement of sisal and the sisal plus glass fibre (E-type, 12 mm) sand-witched between two layers of the plaster slurry. The plaster boards were demoulded after 2.0 hours, dried in the oven at 42 ± 2°C and tested for various properties as per IS : 2542 (Part II)-1976, methods of test for gypsum plaster, concrete, and products, Part II – Gypsum Products. The properties of the plaster-board produced out of unprocessed and the beneficiated phosphogypsum plasters are listed in Tables 6.15 and 6.16.

The data show that the board specimens comply with the requirements of IS:2095 (Part 3) – 1996. The addition of 0.5% glass fibre improves the transverse load and the surface hardness. It can be seen that the transverse load and bulk density (Table 6.16) values are improved on using the beneficiated phosphogypsum plaster over the unprocessed phosphogypsum plaster (Table 6.15).

Table 6.15 Physical Properties of Fibrous Plaster Boards Produced out of Unprocessed PG Plaster

Sl. No.	Fibres (w/w. %)		Bulk density (kg/m³)		Properties Transverse Load (N)		Surface Hardness (mm)	
Sisal Fibres – Glass Fibre (E-Type, 12 mm)			*UR **R		UR R		UR R	
1	2.0	0	1250	1230	360	340	1.0	1.0
2	1.5	0.5	1300	1260	450	410	0.8	1.0
IS:2095 (Part –3) 1996 Limits			N.S.		Min. 340		Max. 5.0 mm	

N.S. = Not specified; *UR = Unretarded, **R = Retarded

Table 6.16 Physical Properties of Fibrous Plaster Boards Produced out of Beneficiated Phosphogypsum Plaster

Sl. No.	Fibres (by wt.%)		Bulk density (kg/m³)		Transverse load (N)		Surface Hardness (mm)	
	Sisal fibre - Glass fibre (E-type) 12 mm long		UR R		UR R		UR R	
1	2.0	0	1280	1250	400	370	0.7	1.0
2	1.5	0.5	1350	1300	500	450	0.8	1.1

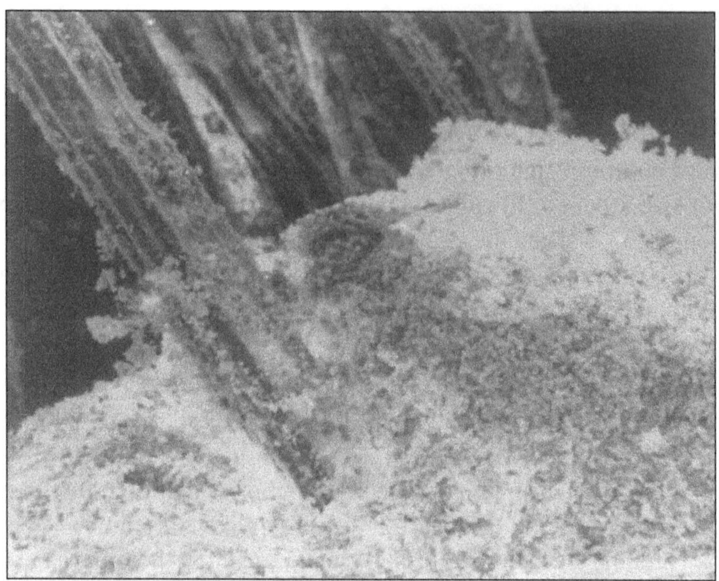

Fig. 6.3 Microstructure of Gypsum plaster-Sisal Fibre Matrix

Figure.6.3 shows microstructure of the gypsum plaster reinforced with the sisal fiber. It can be seen that the fibres are hollow and have rough surface . Sisal fibers in gypsum plaster make intermittent contact with the matrix through recrystallized gypsum crystals.

6.5.4 Thermal Conductivity

Gypsum based building materials are characterized by their excellent thermal insulating properties. With this objective in view, boards of size 30 cm × 30 cm having 2.5 cm thickness were cast using processed phosphogypsum plaster and teased sisal fiber as the reinforcing material at 90% water/plaster ratio. The boards were demoulded after 1 hour of setting and dried in the sun to constant weight. The weight of the boards was noted. These boards were then subjected to the thermal conductivity test as per heat flow method specified in ASTM-C-518–1971. The results obtained are given in Table 6.17. The data

Table 6.17 Thermal Conductivity of Plaster Boards

Sl. No.	Materials	Bulk Density (kg/m^3)	Thermal Conductivity K.Cal/m/h/°C
1.	Gypsum Plaster Board	1100–1200	0.141–0.16
2.	Light Weight Concrete	500–1800	0.3–0.7
3.	Common Clay Brick	1600–1800	0.65–0.70
4.	Asbestos Sheet	900–1000	0.25–0.28

(Table 6.17) show that plaster board has comparatively low thermal conductivity value than the conventional building materials like light weight concrete, common burnt clay bricks and asbestos cement roofing sheets. The low value of thermal conductivity affords good thermal insulation in buildings.

6.5.5 Predesign, Cost Estimates for the Production of Fibrous Gypsum Plaster Boards

Production : Gypsum Plaster Board (Size 120 cm × 60 cm × 12 mm)

Capacity : (*a*) 1,000 boards per day (of three shifts)

(*b*) 3,00,000 boards per year (300 working days)

Capital Investment:

	₹
A. Fixed Capital on Building	
Land 800 sq.m @ ₹1500/m^2	12,00,000.00
Yard improvement (LM)	50,000.00
Building 300 sq.m @ ₹4000/m^2	12,00,000.00
Shed 600 sq.m @ ₹1500/m^2	9,00,000.00
B. Fixed Capital on Plant	
(*a*) Purchased Equipment	
(*i*) Concrete casting tables 60 Nos. @ ₹5000/- each	3,00,000.00
(*ii*) M.S.Moulds: 300 Nos. @ ₹200/- each	60,000.00
(*iii*) Steel racks for drying the boards 50 Nos. (20 boards capacity) @ ₹600/- each	38,000.00
(*iv*) Bucket and drums etc. L.S.	5,000.00
	4,03,000.00
(*b*) Electrical installation 15% of PF	58,700.00
(*c*) Water services and drainage about 5% of PE	19,900.00
(*d*) Instruments & controls @ 5% of PE	19,900.00
(*e*) Laboratory & workshop 10% of PE	39,800.00
(*f*) Engineering and Supervision 10% of PE	39,800.00
(*g*) Contingencies (LS)	60,000.00
Fixed Capital on Plant	6,37,100.00
Total fixed capital (A+B)	39,87,100.00
Working Capital (@ 20% on Fixed Capital)	12,7,420.00
Total Capital Investment (A + B + C)	**41,14,520.00**
Cost of Production	
(Basic: 300 working days per annum)	

Cost of Production:

	₹
1. Raw Materials	
(*a*) Calcined phosphogypsum 3000 tonnes @ ₹1500.00 tonne	45,00,000.00
(*b*) Sisal fibre: 76.0 tonnes @ ₹30,000/- tonne	22,80,000.00
(*c*) Stripping agent and citric acid retarder (L.M.)	1,00,000.00
	68,80,000.00

	₹
2. Utilities	
(*a*) Electrical Power: 10,000 kWH @ ₹4/- per kWH	40,000.00
(*b*) Water: 31000 K.L. @ ₹2.0/KL	62,000.00
	1,02,000.00
3. Labour and supervision (L&S)	
(*a*) Skilled Labour : 18 Nos. @ ₹3000 p.m.	6,48,000.00
(*b*) Mechanic cum operators 3 Nos. 5000 p.m.	1,80,000.00
(*c*) Supervisor cum Manager: 1 No. ₹10000 p.m.	1,20,000.00
	9,48,000.00
4. Maintenance and Repairs (M&R)	
(*a*) Plant @ 6% of B for fixed plant	38,226.00
(*b*) Building @ 2% of Building and Shed cost	42,000.00
	80,226.00
5. Operating Supplies : 10% of M&R	8,023.00
6. Taxes and Insurance of 2% of total fixed capital (A+B)	79,742.00
7. Plant Overheads (@ 5% of B)	31,855.00
8. Depreciation	
Plant: About 10% of item B	63,710.00
Building 2.5% of Building and Shed cost	52,500.00
	1,16,210.00
9. Interest on total capital investment	
(*a*) On fixed capital @ 10% per annum	3,98,710.00
(*b*) On Working capital @ 12% per annum	15,790.00
Total Cost of Production	86,60,056.00
	86,60,056.00
Cost of Production per Board	3,00,000.00
	28.86

C. Profitability:

1. Gross Annual Income (Annual Sales) ₹35/- each board	1,05,00,000.00
2. Annual Cost of Production	86,60,056.00
3. Annual Return (Item-Item 2)	18,39,944.00
4. Return on Investment	

$$\frac{\text{Annual Return} \times 100}{\text{Total Capital Investment } 41{,}14{,}520} = \frac{18{,}39{,}944 \times 100}{41{,}14{,}520} = 44.7\%$$

6.5.6 Casting of Large Size Plaster Boards

A steel frame mold of dimensions 120 cm × 60 cm × 12 mm was laid over the top surface of concrete casting table, which was made smooth by applying the stripping agent. Calcined phosphogypsum was mixed with 70 – 80% water to form thin uniform slurry or slip. The slurry was poured and evenly spread in the mould to a thickness of 4 mm. Teased sisal/sisal plus glass fibre about 2.0% by the weight gypsum plaster was then spread over the surface. The fibre near the

Fig. 6.4 Plaster Boards from Phosphogypsum Plaster

edges was pressed into the plaster slurry by hand to provide extra reinforcement near the edges. The remaining slurry was then poured, labelled and allowed to set. The set plaster boards were demoulded after two hours and dried in steel racks in air to constant temperature. Typical photographs of gypsum plaster boards are shown in Fig. 6.4.

These gypsum boards are quite common in Australia and UK where casting of gypsum sheets is carried out manually earlier but in modern time the casting is done in factories. It is still the common practice in India and subcontinent by the small scale entrepreneurs to produce gypsum plaster boards for ceiling or partitions purposes in auditorium, hospitals, cinema theatres, hotels, etc. Some of the photographs of the fibrous gypsum boards cast by manual and at factory level using sisal fibre are shown in Figs. 6.5 to 6.9.

6.6 Glass Reinforced Gypsum Binder Composites

Gypsum plaster, like other inorganic cements, is strong in compression but weak in tension and has low impact strength. These brittle characteristics prevent effective use of high compressive strength in structural applications. Some improvement has been described above wherein sisal, coir or mixture of sisal with little glass fibres have been blended with gypsum plaster/binders. The properties of calcined gypsum and building products containing such fibres are shown in Table 6.18. Glass reinforced gypsum (GRG) gives high impact and tensile values. The strength properties of such composites can approach that of sheet timber, possess high fire resistance than gypsum plaster boards but at the same time they have higher densities also to cause fixing problems[18].

Fig. 6.5 Spread of Gypsum Plaster Slurry

Fig. 6.6 Teased Sisal Fibre being Spread Over a Casting Table on Unset Plaster

Fig. 6.7 Sisal Fibre Being rolled into Unset Plaster

Fig. 6.8 Machine Made fibrous Plaster Board Sisal Fibre Being Spread Over a Casting Table Covered with Unset Plaster

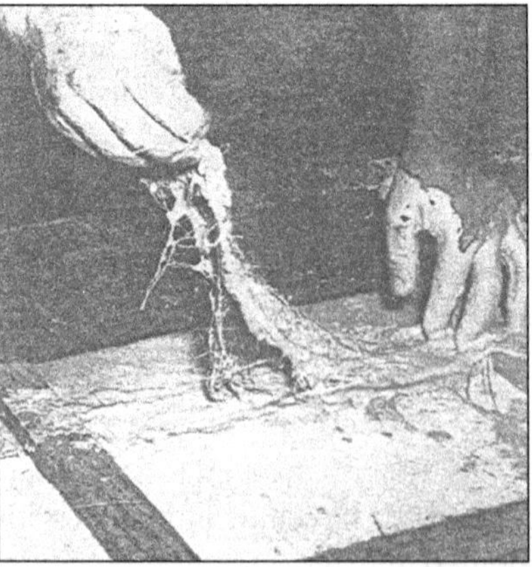

Fig. 6.9 Scrim Being Applied to the Back of Joint in a ceiling

Table 6.18 Properties of Calcined Phosphogypsum and Its Products

Materials/ Product	Properties			
	Density g cm^{-3}	Compressive strength (N mm^{-2})	Transverse load (N)	Thermal conductivity (Kcal m^{-1} $h^{-1}{}^{\circ}C^{-1}$)
β-hemihydrate	1.0–1.1	8.1–13.1	—	—
α-hemihydrate	1.3–1.4	22–30.0	—	—
β-anhydrite	1.6–1.7	30–33.3	—	—
Fibrous plaster board	1.0–1.1	—	350–500	0.14–0.17
Building blocks	0.9–0.95	2.5–3.5	—	0.13–0.15
Slotted tiles	0.42–043	1.5–2.0	—	—

Table 6.19 Physical Properties of E-Type Glass Fibre

Properties	Values
Diameter of the fibre filament, micro metre	8–10
Numbers of filaments in strand	204
Tensile strength of glass fibre, MPa	1750
Young modulus of glass fibre, MPa	6890–7600

Generally E-type glass fibre is used for reinforcing the gypsum plaster. The Properties of glass fibres are given in Table 6.19.

There are two main methods of reinforcing gypsum with fibre materials. One method is to concentrate fibres in the tensile zone of the resulting structural element so as to match the external tensile force and use the matrix to match the external compressive force. The other method is to disperse the glass fibres uniformly in the matrix so as to form a homogeneous mixture. This ensures a high degree of stress distribution by the fibre which then acts as a crack arrester. This method was selected to produce GRG using gypsum binder and E-type glass fibres.

The GRG composites were cast by reinforcing the chopped glass fibre of different sizes and lengths (Optimization) in between gypsum binder paste followed by applying suction for 15 minutes to remove extra water from the matrix. The composites were then demoulded from the casting table, cured under 90% RH at 27 ± 2°C for a period up to 28 days. The GRG composites were tested for various properties (Table 6.20).

The increase in flexural strength (F.S.) of GRG composites is always required without effecting the density of the composites and some efforts were made in this direction. The effect of consolid - 444 polymeric additive was studied on the

Table 6.20 Properties of Glass Fibre Reinforced Gypsum Binder Composites

Property	GRG Binder Composites	Plain Plaster Composites
Bulk density, g/cm^3	1.628	1.20
Consistency, %	65.00	81.00
Flexural strength, MPa		
1 day	10.70	4.96
3 days	12.17	4.97
7 days	13.21	4.98
28 days	22.00	4.96
Tensile strength, MPa (28 days)	18.00	2.75
Impact strength, N/mm^2 (28 days)	18.60	10.20
Thermal conductivity, Kcal/m h°C	0.09	0.12

F.S. of two binder composites - A2 (F.S. at 1, 3, 7, 28 days : 10.1, 23.1, 28.6 and 35.0 MPa and B2 (F.S. – at 1, 3, 7 and 28 days : 13.7, 19.0, 20.9 and 21.3 MPa) (Fig. 6.10). The properties of binder composites suggest its use as an alternative to timber in door panels, partitions, ceilings, cup boards, etc[19].

The durability of gypsum binder composites was checked by immersing the gypsum composites in water for a period of 28 days. After 3 days of immersion gypsum binder composites have a much lower water absorption (17%) than the composite based on plain gypsum plaster (35%), indicating the better water resistance of the former compared with the latter[20]. The effect of alternate wetting and drying and heating and cooling cycles at 27–60°C on gypsum binder composites has been studied which indicates that these composites can be used under 50°C safely. Beyond 50°C the strength of gypsum binder composites drops at faster rate under wetting and drying cycles than the heating and cooling cycles. The decaying of the glass fibres embedded in the gypsum composites was also studied. The results showed higher surface etching and loss in weight of the glass fibres under wetting and drying cycles than the heating and cooling cycles.

6.6.1 Microstructure at the Interface

The microstructure of glass fibre (E-type) reinforced composites based on slag based gypsum binder exposed to water, accelerated ageing *i.e.,* alternate wetting and drying cycles at 27 and 60°C and to natural weathering are reported. Micrographs of the fibre and the fibre/matrix interface are shown in Fig. 6.11 (a–d). These are typical of all the samples stored in water, exposed to wetting and drying cycles and natural weathering. Indeed it is significant that few differences between wet and dry stored 28 days old samples have been observed

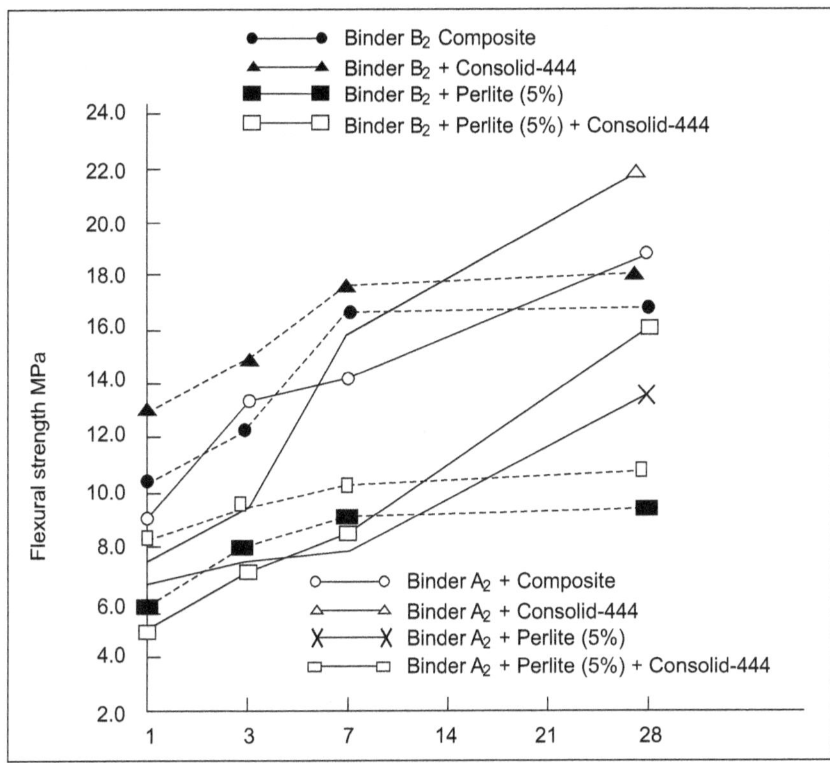

Fig. 6.10 Effect of Curing Period on Flexural Strength of Gypsum Binder Composites

with gypsum binder composite Fig. 6.11(*a*) shows the filling of spaces between the filaments of a glass fibre strand with a dense crystallized gypsum binder products cured in water. A close examination of fibre shows that no visible deterioration of the glass fibre occurred.

Figures 6.11 (*b*) and (*c*) show microstructure of glass fibre exposed to 50 alternate wetting and drying cycles at 27°C and 60°C. It can be seen from Fig. 6.11 (b) (27°C) the crystallized gypsum binder is tenaciously bonded to the glass fibre at the interface. While Fig. 6.11(*c*) (60°C) clearly shows deposition of matter at the surface and in between glass fibre in an irregular manner showing cracking of the matrix leading to reduction in the mechanical properties due to weakening of the fibre/matrix bond.

Figures 6.11(*d*) shows microstructure of composites exposed to natural weathering at 28 days showing an intermittent point contact between glass fibre and the hardened gypsum binder phases in a haphazard manner resulting in a discontinuous and irregular interfacial bond.

6.6.2 Durability of Glass Fibre

Microstructure of glass reacted with the aqueous extract of gypsum binder at 27°C, 40°C, 50°C and 60°C are shown in Figs. 6.12–6.15. It can be seen from the Figs. 6.12

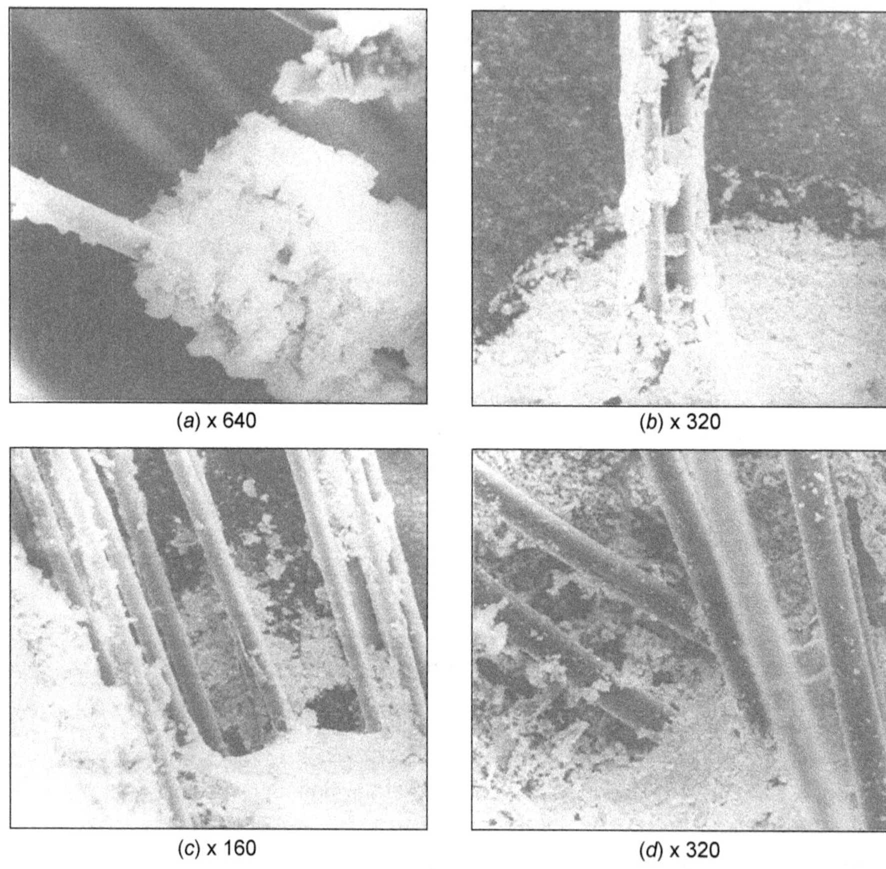

(*a*) x 640 (*b*) x 320

(*c*) x 160 (*d*) x 320

Fig. 6.11 Micrographs of Glass Fibre/Matrix Interface of Gypsum Binder Composites: (*a*) Cured in Water (28 days); (*b*) Exposed to 50 Alternate Wetting and Drying Cycles at 27°C; (*c*) Exposed to 50 Alternate Wetting and Drying Cycles at 60°C; (*d*) Exposed to Natural Weathering (28 days)

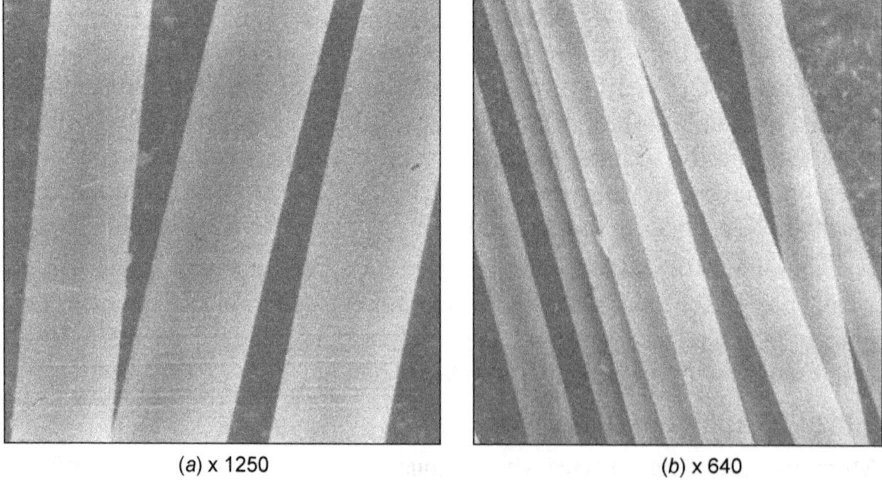

(*a*) x 1250 (*b*) x 640

Fig. 6.12 Microstructure of Glass Fibres at 27°C: (*a*) 3 Days and (*b*) 7 Days

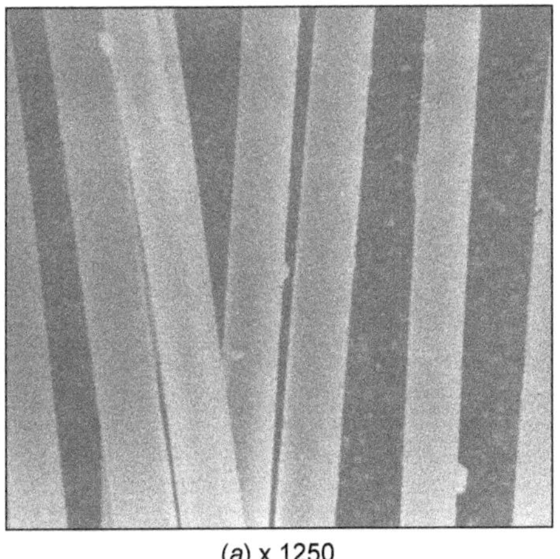

(*a*) x 1250

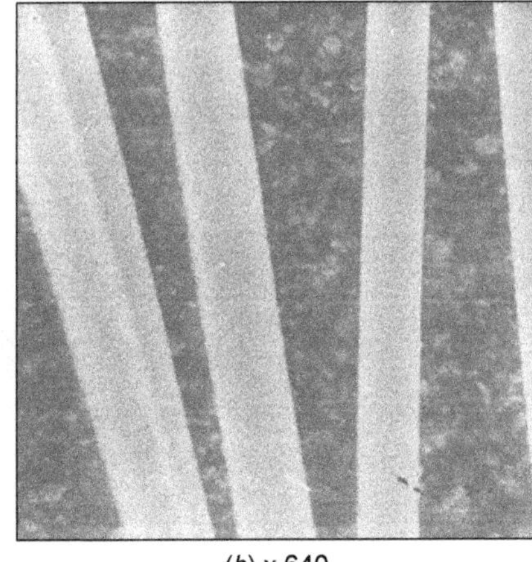

(*b*) x 640

Fig. 6.13 Microstructure of Glass Fibres at 40°C: (*a*) 3 Days and (*b*) 7 Days

(*a*) and 6.12 (*b*) (3 and 7 days at 27°C) that euhedral fibres showing no pitting or deterioration of surface are observed, little deposition of material is formed on the glass surface when fibres reacted with binder extract for 3 and 7 days at 40°C [Figs. 6.13(*a*) and 6.13(*b*)]. No indication of damage is visible. When glass fibres are subjected to similar treatment at 50°C, evidence of attack of alkalies is easily seen at 7 and 28 days of reactions [Figs. 6.14(*a*) and 6.14(*b*)]. When the glass fibres are reacted with the aqueous extract at 60°C, the attack on glass fibre surface is visible at 7 days [Fig. 6.15(*b*)]. It is quite clear that a reaction film had formed on the surface of the fibre which subsequently detached itself from the bulk. As a consequence of surface degradation, glass fibres become weaker with time when placed in binder environment and the possibilities exist that the submicroscopic etch pits provide locations for nucleation and growth of crystalline phases in the interfacial zone is important, however, the improved properties of glass reinforced gypsum binder composites are to be retained. One way of achieving this objective is by using the composite in temperature not exceeding 50°C.

The weight loss of glass fibres reacted with binder extract up to a period of 3 months at 27°C to 60°C are plotted in Fig. 6.16. It can be seen that with the increase in temperature, the weight loss of glass fibre is increased and maximum weight loss in fibre weight is visible at 60°C. The etching of the glass fibre and weight loss at 50°C and 60°C is directly responsible for the fall in the strength of gypsum composites when subjected to alternate wetting and drying cycles.

6.6.3 Gypsum Fibre Board

Gypsum Fibre Board (GFB) is a new building material mainly applicable at present for dry, interior building work. The basic materials for the manufacture of GFB are high-purity gypsum (85% $CaSO_4.2H_2O$) and waste paper. In these boards, the cellulose fibres produced from the waste paper are intimately embedded in the gypsum mass. By the homogeneous distribution throughout

Fig. 6.14 Microstructure of Glass Fibres at 50°C: (*a*) 3 Days and (*b*) 7 Days

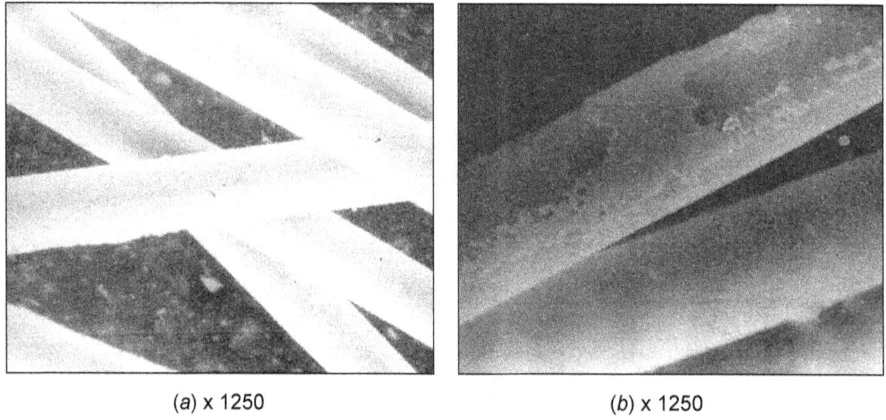

Fig. 6.15 Microstructure of Glass Fibres at 60°C: (*a*) 3 Days and (*b*) 7 Days

the complete board, the cellulosic fibres become a reinforcement and bind the gypsum composite together and make it elastic. Sometimes a surface coating/treatment is applied to the board which increases its mechanical strength significantly. Siemkel Kemp[21–22] a German company, has brought the manufacturing process to industrial maturity. At the beginning of industrial production, the dimensions of the boards were 1500 × 1000 × 15 mm and these were later increased to 6000 × 2500 × 10 mm. A GFB plant [Wurtex (Process)] with a capacity of 1575 m^2 h^{-1} of board of 10 mm thickness is working at present in the Netherlands.[23]

GFB is water and fire resistant and possesses excellent sound insulation properties. Laminated GFB with plastic decorative sheets and wood veneer can be produced. These boards can be used for partition walls inside large houses and multistoried buildings. GFB is not available in India at present but

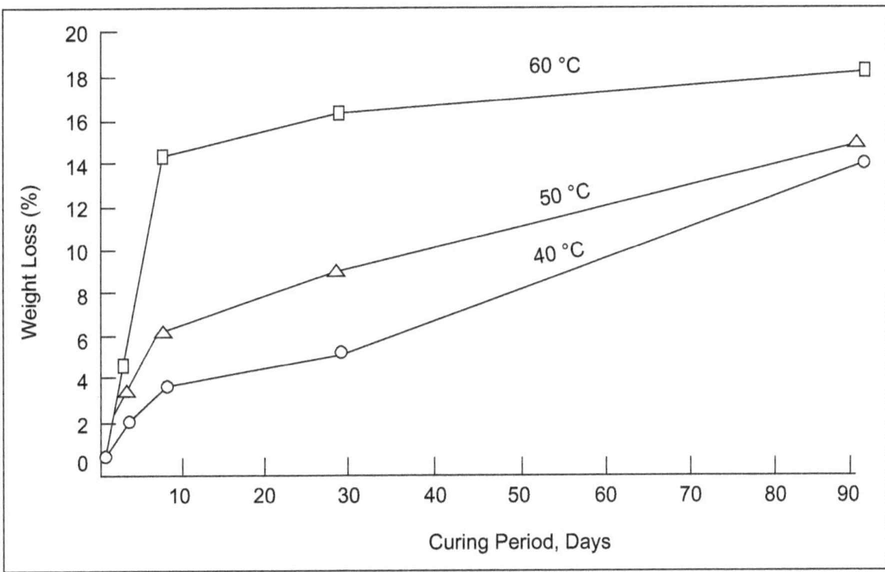

Fig. 6.16 Effect of Immersion on the Weight Loss of Glass Fibres in Aqueous Gypsum Binder Extract at Different Temperatures

the Bureau of Indian Standards has taken the initiative to formulate a standard specification of these boards for the consumer in the near future.

Gypsum coated with paper liners are also popular. These boards are produced as per IS : 2095 (Part-1) and are well known for thermal insulation . fire resistance and acoustic properties. The properties of these boards are shown in Table 6.21.

6.7 Gypsum Bonded Particle Board

Gypsum bonded particle board is being produced in Scandinavian[24] plants that utilizes the semi-dry process developed by KOSSATZ. In the conventional manufacturing processes of gypsum bonded building boards, gypsum plaster is utilized in a highly viscous or liquid state (water/gypsum ratio 0.7 to 0.9).

Table 6.21 Properties of Gyp Board

Edges	Square	Tapered	Bevelled	Rounded
Thickness (mm)	9.5	2.5	15.0, 19.0	23.0, 25.0
Width (mm)	619	914	1219	—
Length (mm)	610	1220	2438	—
Weight (kg/m^2)	8.0	10.5	12.80	20.0
Thermal resistance (m^2 K^{-1}W^{-1})	0.06 (9.5 mm)	0.08 (12.5)	0.09 (15.0 mm)	—
Fire protection	Fire propagation conforms to BS 476 Part 6, 1981, Surface spread of flame conforms to BS 476 Part 7 1971			

Theoretically a water/gypsum ratio in between 0.16 and 0.19 is sufficient to obtain the complete reaction of hemihydrate to dihydrate. In this process the wood furnish flakes, particles or fibres provides sufficient moisture for the binder to set. The properties of gypsum bonded particle boards produced by the semi-dry process are comparable with or superior to gypsum bonded board products, such as plaster board and gypsum fibre boards (Table 6.22).

6.7.1 End-use Characteristics and Applications for Gypsum Bonded Particle Boards

Gypsum-bonded particle boards provide a surface for a variety of surface systems (*e.g.*, wood veneers, decorative papers and plastic foils) and satisfactory bond qualities can be obtained with commercially available adhesives.

Gypsum-bonded particle boards perform better than other wood composite boards when exposed to fire. This can be ascribed to the 20.0% water of crystallization present in the gypsum. On conflagration, the water evaporates and forms a vapor which acts as a barrier in between the flame and the board. This property opens up possibilities for the use of gypsum bonded particle boards as interior wall panels in commercial buildings and in schools, hospitals and hotels.

GRG has some of the properties of timber and can be sawn, drilled, screwed and nailed. It can also be polished to look like wood but with almost identical

(*a*) (*b*)

Fig. 6.17 Phosphogypsum Door Shutters – Gypsum Panels Encased in (*a*) MS section and (*b*) MS Angle Frames

Table 6.22 Properties of GFB Binder Composites and Some Conventional Building Materials

Sl. No.	Materials	Property				
		Bulk density	Flexural strength	Impact strength	Water absorption	Swelling (24 h)
1	Low grade gypsum	1.2–1.3	4–5	1.0	35	1.0
2	Autoclaved gypsum	1.3–1.4	7–10	1.5	32	0.9
3	GRG (commercial)	1.6–1.8	15–30	12–20	30	0.5
4	GRGB (CBRI)	1.5–1.6	20–25	16–18	16	0.4
5	Gypsum bonded particle board	1.1–1.2	6–10	—	—	2.5
6	Gypsum plaster board	0.8–0.98	3–8	—	35	2.0
7	Gypsum fibre board	1.1–1.3	5–8	—	—	2.0
8	Particle board	0.8–2.2	14	4.5	70	26
9	Asbestos cement	2.1	18–23	2.0	—	0.60
10	Glass-reinforced cement (GRC)	2.1	30–50	15–30	—	—
11	Cement-bonded particle board	1.1–1.3	9–15	—	—	2.0
12	European resin-bonded particle board	0.65–0.75	12–24	—	—	11–15
13	Polycoir board	0.9–1.15	48–49	—	2.2	2.4
14	Rice husk particle board	0.8–0.83	12–13	—	16	13.5–14.5

mechanical properties in all directions in contrast to timber which is very much weaker across the fibre direction. As a result much wider thinner GRG boards may be produced compared with timber panels, bringing substantial savings in costs. GRG can be used in door panels, structural partitions, cupboards, furniture, etc.

6.7.2 Gypsum Door Shutters

Door shutters (size 2000 × 900 × 30 mm) produced from sisal fibre reinforced gypsum binder boards (SRGB) panels[25] encased in mild steel (Fig. 6.17) and aluminum frames have been evaluated for various properties as per BIS : 4020–1967 and found to comply all the requirements. The cost of door shutters has been worked out to be ₹1750/- for the mild steel framed door stutters and ₹3400/- for the aluminium framed door shutters respectively. The cost of the SRGB panels used in the door shutters lies in the range of 3 to 6 per cent of the total cost of

Table 6.23 Cost of SRGB Panels *vis-à-vis* Traditional Building Materials

Materials	Approx. Cost in ₹ (per.sq.m.)
Plywood 4 mm, 6 mm, 8 mm, 12 mm 19 mm	106, 150, 167, 213, 366
Roofing sheet, Ac Sheet (4 mm thick) 44' x 4', 6' x 4', 8' x 4'	130, 200, 260
Particle board (9 mm, 12 mm, 18 mm, 25 mm)	70, 140, 175, 250
Gyp board (12 mm thick)	—
Plain: 595 mm x 595 mm x 12 mm	70.0
Gyp board: 9.5 mm,12.5 mm,	154, 144
GRG (12 mm)	190
Thermo plastic sheets (1 mm, 2 mm, 3 mm, 4 mm, 5 mm, 6 mm)	70,140, 190, 260, 330, 380
PVC Compact sheet (1 mm, 2 mm, 3 mm, 4 mm, 5 mm)	150, 240, 330, 440, 530
SRGB (Based on water resistant gypsum binder)	70

the normal door shutter. In spite of the high cost of the metallic frames, the door shutters have the following specific advantages over the conventional door shutters.

- Total elimination of timber
- Water proof
- Termite proof
- Fire resistance
- Waste utilization
- Pollution abatement

The major thrust on actual utilization of these door shutters can only be realized if testing and evaluation of these shutters can be adopted by the standards other than BIS : 4020–1967 or equivalent due to brittle nature of the GRG/SRGB panels. Central Glass & Ceramic Research Institute (CGCRI), Kolkata (India), a CSIR Laboratory possesses the technology for manufacturing of GRG panes/boards through NRDC. These GRG boards are similar to the boards developed by Majumdar and his coworkers[26–29].

The cost of various building materials *vis-à-vis* gypsum panels is given in Table 6.23.

6.8 Gypsum Blocks

To arrive at optimum strength and density characteristics of the gypsum blocks, the phosphogypsum plaster produced out of beneficiated and the unbeneficiated plaster, β-hemihydrate was mixed with 60, 70 and 80% water by the weight of the plaster and the uniform slurry was made by thorough mixing either

Table 6.24 Properties of Phosphogypsum Plaster Cubes

Plaster Design	Consistency (%)	Compressive strength (MPa)	Bulk density (Kg/m^3)	Porosity		
				2 hr	8 hr	24 hr
UB	60.0	6.7	1190	30.1	34.2	35.1
B	60.0	7.2	1200	31.8	36.4	37.8
UB	70.0	4.7	1090	35.2	35.1	38.1
B	70.0	5.6	1100	36.3	36.7	37.6
UB	80.0	3.92	1040	40.20	38.6	39.9
B	80.0	4.2	1080	41.42	39.9	40.6

UB = Unbeneficiated Plaster
B = Beneficiated Plaster

manually or by a stirring mechanism. The gypsum slurry was poured into 5 cm × 5 cm × 5 cm cube moulds tempted to remove any void/air bubbles. The superfluous material is removed with spatula to give smooth top surface. The cubes were demoulded after 2.0 hours, dried at 42 ± 2°C and tested for compressive strength, bulk density and porosity values. The results are listed in Table 6.24.

Fig. 6.18 Blocks from Phosphogypsum

Table 6.25 Properties of Phosphogypsum Blocks

Sl. No.	Property	Value		IS : 2849–1983
		Unbeneficiated	Beneficiated	
1	Visual Inspection	Passes	Passes	Blocks should be sound, free from cracks & imperfection
2	Bulk density, kg/m^3	1031	1066	N.S.
3	Compressive strength, MPa	1.1	1.3	Min. 0.5
4	Non-combustibility	Passes	Passes	Cause the temperature of furnace thermocouple to rise by more than 50°C above the initial furnace temperature

The data depicts that the plaster produced out of beneficiated phosphogypsum possesses higher values of strength, bulk density and the porosity than the plaster made from the unbeneficiated phospho plaster.

6.8.1 Casting of Large Size Blocks

Based on the data generated by 5 cm × 5 cm × 5 cm cubes, 40 × 20 × 10 cm blocks were cast at 70–80% consistency to get gypsum blocks of lighter density. The dimensions and tolerances of the blocks are within the limits of IS : 2849–1983. The properties of gypsum blocks are given in Table 6.25.

It can be seen that, the gypsum blocks comply with the requirements laid down in IS : 2849–1983. The gypsum blocks produced out of beneficiated plaster showed better strength and density properties than the blocks produced from the unbeneficiated phospho plaster. The gypsum blocks are recommended for internal use only. Figure 6.18 shows gypsum blocks produced from phosphogypsum.

6.8.2 Gypsum Bricks

Attempts were made to produce gypsum bricks from the calcined phosphogypsum. The gypsum bricks of dimensions 19.5 × 9.5 × 10 cm were produced from unbeneficiated and the beneficiated phospho plaster at 60% consistency. The bricks were cast by pouring gypsum plaster slurry into moulds of dimensions mentioned above, flushing, demoulding and drying the bricks at 42 ± 2°C. The gypsum bricks were tested for compressive strength, bulk density and water absorption. The results are reported in Table 6.26. A combined photograph of gypsum building products is shown in Fig. 6.19.

Table 6.26 Properties of Gypsum Bricks

Sl. No.	Property	Value		IS : 1077 limits
		Unbeneficiated	Beneficiated	
1	Compressive Strength, MPa	5.2	6.5	Min. 3.5
2	Bulk density, kg/m^3	11.80	12.24	N.S.
3	Water absorption, % (24 hrs)	31.8	29.5	Max. 20.0
4	Porosity	37.84	36.21	N.S.
5	Drying shrinkage,%	0.10	0.08	Max. 0.15
6	Efflorescence	slight	slight	Not more than moderate upto class 10 and slight for higher classes

6.8.3 Prefabricated Gypsum Building Components

Prefabricated gypsum products are manufactured in large quantities: plaster boards, partition panels, ceiling tiles and fibre-reinforced boards. These are light, porous, dry, and non-brittle products possessing excellent workability. β-hemihydrate is the starting material for all these products as it sets faster and satisfy building industry requirements for certain properties of finished products. The manufacturing of gypsum products are normally built close to gypsum source.

Gypsum plaster boards are large thin panels of gypsum plaster covered with card board of density of 750–950 kg/m^3. They are produced by feeding β hemi-hydrate

Fig. 6.19 Gypsum Building Products Produced from Phosphogypsum

Table 6.27 Properties of Gyp Board (Saint Gobain Gyproc India Ltd., Jind, Haryana)

Edges	Square	Tapered	Beveled	Rounded
Thickness (mm)	9.5	12.5	15.0–19.0	23.0–25.0
Width (mm)	619	914	1219	—
Length (mm)	610	1220	2438	—
Weight (kg m^{-2}) Thickness	7.5 (9.5 mm)	9.5 (12.5 mm)	11.92 (15.0 mm)	19.8 (23.0 mm)
Thermal resistance (m^2 K^{-1} W^{-1})	0.07 (9.5 mm)	0.07 (12.5 mm)	0.08 (15.0 mm)	0.09
Fire Protection	Fire propagation index of performance a(I) not exceeding 12 and a sub index (i) not exceeding (both sides), class I (both sides). When each side tested separately to BS-476: Part 6, 1981. Surface spread of flame when tested to BS-476: Part 7, 1971.			

(*a*)

(*b*)

(*c*)

Fig. 6.20 Continuous Process for Manufacturing Paper Coated Gypsum Plaster Boards showing (*a*) Continuous casting table with drier (*b*) Continuous casting table with cutter and stacking and (*c*) Paper rolls for lining

into continuous mixer from controlled feeding devices, mixing it continuously with water and additives – adhesive to get uniform rapid setting gypsum slurry. This slurry is spread on a continuous sheet of cardboard of thickness 0.6 mm and passed over a moulding platform to be cast into a completely encased strip, 1.20–1.25 m wide and 9.25 to 25 mm thick (Fig. 6.20). This strip of plaster-board is cut into separate panels. These panels with one third of their weight are immediately dried in a continuous tunnel drier heated directly with steam or gas or oil. The finished plaster boards, consisting of gypsum core encased cardboard, is considered to be a laminated building material.

The properties of gyp boards produced at M/s Saint-Gobain Gyproc India Ltd. (Jind, India) are given in Table 6.27.

Many different types of gypsum plaster-boards are manufactured, depending on their intended use. Distinctive features are size, edge configuration, weight, water resistance, structural behaviour, and strength, Gypsum plaster boards with specific fire-resistant properties incorporate fiber glass. The non-combustible gypsum lightweight board is a recent development, making use of a laminate of glass fiber mat with a lined-on glass silk scrim to replace the card board[30].

Modern plants for the manufacture of plasterboards have a capacity of about 20×10^6 m^2 per year and an annual consumption of 150000 t of hemi-hydrate plaster.

Plasterboard is used in interior finishing, *e.g.,* for ceiling and wall panelling. It is screwed to wooden or metal frames or pasted on masonry or concrete with special building plaster (adhesive plaster, the German Ansetzgips). The joints are covered and finished with special paper and joint filter to form a smooth surface. Plasterboard can also be used for the construction of dismountable partitions and lightweight dividing walls having various characteristics (weight, sound proofing, fire resistance) in concrete or steel framed building as well as in prefabricated dwellings. Plaster board undergoes further processing in factories into special-sized panels, coffers, or lightweight laminated panels with intermediate layers of polystyrene or polyurethane, the last called insulating board. Multilayered plas-terboards are used for dry floorings and for the construction of lift shafts. Factory made tiles with decorative surface finishes such as plastic sheet or special coats of paint or else an aluminum foil to prevent water vapor transmission are available.

Gypsum partition panels consist of set gypsum plaster. To produce them β-hemihydrate plaster is mixed with water (water-to-plaster ratio 0.9–1.0) and the slurry, which sets quickly, is poured into molds. After 5–8 min the panels are taken out of the molds and dried[31].

The standard size of gypsum partition panels (Fig. 6.21) is 500 × 666 mm, with a thickness of 60, 70, 80, or 100 mm and a density of 700–900 kg/m^3 . They are used in interiors as lightweight dividing walls, the tongue and groove joints being bonded with joint plaster. The partition walls can be single or multilayered. Characteristic features are low weight, average to good sound insulation, and excellent fire resistance.

Fig. 6.21 Manufacturing of Gypsum Partition Panels (Size: 500 x 666 mm, thickness: 60 to 80 mm)

Gypsum ceiling tiles are produced by mixing β-hemihydrate, water, and small amounts of glass fiber and pouring the slurry into rubber molds. The molds allow for designs of individual choice.

Ceiling tiles are used as decorative tiles, ventilation tiles, heating tiles, or sound-proofing tiles with mineral wool bonded on the back. They are screwed to a base frame and fitted into the framework. They are normally 625 × 625 mm in size and weigh between 10 and 20 kg/m^2.

Gypsum fiber board and fiber-reinforced gypsum elements are another group of gypsum building components manufactured from hemihydrate plaster and paper fib, glass fibers, or another type of fiber material[32–36]. Glass fiber, as mat or web, can be incorporated[37]. The proportion of fiber in these components, evenly distributed in the plaster, ranges up to 15 wt%. These fiber boards have an apparent density between 800 and 1200 kg/m^2.

Factories for the manufacture of gypsum partition panels and ceiling tiles have capacities up to 10,00000 m^2/a, consuming about 50,000 t/a of hemihydrate plaster, or else gypsum fiber boards up to 5×10^6 m^2/a, consuming the same amount of hemihydrate.

6.8.4 Water Proof Gypsum Cabinet

This is new concept in gypsum industry wherein a water repellent gypsum prepared by mixing the calcined gypsum /plaster formed at 170–180°C, with cement, fly ash, SiO_2 and some water miscible polymeric compounds in water forming uniform slurry and filling in the moulds of cabinet size. The gypsum slurry is reinforced with 3 to 5 mm galvanized steel rods in cross mesh and vibrated. The gypsum mass during setting and hardening is steamed cured from outside to accelerate strength development in the component through pozzolanic reaction. The units are demoulded after one to two hours and then dried at 110–120°C in air circulated ovens. The cabinet is a structure of combined bathroom and a toilet. The cabinet is produced in the factory and then lifted and transported on trucks or rail carriage to the construction site for use. These units are produced in Baltic countries - Latvia, Estonia and Lithuania[38].

6.8.5 Polymerized Gypsum

This is a new class of water-resistant high strength materials used in buildings. Gypsum impregnated with vinyl, polyurethane, polyester or epoxy resins and polymerized is sometimes are termed as Gypsum Polymer Composite (GPC). A lustrous material with low water absorption is obtained and its use in decorative furniture is well known in USA and UK. A material developed in the UK based on glass fiber, a thermosetting resin and α-hemihydrate gypsum shows a remarkably good performance in outdoor exposure. The material is very expansive and cannot compete with concrete but interesting if compared with some high performance fibre reinforced cement composites. Recently, a material called 'Forton-Jasonite polymer-modified glass fibre reinforced gypsum, comprising of 49% m/m α-hemihydrate gypsum, 22% m/m polymer, 12% glass fibre and 17% water has been produced in Netherlands by Bijen and Plas[39–40]. The glass bundles are well bonded to the matrix due to the polymer. In tensile failure a part of the fibres fractures while for another part matrix failure occurs through shearing. The main features of material are high tensile strength and good weather ability. The latter allows outdoor application whereas traditional applications of gypsum are inside buildings. Typical characteristics are a modulus of rupture of 75 N/mm^2, a tensile strength of 30 N/mm^2 and a strain capacity of 1.8%. Other interesting properties are high impact resistance and good performance in ballistic tests. At present the material is used for cladding panels, window frames, imitations of natural stones, etc.

Typical photographs of the polymerized glass reinforced gypsum are shown in Figs. 6.22 to 6.24.

The effect of addition of epoxy formulation on the properties of the gypsum plaster has been studied and reported by Singh[41]. The epoxy composition was prepared by mixing equal proportion of resin-diglycidyl ether of bisphenol A, hardener-trimethyl hexamethylenediamine and small quantity of digylcidyl ether at viscosity of poise 70. Different percentages of the epoxy formulation were added to the b-hemihydrate plaster at 50% consistency at 30°C. The properties are listed in Tables 6.28 and 6.29.

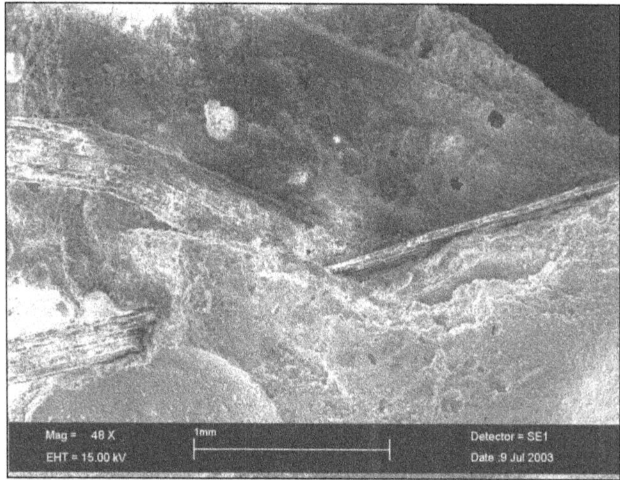

Fig. 6.22 Microstructure of PGRG

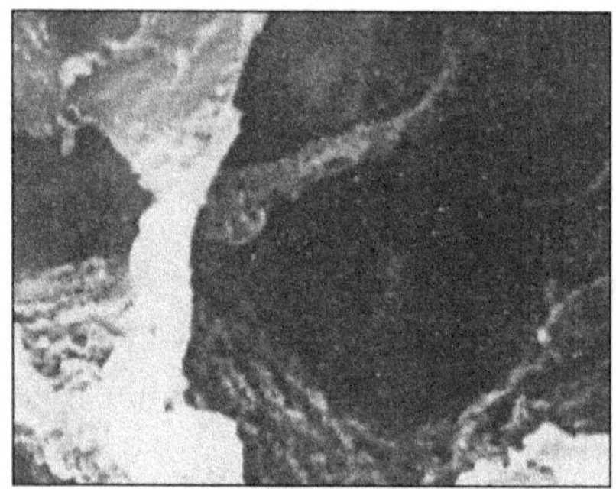

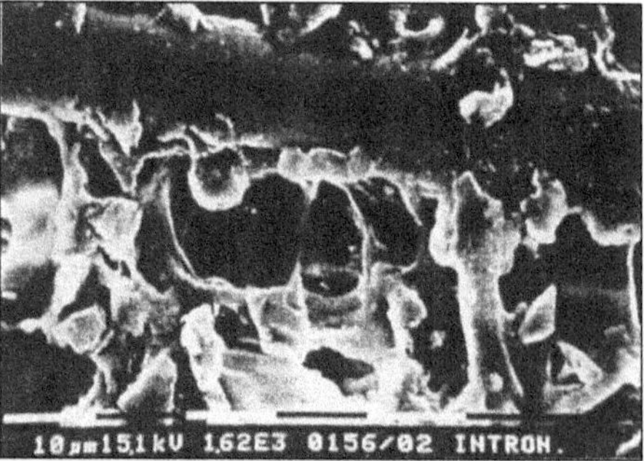

Fig. 6.23 Photographs of polymer structure of PGRG

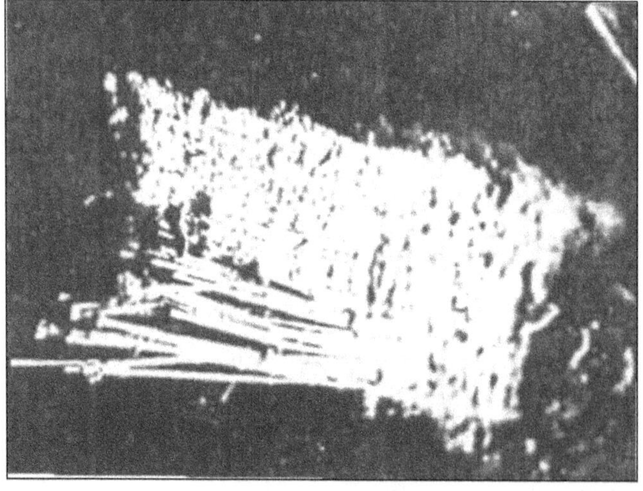

Fig. 6.24 Glass fibre strand pulled out gypsum matrix

Table 6.28 Properties of Gypsum Plaster in Presence of Different Concentration of Epoxy Formulation

Epoxy formulation (%)	Compressive strength, MPa			Bulk density, (g/cc)		
	3 day	7 day	28 day	3 day	7 day	28 day
3	12.8	13.3	16.7	1.28	1.32	1.33
5	16.3	17.4	19.8	1.39	1.42	1.49
7	13.8	14.5	18.0	1.26	1.29	1.32

Table 6.29 Water absorption and Porosity of Gypsum Plaster in Presence of Different Concentration of Epoxy Formulation

"Epoxy formulation (%)"	Water absorption (%)			Porosity		
	2 hr	8 hr	24 hr	2 hr	8 hr	24 hr
3	21.78	21.80	21.80	21.8	26.25	26.37
5	9.50	9.67	9.70	13.77	14.02	14.06
7	16.76	16.8 0	16.82	22.45	22.51	22.54

The bonding of polymeric material with gypsum matrix is shown in Fig. 6.25.

With the addition of EPI epoxy polymer, there is increase in compressive strength and bulk density of the plaster with increase in curing period. At 5% of the epoxy, highest compressive strength and maximum reduction in water absorption and porosity values was observed. In polymerized gypsum polymer particles can fill up the interstices between the gypsum plaster particles and modifies the matrix leading to decrease in water absorption and improve

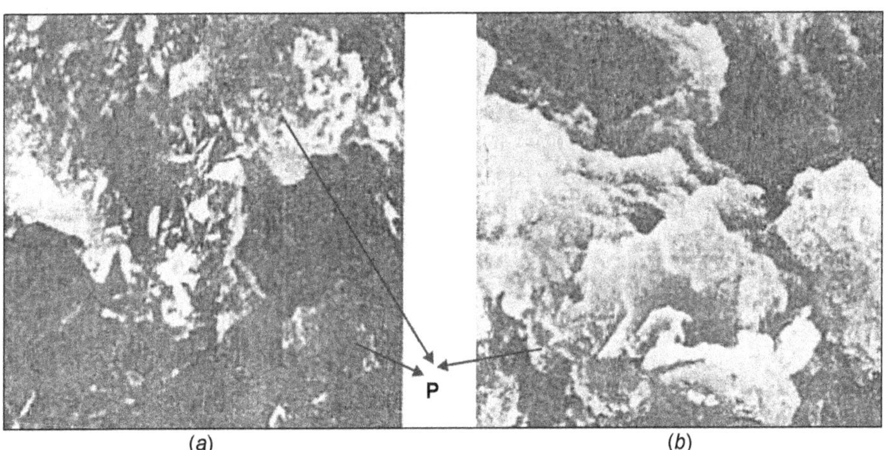

Fig. 6.25 SEM of Polymerized Gypsum plaster cured for (*a*) 7 Days (*b*) 28 Days (P = Polymer)

in compressive strength of the plaster. The treated gypsum plaster has great potential for use in external situations and damp environment.

6.9 New Plaster Compositions from Gypsum

6.9.1 Development of Multiphase Plaster

Gypsum is an accomplished building material known for rapid setting, medium strength and excellent performance under dry climatic conditions. The industrial importance of gypsum is due to its ability to form hemihydrate plaster at much lower temperature than that for lime or cement manufacture. Over 95 per cent of the calcined gypsum produced world over is β-hemihydrate plaster. Multiphase plaster is a versatile variety of gypsum plaster composed of β-hemihydrate and anhydrite III in range 30%–60% as well as $CaSO_4$ II. It is called "Putzgips" in Germany and "plaster de construction" in France. Multiphase plaster makes up about 1/3rd of the world's building plaster production. The β-hemihydrate or plaster of Paris made at 150 - 160°C sets quickly and requires retardation whereas anhydrite plaster made at 800 - 850°C requires fine grinding and activation. The plaster of Paris on setting and hardening produce low density and medium strength while anhydrite plaster sets slowly and attains maximum strength at 28 days of curing. Multiphase plaster has quick initial setting and gradual final setting and these plasters develop quite high strength (15–20 MPa).

Most of the work reported on multiphase plaster is patented and little information is available. Moreover, multiphase plasters have been produced using natural gypsum so far. Investigations were, therefore, undertaken to produce multiphase plaster which does not require either retardation or activation and gives properties required for plastering and other applications. The multiphase plaster (MPP) was produced by heating the waste gypsum of phosphoric acid (phosphogypsum), hydrofluoric acid (fluorogypsum) and H-acid (intermediate dyes) origin at different temperatures between 300–1000°C to get a mixture of hemihydrate, soluble and insoluble anhydrite plasters[42–43]. The MPP made from waste phosphogypsum was tested and evaluated for various physical properties as per relevant Indian standards. The results are listed in Table 6.30.

Data show an increase in compressive strength, bulk density and chemically combined water with increase in curing. Addition of activator (mix. C) does not enhance strength appreciably. The attainment of strength is remarkably higher than the POP. The increase in chemically combined water confirms conversion of various phases of gypsum plaster into the hardened dihydrate gypsum. The Scanning Electron Microscopy (SEM) of the hydrated MPP is shown in Fig. 6.26 (*a*) at 7 days, (*b*) at 28 days. At 7days subhedral to euhedral prismatic crystals of irregular boundaries are formed while at 28 days, euhedral prismatic, lath and fibrous needles shaped crystals with uniform stacking with cohesiveness are crystallized.

Table 6.30 Properties of Multiphase Plaster

*Mix Compn.	Cosistency (%)	Setting time (Minutes)	Compressive strength (MPa)			Bulk density (g/cc)			Chemically combined water (%)		
			3d	7d	28d	3d	7d	28d	3d	7d	28d
A.	46.0	25.0	5.6	6.8	15.6	1.0	1.1	1.18	5.5	7.0	14.7
B.	40.0	10.0	5.5	6.0	16.8	0.9	1.0	1.06	6.9	10.2	16.9
C. (a)	38.6	15.0	6.0	6.8	17.8	1.1	1.1	1.20	7.5	8.6	18.0
POP	56.0	8.0	10.8 (24 hrs)			1.10 (24 hrs)			18.8 (24 hrs.)		

*Mixes A, B, C: 65 % Anhydrite plaster (Prepared at different temperatures) plus Soluble anhydrite and Hemihydrate plasters, a: Activator

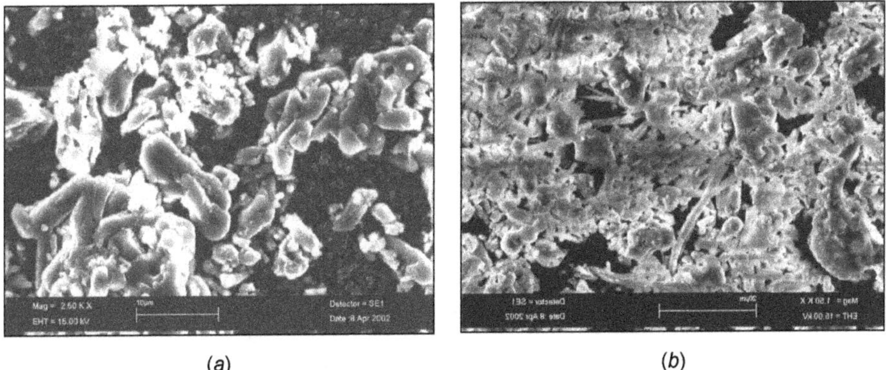

(*a*) (*b*)

Fig. 6.26 SEM of MPP Hydrated for (*a*) 7days and (*b*) 28 days

6.9.2 Applications of Multiphase Plaster

The multiphase plaster can be used in plastering both for finish coat as well as base coat in buildings. The highly glossy surface of MPP may replace "Neeru" (a lime based material) or commercial putties used over coarse plastering in cosmopolitan cities. Multiphase plaster may be blended with the lightweight exfoliated vermiculite, perlite or any compatible aggregate for making lightweight thermal insulative plasters. The MPP has been found eminently suitable for making fibrous plaster boards as per the Indian Standards IS : 2095-(Part-3)–1990. Many entrepreneurs have shown keen interest to adopt this technology. The process know-how has been commercialized.

6.9.3 Waste Gypsum from Intermediate Dye Industries

This is another variety of by-product gypsum popularly known as H-acid gypsum, is produced in India to the tune of 0.5 million tonnes per annum from the dye intermediate by the neutralization of free sulphuric acid with limestone. The H-acid gypsum is an intermediate for the manufacture of different dyes, in the process of sulphonation of naphthalene with oleum 65% and sulphuric acid followed by nitration with nitric acid. The nitro mass is reacted with limestone to remove the unreacted sulphuric acid present in it as per the following reaction:

$$H_2SO_4 + CaCO_3 + H_2O \longrightarrow CaSO_4.2H_2O + CO_2$$

This treatment produces the waste gypsum which is filtered, washed and disposed off as land fill. The filtrate is reduced by reduction with iron and HCl to give an amine which is reacted with caustic soda to form hydroxy amine called H-Acid. The chemical name of H-Acid is 1-Amine-8-hydroxy-3,6-disulphonic acid. The H-Acid gypsum samples were collected from two plants from Amal Rasayan Limited, Atul, Gujarat and the Zenith Chemicals Limited, Thane, Maharashtara. The samples were analyzed for their chemical constituents as per IS :1288–1986 (Table 6.31). They were also evaluated by differential thermal analysis and scanning electron microscopy (SEM)[44].

1. Characterization of H-Acid Gypsum

The results of chemical composition of the gypsum samples are shown in Table 6.31. Data show that gypsum contains impurities of organic matter and the unreacted carbonate particles. The organic matter may affect the setting and colour of the calcined gypsum, it is important to beneficiate gypsum samples to get the product with suitable properties fit for making building or ceramic grade plasters and the gypsum components.

Table 6.31 Chemical Composition of H-Acid Gypsum

Constituents %	H-acid gypsum Samples	
	*A	*B
SiO_2 + insoluble in HCl	3.4	1.78
CaO	28.14	29.45
$R_2O_3(Al_2O_3 + Fe_2O_3)$	6.9	5.2
MgO	2.2	2.1
SO_3	40.20	41.30
Organic matter	1.2	0.36
$CaCO_3$	2.0	1.67
Combined water	18.70	18.20
pH	6.96	6.97

*A = Amal Rasayan Ltd., Atul, B = Zenith Chemicals Ltd., Thane

Table 6.32 Properties of Beneficiated H-Acid Gypsum

Constituents, %	H-Acid Gypsum	
	A	B
SiO_2 + insoluble in HCl	2.89	1.45
$R_2O_3(Al_2O_3 + Fe_2O_3)$	4.26	3 .6
SO_3	42.2	41.40
Organic matter	0.80	0.11
pH	7.96	8.0

2. Beneficiation of H-Acid Gypsum

The gypsum samples were thoroughly churned with water for 15–20 minutes in the mixer, vacuum filtered and washed with water and then dried in the drier at 100–110°C to the moisture content below 5%. The beneficiated gypsum samples were tested for the residual impurities.

The level of impurities present in the benficiated H-acid gypsum are shown in Table 6.32. The results show reduction in the organic matter, silica, R_2O_3 contents and an increase in the SO_3 content. The pH of the gypsum samples increased thus indicating an improvement in the beneficiated gypsum.

The SEM of the H-Acid gypsum are shown in Fig.6.27. It can be seen that the unbeneficiated H-acid gypsum sample A [Fig. 6.27 (*a*)] is mainly comprised of prismatic crystals along with the presence of lath and twinned small anhedral to subhedral prismatic crystals coated with the attrited gypsum particles. Occasionally formation of rounded bodies comprising calcium carbonate interspersed with fused organic matter can be seen. The gypsum sample after beneficiation [Fig. 6.27 (*b*)] shows agglomerated euhedral prismatic crystals in predominance in association with the rounded bodies of calcium carbonate in comparatively lesser quantity than the unprocessed gypsum sample [Fig. 6.27 (*a*)]. The presence of reduced

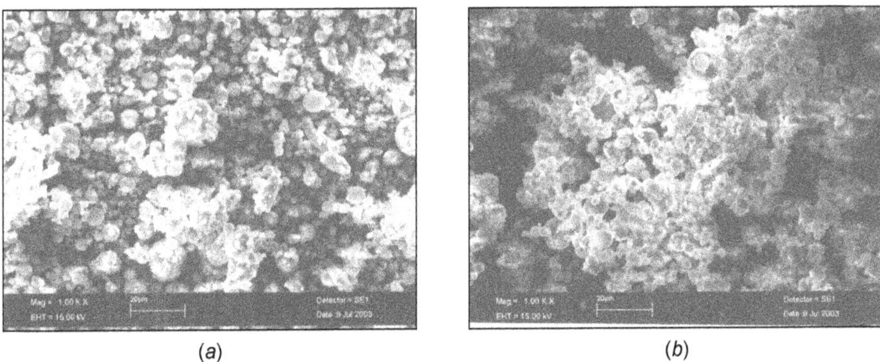

(*a*) (*b*)

Fig. 6.27 SEM of H-Acid Gypsum Sample 'A' Before beneficiation (*a*) and After beneficiation (*b*)

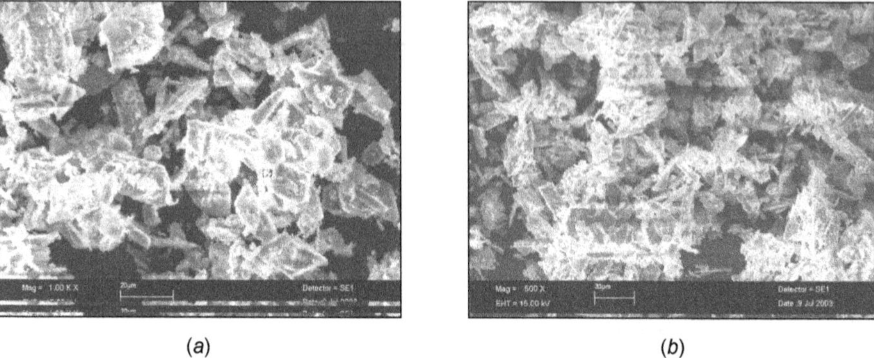

(*a*) (*b*)

Fig. 6.28 SEM of H-Acid Sample 'B' Before Beneficiation (*a*) and After Beneficiation (*b*)

quantity of calcium carbonate and the euhedral prismatic crystals indicate removal of the impurities from the gypsum sample.

The SEM of H-acid gypsum sample 'B' [Fig. 6.28 (*a*)] shows presence of anhedral to subhedral prismatic and lath shaped crystals with occasional twinning and agglomeration. The agglomeration is due to presence of organic matter and the presence of fine gypsum crystals. The beneficiated gypsum sample [Fig. 6.28 (*b*)] shows appearance of euhedral prismatic crystals in majority showing uniform stacking. The absence of agglomeration and formation of euhedral prismatic crystals confirm decrease in the impurities in the beneficiated gypsum sample than the unprocessed gypsum.

3. Properties of Gypsum Plaster

The properties of gypsum plaster produced by calcining unprocessed and beneficiated H-acid gypsum are listed in Table 6.33. It can be seen that plaster sample 'B' possesses better setting and strength properties than the plaster sample 'A' probably due to higher purity and the low consistency values of the plaster in

Table 6.33 Properties of Gypsum Plaster Produced by Calcining H-Acid Gypsum

Sample des.	Properties						
	Consistency (%)	Setting time (Minutes)		Bulk density (kg/m^3)		Compressive strength (MPa)	
		UR	R	UR	R	UR	R
A							
Unprocessed	67.0	7.0	25.0	1240	1210	9.20	8.50
Beneficiated	62.0	10.0	35.0	1250	1225	11.60	10.60
B							
Unprocessed	65.2	6.0	28.0	1280	1225	11.20	9.50
Beneficiated	61.0	10.0	35.0	1250	1225	12.6	11.44

the former than the latter plaster samples. However, both the plasters complied with the requirements laid down in IS : 8272–1981, Specification of gypsum plaster for use in manufacture of fibrous plaster boards and IS : 2547 (Part 1)–1976, Specification for gypsum building plaster.

4. Preparation and Testing of Plaster Products

Fibrous Gypsum Plaster Boards

The properties of fibrous gypsum plaster boards (FGPB) prepared and tested as per procedures explained earlier in the chapter are listed in Table 6.34. It can be seen that the properties conformed to the requirements of IS : 2095 (Part–3) - 1996, Gypsum plaster boards - specification, Part 3-Reinforced gypsum plaster boards. The plaster boards have lower thermal conductivity than the conventional asbestos cement sheets generally used as false ceiling, there by facilitating the better insulation of the gypsum boards. The boards are known for their excellent fire resistant and acoustic properties and for their use in false ceiling, light-weight partitions and cladding.

Gypsum Building Blocks

The properties of gypsum blocks are listed in Table 6.35. Data shows the gypsum blocks possess bulk density in the range 1,000–1,100 kg/m^3 and the compressive strength 2.5–3.0 MPa. The compressive strength conformed to the

Table 6.34 Properties of Fibrous Gypsum Plaster Boards Made from Gypsum Sample 'A'

Characteristics	Results		
	Asbestos-cements	FRGB	IS:2095–1996 limits
Transverse strength, N	500–550	500–600	Min.340
Bulk density, kg/m^3	1600–1700	1000–1100	—
Deflection, mm	—	1.5	Min.19.0
Hardness, mm	Passes	Passes	Max. 5 mm
Thermal conductivity, (Kcal/m/hr/°C)	0.24–0.30	0.12–0.14	—

Table 6.35 Properties of Gypsum Building Blocks Made from Gypsum 'A'

Property Studied	Results	IS : 2849–1983 Limits
Compressive Strengrth, MPa	2.2–2.5	—
Bulk Density, kg/m^3	1150–1180	N.S.
Water absorption, %	35–40	N.S.

N.S. = Not Specified

requirements laid down in IS : 2849–1983, Specification for non-load bearing gypsum partition blocks (Solid & hollow type). The values of the bulk density and water absorption have not been specified. Gypsum blocks are lightweight, fire resistant building components suitable for use as lightweight partitions for non-load bearing walls. The gypsum blocks should not be used in the exposed situations being soluble in the water. The gypsum blocks may find wide applications if the water-resistance properties are imparted to them.

In this regard, the addition of ground granulated blast furnace slag (GGBS) or fly ash along with other suitable additives may improve weather – resistance of these blocks at economic cost.

References

1. Singh Manjit, Phosphogypsum based Composite Binders, Journal of Scientific & Industrial Research, October 2001, Vol. 60, pp. 812–817.
2. Singh Manjit and Garg Mridul, Cementitious Binder from Fly ash and Other Industrial Wastes, Cement and Concrete Research (U.S.A.), 1999, Vol. 29, pp. 309–314.
3. Singh Manjit and Garg Mridul, Durability of cementitious binder derived from industrial waste, Materials and Structure J., (France) RILEM, December 1997, Vol. 30, pp. 607–612.
4. Singh Manjit, Garg Mridul, Phosphogypsum - Fly ash Cementitious Binder– Its Hydration and Strength Development, Cement and Concrete Research (U.S.A.), 1995, Vol. 25, No. 4, 1995, pp. 752–758.
5. Singh Manjit & Garg Mridul, Investigation of durable gypsum binder for building materials, Construction and Building Materials (U.K.), March 1992, Vol. 6, No. 1, pp. 52–56.
6. Singh Manjit & Garg Mridul, Studies on the formation of cementitious compounds using phosphogypsum and fly ash, 2nd Annual General Meeting of MRSI, NPL, New Delhi, 9–10 Feb. 1991.
7. Singh Manjit & Garg Mridul and Rehsi, S.S.. Water resistant binder from waste phosphogypsum, International Congress CIB 89 Quality for Building Users Throughout the World, Paris (France), Theme 11, Vol. 11, June 19–23, 1989, pp. 339–352.
8. Singh Manjit & Garg Mridul, Development of cementitious properties in phosphogypsum flyash system, 9th International Congress on Chemistry of Cement, New Delhi, Vol. IV, Theme 111, 23–28 No. 1992, pp. 489–94.
9. Colak, A., The long term durability performance of Gypsum - Portland Cement-Natural Pozzolana Blends, Cement & Concrete Research, 2002, Vol. 32, No. 1, pp. 109–115.
10. Singh Manjit, Phosphogypsum - cement - pozzolana binder for use in construction work, Intl. Conf. on Low Cost Housing for Developing Countries, CBRI, Roorkee, March 8–12, 1985, Vol. 1, pp. 233–237.
11. Singh Manjit, Verma, C.L. & Garg Mridul, Development of New Eco-friendly building materials from by product Gypsum, National conference on building materials, the Emerging Technologies in building materials, Hitex, Hyderabad, December 6, 2003.
12. J,C. Pearson, Measurement of bond between bricks & mortar, Proc. A.S.T.M, 1943, Vol. 43, p. 857.
13. Singh Manjit, (Unpublished), Masonry cement from Phosphogypsum.
14. Singh Manjit, Rehsi, S.S. and Taneja, C.A., Beneficiation of phosphogypsum for use in building materials, National Seminar on Building Materials-Their Science and Technology, New Delhi, 15–16 April 1982, IIA(I), pp. 1–5.

15. Singh Manjit & Garg Mridul, Gypsum based fibre reinforced composite - an alternative to timber, National Symposium on Substitute of Wood in Buildings, C.B.R.I., Roorkee, 12–13 Dec., 1991.
16. Singh Manjit & Garg Mridul, Gypsum binders and fibre reinforced gypsum products, Indian Concrete Journal, August 1989, Vol. 63, No. 8, pp. 387–392.
17. Singh Manjit & Garg Mridul, Gypsum based fibre reinforced composites, An Alternative to timber. Construction and Building Materials (U.K.), 1994, Vol. 8, No. 3, pp. 155–160.
18. Singh Manjit & Garg Mridul, Glass fibre reinforced water resistant gypsum based composite, Cement Concrete Composite, (U.K.), 1992, Vol. 14, No. 1, pp. 23–32.
19. Singh Manjit, Garg, M, Verma, C.L. and Kumar. R., Possibilities of Utilizing Phosphogypsum as building material. National Seminar on 'Cement & Building Materials from Industrial Wastes, July 24–25, 1992, Hyderabad, pp. 86–93.
20 Singh Manjit & Garg Mridul, Microstructure of glass fibre reinforced water resistant gypsum binder composites, Cement and Concrete Research (U.S.A.), 1993, Vol. 23, No. 1, pp. 213–220.
21. Kraemer, E.F., The Vidin gypsum fibre board plant, Bulgaria, Zement-Kal-Gips, 1990, Vol. 43, pp. 330–333.
22. J. Bold, Kaiserslauteru, A New Method of Producing Gypsum fibre Board at a Dutch Gypsum Plant, Zement-Kal-Gips, 1989, Vol. 5, pp. 255–258.
23. Lamper Karsten, Thomas, Hilbert and Gurizerodt, Helge, Development of Gypsum bonded particle board Manufacture in Europe, Forest Res. Technol., 1990, Vol. 40, pp. 37–49.
24. Kossatz, G, Verfahren Zum Herstellen Venn Gypsbanilen Insbesondene Gypsplaten, De-AS 2919311, Federal Republic of Germany, 1979.
25. Singh Manjit, Verma, C.L., Garg Mridul, Handa, S.K. & Kumar R., Studies on Sisal fibre Reinforced Gypsum Binder for Substitution of Wood. Research and Industry, March, 1994, Vol. 39, No. 3, pp. 55–59.
26. Majumdar, A.J., Glass fibre reinforced cement and gypsum products,, Proc. R. Soc 319, (London 1970), pp. 69–78.
27. Walton, P.L. and Majumdar, A.J., Fraction Energy of Plain and Glass Reinforced Gypsum Plaster, J. Mater. Sci., 1977, Vol. 12, pp. 831–836.
28. Ryder, J.F., Glass Fibre Reinfornced Gypsum Plaster, Pro., International Building Exhibition Conference on 'Properties for Fibre Reinforced Construction Materials', London, (Building Research Station, Watford 1972), pp. 69–89.
29. Neuhauser, G., Bundesbaublatt, Vol. 31, 1982, pp. 566–569.
30. Aeppli, E., Eurugypsum, Stockholm, 1972.
31. Knauf, N.W., DE 1104419, 1957. h
32. FERMA Gesellschaft fur Rationelle Fertigbaumethoden und Maschinenenanlage GmbH & Co., CH 505674, 1969.
33. Ali, M.A. and Grimmer, F.J., Mechanical Properties of Glass Fibre Reinforced Gypsum, J. Matr. Science, 1969, Vol. 4, p. 389.
34. Kazimir, J, Tonind. Ztung. Keram, Rundscch., 1967, Vol. 91, pp. 22–25.
35. Singh Manjit, Recent Advances in the Utilization of Waste Fluorogypsum, Global Gypsum Magazine (U.K.), No. 2011, pp. 18–29.
36. Kossatz, G, and Lemper, K., Holz. Roh. Werkst., 1982, Vol. 40, pp. 333–337.
37. Kossatz. G., Baustoffindustrie, 1966, Vol. 9, pp. 1–5.
38. Sauriesu Puvmateriaulu Kombinates, Izstradajumu Katalogs, Riga, Latvia, 1988.
39. J. Bijen, J. and Van der Plas, C., Polymer Modified Glass Fibre Reinforced Gypsum, Polymeergmodi-ficeered Glascezelversterkt Gips, Dec., 1990, CUR Report, pp. 90–98.

40. Bijen, J. and Jacobs, M., Properties of Glass Reinforced polymer Modified Cement, J. Mat. Constr, Vol. 15, No. 89, pp. 446–452.
41. CBRI Biennial Report, 2004–2006, pp. 23–24.
42. Singh Manjit & Garg Mridul, Development of Multiphase Plaster from Waste Gypsum, Indian Concrete Journal, May 2003, Vol. 77, No. 5, pp. 1086–1089.
43. Singh Manjit & Garg Mridul, Multiphase Plaster from Waste Gypsum – A New Preposition, National Seminar on Recent Trends in Building Materials, Bhopal, 26–27 Feb., 2004, pp. 40–45.
44. Singh Manjit & Garg Mridul, Recycling of Waste Gypsum from Intermediate Dye Industries for the Production of Value Added Building Materials, 8th International Seminar on Cement and Building Materials, 18–21 No. ember 2003, New Delhi.

7

Anhydrite Plaster Based Building Products

INTRODUCTION

Anhydrite ($CaSO_4$) as such, does not combine or react with water. On addition of suitable accelerators, generally soluble sulphates, it gets hydrated and forms a cementing materials. Its curing in high humidity is essential for the development of strength. Besides, high compressive strength and density, its chief advantages are resistance to moisture attack, and cracking. Its capacity of receiving polish is put to advantage in making ornamental plaster and imitation marble.[1–5] Anhydrite is available in nature also. Although natural anhydrite, available in Germany and Russia, has been used as chief cementing material in these countries[6], there is no mention of the use of by-product anhydrite in India. It is produced artificially both from natural mineral as well as by-product phospho or any other variety of gypsum.

On calcining gypsum or phosphogypsum (PG) at elevated temperature, anhydrite is formed. When phosphogypsum is heated then phosphoanhydrite is formed. On heating PG, the phosphate compounds present as impurity in the phosphogypsum are converted into water insoluble inert form and the fluoride gets volatilized partially. In the present chapter, the effect of different chemical accelerators on various properties of phosphoanhydrite has been studied for the development of suitable plasters.

7.1 Preparation of Phosphoanhydrite

For the preparation of phosphoanhydrite, phosphogypsum samples Coromondal Fertilizers Ltd., Vizag (CFL) and Southern Petrocheical Industries Corporation (SPIC) Tuticorin were calcined in the electric furnace at the temperature of 500°C, 600°C, 700°C and 800°C, 900°C and 1000°C for 4 hours. The anhydrite samples were cooled and ground to a fine powder.

7.1.1 Properties of Anhydrite Plaster

The water soluble impurities of P_2O_5 and F can be easily removed from PG by washing with water. However, the P_2O_5 occluded, *i.e.*, $CaHPO_4.2H_2O$ in the gypsum crystals is rather difficult to be removed due to its low solubility 0.0286 g/100 cm^3 water at 35°C. The strength of cement and calcined gypsum are adversely affected by slow dissolution of $CaHPO_4.2H_2O$[7–8]. This was confirmed by obtaining the absence of phosphate in the extract prepared by shaking the anhydrite prepared from PG in saturated lime water for 4 hours. The phosphate content does not change only the organic matter is volatilized to large extent. The results showed that stable anhydrite is formed between 800–1000°C. The other properties such as density, specific surface area are listed in Table 7.1. Data showed that with enhancement in calcination temperature, the values of specific gravity, pH and specific surface area increased and that of $CaSO_4.2H_2O$ content reduced. The increase in pH value beyond 800°C is probably due to formation of free lime from the partial decomposition of gypsum. The loss on ignition of anhydrite at 800°C, 900°C and 1000°C was within the maximum specified value of 2.0% laid down in IS : 2547(Part-1)–1976[9], thereby, establishing the formation of stable anhydrite content. It is interesting to note here that anhydrite cement produced at 1000°C is less energy intensive than the Portland cement and other building materials like wood particle board, quick lime, mild steel, PVC, aluminium, etc.

X-ray diffraction patterns of unprocessed phosphogypsum and the phosphoanhydrite obtained by firing at 500°C (*a*), 600°C (*b*), 700°C (*c*), 800°C (*d*), 900°C (*e*) and 1000°C (*f*) are shown in Fig. 7.1. It can be seen from X-ray diffraction patterns that compared to unprocessed phosphogypsum, shifting of the most intense lines/reflections takes place in case of anhydrite cement obtained at 500°C, 600°C and 700°C. As the calcination temperature is increased, no shifting in lines occurs, which shows the absence of secondary rearrangement of the gypsum crystal lattice. An increase in the intensity of some reflections and decrease in others reflections are found. Major

Table 7.1 Characteristic Properties of the Anhydrite Produced from Phosphogypsum

Temperature of Calcination (0°C)	Properties			
	Density (kg/m^3)	Specific surface Area Blaine (m^2/kg)	pH	Loss on ignition (%)
500	2650	260	7.1	3.0
600	2700	266	7.1	2.5
700	2780	270	7.1	2.3
800	2800	275	7.2	2.0
900	2810	280	7.3	1.8
1000	2900	288	7.3	1.6

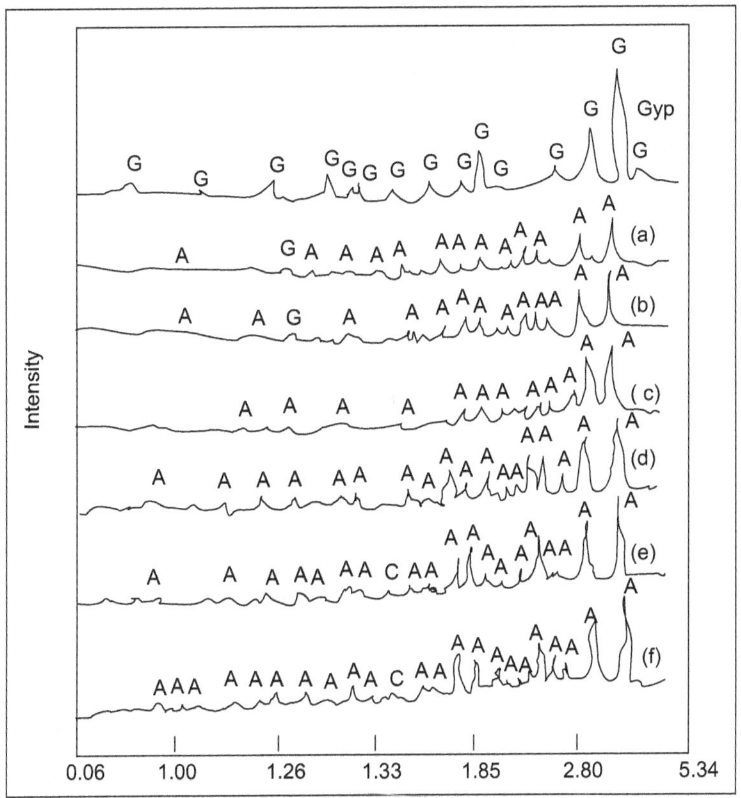

Fig. 7.1 X-ray Diffractograms of Phosphogypsum Calcined at Different Temperatures [G-Gypsum ($CaSO_4.2H_2O$), Anhydrite-($CaSO_4$) and C-lime (CaO)]

reflections were noticed at 3.549, 2.849, 2.3298 and 2.2115 Å which confirm the conversion of phosphogypsum into anhydrite of stable configuration.

The microstructure of fractured surface of anhydrite cement produced at 500°C, 600°C, 700°C, 800°C, 900°C and 1000°C are shown in Fig. 7.2 denoted by (*a*), (*b*), (*c*), (*d*), (*e*) and (*f*) respectively. At 500°C, 600°C and 700°C [Fig. 7.2 (*a*)–(*c*)], anhedral to euhedral prismatic and rhombic shaped crystals of irregular boundaries and stacking are formed. The formation of small-size pseudoamorphous particles has largely enhanced. The anhydrite produced at 800°C, 900°C and 1000°C [Fig. 7.2(*d*)-(*f*)] shows appearance of euhedral to anhedral prismatic and rhombic shaped crystals of platy structure binding the smaller anhydrite grains. The pseudoamorphous gypsum crystals break down into micro-crystals with the increase in calcination temperature. These results corroborate the findings of Berezovskii[10]. The pseudoamorphous crystals may be attributed to the crystals of soluble anhydrite (γ-$CaSO_4$) which are transformed to anhydrite with the increase in temperature. At 900°C [Fig. 7.2(*e*)], the cavities and irregular boundaries with variable stacking of anhydrite crystals can be seen. However, the size of prismatic crystals increased in association with

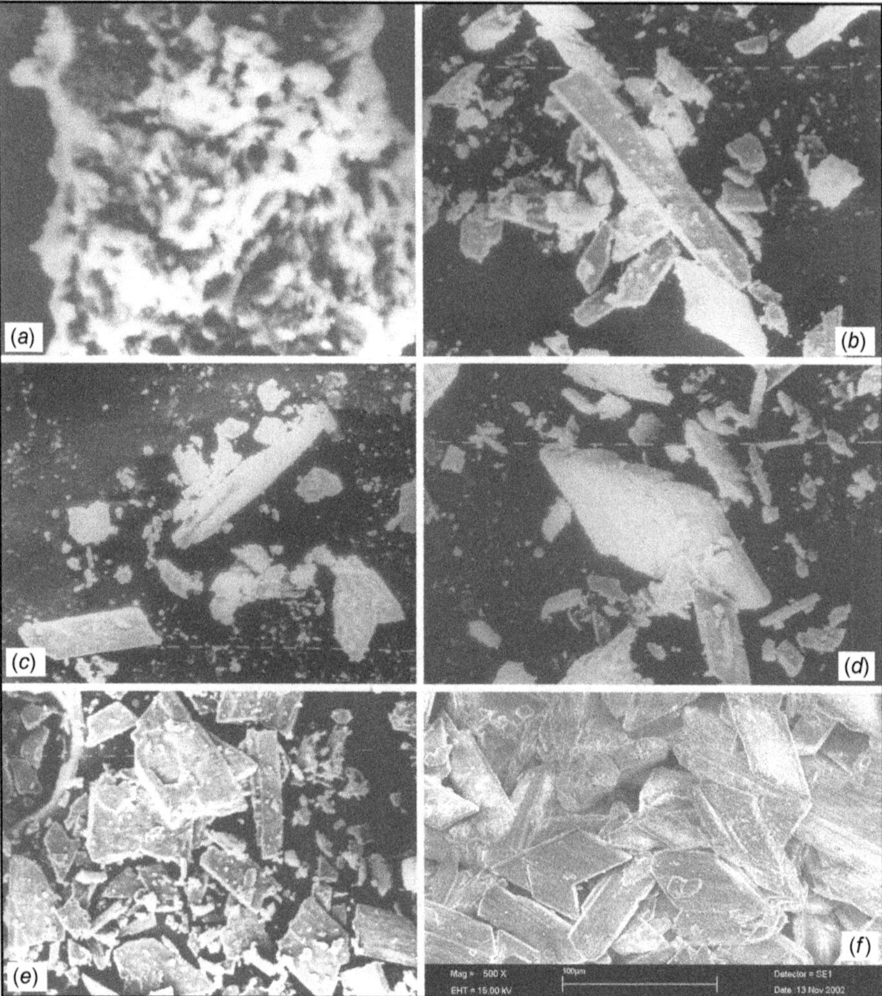

Fig. 7.2 Microphotographs of Anhydrite Cement Produced at Different Temperatures (*a*) 500°C, (*b*) 600°C, (*c*) 700°C, (*d*) 800°C, (*e*) 900°C and (*f*) 1000°C (magnification × 1200)

occasional formation of rhombic shaped crystals with sharp boundaries at 1000°C [Fig. 7.2(*f*)]. The formation of euhedral prismatic crystals and transformation of pseudamorphous crystals of soluble anhydrite into micro crystallites having sharp edges and uniform stacking could be the deciding factor for the optimum strength development in phosphoanhydrite cement.

7.1.2 Physical Testing and Evaluation of Anhydrite Cement/Plaster

The phosphoanhydrite plaster was tested and evaluated for properties like fineness and setting time as per IS : 2542 (Part 1)–1978[11]. The compressive Strength of anhydrite plaster was tested on 5 cm × 5 cm × 5 cm cubes of neat plaster at normal consistency using the optimum doses of the accelerators (previously determined on the basis of the specified setting time) dissolved in water. The cubes were cured at 27 ± 2°C and relative humidity of more than 90°% in sealed containers. After different periods of curing the cubes were dried at 42° ± 2°C to constant weight, cooled to ambient temperature and tested for compressive strength. The effect of different accelerators dissolved in water on the setting of neat plaster was examined. The mineralogical studies of anhydrite plaster/cement were determined by X-ray diffraction (Phillips diffractometer, Netherlands), scanning electron microscope (SEM model 508, Philips). The strength development of anhydrite cement was evaluated with differential thermal analysis (Stanton Redcroft, UK) and X-ray diffraction. The chemically combined water was determined as per the ASTM Srandard: C 471–76.

For determining efflorescence each, sample of anhydrite plaster was mixed with 0.1% brown iron oxide pigment to impart deep red colour to the plaster in order to distinguish white patches of efflorescence caused by the addition of accelerators. Samples of 10 cm diameter and 0.5 cm thickness were cast and left for drying at room temperature and the efflorescence was observed.

The effect of chemically combined water in the set specimens containing different doses of accelerators was studied by heating the ground samples at 400°C to constant weight and determining the loss in weight.

To study the effect of temperature on hydration and development of strength, three set of samples of anhydrite plaster were cast at normal consistency into 5 cm. cubes and cured at 27°C, 35°C and 40°C and relative humidity of more than 90%.

To study the durability of phosphoplaster, 5 cm set plaster cubes were cured for 28 days in air at 27°C and dried at 42 ±2°C for 3 days. Then they were subjected to alternate heating and cooling cycles. One cycle of alternate heating and cooling consisted of heating at 42°C for 6 hours and cooling at 27°C for 18 hours.

7.1.3 Assessment of Anhydrite Plaster

1. Fineness

The results of fineness are given in Table 7.2. It can be seen from the results that phosphoanhydrite plaster made at different temperatures complied with the specified fineness values.

2. Effect of Accelerators on Setting Time

The effect of different accelerators on the setting characteristics of DC and HDS phosphoanhydrite plasters is shown in Tables 7.3 and 7.4. The maximum limit of setting time specified is 6 hours. The data obtained showed that the plaster complied with the specified setting time when K_2SO_4, K_2SO_4, + $ZnSO_4$, Na_2SO_4, Na_2SO_4, + $FeSO_4$, $NaHSO_4$, were added in optimum quantities. Since

Table 7.2 Fineness of DC and HDS Phosphoanhydrite Plasters

Sl. No.	ASTM Sieve Nos.	CFL phosphoanhydrite			HDS Phosphoanhydrite			ASTM
		650°C % Passing	750°C % Passing	850°C % Passing	650°C % Passing	750°C % Passing	850°C % Passing	Limits
1	14 (1.405) mm	Passes com-pletely	Passes com-pletely	Passes com-pletely	Passes com-pletely	Passes com-pletely	Passes com-pletely	Shall pass com-pletely
2	40 (0.222) mm	99.4	99.0	99.45	99.60	99.7	99.18	Shall pass com-pletely
3	100 (150 micron)	97.0	99.5	98.0	96.0	98.4	98.60	Shall pass com-pletely

the phosphoanhydrite plasters were made at different temperatures, the optimum quantity of accelerators used was also variable. As shown in Table 7.3, the optimum quantity of K_2SO_4, Na_2SO_4, $NaHSO_4$, was in the range of 2.0 – 3.0% and for combinations like K_2SO_4, + $ZnSO_4$, Na_2SO_4,+ $FeSO_4$, $NaHSO_4$, + $FeSO_4$ and $Al_2(SO_4)_3$. $16H_2O$+ K_2CO_3, in the range of 1.5 – 1.8 + 0.2 – 1.5% respectively. The optimum quantity of accelerator used, was found to be lower in case of phosphoanhydrite made at 750°C.

Similar type of setting behaviour was observed in the case of SPIC phosphoanhydrite sample (Table 7.4). The optimum quantity of accelerator fell in the range 2.0 – 2.5% when K_2SO_4, Na_2SO_4 and $NaHSO_4$ were used independently. On using the combinations of K_2SO4, + $ZnSO_4$, Na_2SO_4, + $FeSO_4$ and $Al_2(SO_4)_3.16H_2O$ + K_2CO_3 accelerators the optimum doses was found to be in the range 1.0 – 1.7, 0.3 – 1.7% and 0.5 – 1.5% respectively.

3. Effect of Accelerators on Development of Strength

The effect of optimum quantity of different accelerators on compressive strength of phosphoanhydrite was determined on the basis of attaining specified setting time of 6 hours. The data in Tables 7.5 to 7.8 showed that all the mixes comply with the minimum specified value of 17.5 MPa. The strength development was found to be higher in case of phospoanhydrite plasters made at 1000°C in both the samples DFL and SPIC respectively. Highest compressive strength was obtained when combination of Na_2SO_4 and $FeSO_4$ accelerators were used. Strength-wise SPIC phosphonanhydrite plaster has shown better performance than the DFL anhydrite plaster. This may be attributed to the better quality of the SPIC than the DFL phosphogypsum sample. The strength development of hardened anhydrite was supplemented by X-ray diffraction and differential

Table 7.3 Setting Time of CFL Phosphoanhydrite Plaster with Different Accelerators

Accelerators	Setting time (Hrs.)			
	650°C	750°C	850°C	1000°C
K_2SO_4(1%)	9.5	8.5	8.0	6.0
K_2SO_4(2.0%)	7.0	6.5	7.0	5.0
K_2SO_4(2.5%)	6.5	5.2	5.6	4.6
K_2SO_4(3.0%)	6.0	11.0	—	—
K_2SO_4(1%) + $ZnSO_4$(1%)	12.0	7.5	12.0	10.0
K_2SO_4(1.5%) + $ZnSO_4$(0.5%)	7.0	5.5	8.5	6.5
K_2SO_4(1.7%) + $ZnSO_4$(0.3%)	6.5	6.5	6.0	5.0
Na_2SO_4(2%)	7.0	5.0	6.2	4.8
Na_2SO_4(3%)	6.0	—	5.5	4.5
Na_2SO_4(1.7%) + $FeSO_4$(0.3%)	8.5	6.0	6.0	3.3
$NaHSO_4$(1%)	—	6.5	6.5	4.5
$NaHSO_4$(2%)	7.5	5.5	5.5	4.2
$NaHSO_4$(3%)	7.0	—	5.5	4.0
$NaHSO_4$(1.8%) + $FeSO_4$(0.4%)	6.5	6.5	8.0	6.0
Potash alum (2%)	8.0	7.0	7.0	6.0
Potash alum (3%)	7.0	6.5	7.0	5.5
$Al_2(SO_4)316H_2O$(2%)	48.0	42.0	46.0	31.0
$Al_2(SO_4)316H_2O$(3%)	45.0	41.0	43.0	28.0
$FeSO_4$(3%)	52.0	50.5	40	30.0
$FeSO_4$(5%)	52.0	60.0	51.0	40.0
$Al_2(SO_4)316H_2O$(1.5%) + $K_2C\ O_3$(1.5%)	6.5	6.0	6.0	5.0
$(NH_4)_2SO_4$				
1%	—	—	—	4.6
2%	—	—	—	4.2
3%	—	—	—	4.0
$Ca(OH)_2$(3%)+$CaCl_2$ (0.5%)+Na_2SO_4(2%)	—	—	—	3.3
$Ca(OH)_2$(3%)+$CaCl_2$ (0.5%)+Na_2SO_4(2.5%)	—	—	—	4.2
K_2CrO_7(1%)+K_2SO_4 (1%)	—	—	—	3.3

Table 7.4 Setting Time of SPIC Phosphoanhydrite Plaster with Different Accelerators

Accelerators	Setting time (Hrs.)			
	650°C	750°C	850°C	1000°C
K_2SO_4(1%)	8.0	—	—	—
K_2SO_4(2.0%)	6.5	6.0	6.0	4.0
K_2SO_4(2.5%)	6.5	—	—	—
K_2SO_4(1.5%) + $ZnSO_4$(0.5%)	8.5	—	—	—
K_2SO_4(1.0%) + $ZnSO_4$(0.5%)	—	6.0	7.0	5.5
Na_2SO_4(2.0%)	7.0	5.5	6.0	4.5
Na_2SO_4(2.5%)	6.0	—	—	—
Na_2SO_4(1.7%) + $FeSO_4$(0.3%)	6.0	—	—	—
Na_2SO_4(1.0%) + $FeSO_4$(0.5%)	—	5.0	5.5	3.5
$NaHSO_4$(2.0%)	—	6.0	7.0	4.2
$NaHSO_4$(2.5%)	8.0	—	—	—
$Al_2(SO_4)_3 16H_2O$(1.5%) + K_2CO_3(1.5%)	8.0	—	—	—
$Al_2(SO_4)_3.16H_2O$(1.5%) + K_2CO_3(0.5%)	—	6.0	6.5	5.0
$(NH_4)_2SO_4$	—	—	—	—
1%	—	—	—	4.0
2%	—	—	—	4.0
3%	—	—	—	3.5
$Ca(OH)_2$(3%)+$CaCl_2$ (0.5%)+Na_2SO_4(2%)	—	—	—	3.0
$Ca(OH)_2$(3%)+$CaCl_2$ (0.5%)+Na_2SO_4(2.5%)	—	—	—	4.0
$K_2Cr_2O_7$(1%)+K_2SO_4 (1%)	—	—	—	3.1

thermal analysis. The results are plotted in Figs. 7.3 and 7.4 respectively. X-ray diffractions (Fig.7.3) show reflections for the formation of glauberite ($Na_2SO_4.CaSO_4$), ferrinatrite ($Na_3FeSO_4.3H_2O$), gypsum ($CaSO_4.2H_2O$) and unconverted anhydrite ($CaSO_4$). The intensity of ferrinatrite and gypsum reflections increased while those of glauberite and anhydrite reduced with the increase in hydration of the anhydrite. Differential thermograms of the hardened anhydrite cement (Fig. 7.4) show an increase in the intensity of double dehydration endotherms of gypsum obtained at 130–150°C and 170 – 190°C and that of inversion of γ-$CaSO_4$ and $\beta CaSO_4$ at 360–380°C. The increase in the intensity of endotherms

Table 7.5 Compressive Strength of CFL Phosphoanhydrite Plaster with Different Accelerators

Sl. No.	Accelerators	Compressive Strength MPa											
		650°C				750°C				850°C			
		3d	7d	28d	90d	3d	7d	28d	90d	3d	7d	28d	90d
1	K_2SO_4 (2.5%)	—	—	—	—	7.24	15.4	21.8	22.3	8.0	13.8	19.8	20.8
2	K_2SO_4 (3.0%)	5.6	12.4	12.6	18.2	—	—	—	—	—	—	—	—
3	K_2SO_4 (1.5%)+ $ZnSO_4$ (0.5%)	—	—	—	—	6.5	13.2	18.8	20.0	—	—	—	—
4	K_2SO_4 (1.7%)+ $ZnSO_4$ (0.3%)	—	—	—	—	—	—	—	—	4.5	12.9	12.8	18.8
5	Na_2SO_4 (2.0%)	—	—	—	—	12.2	19.2	26.8	27.4	—	—	—	—
6	Na_2SO_4 (3.0%)	7.8	15.0	19.7	21.3	—	—	—	—	11.4	16.8	20.9	22.0
7	Na_2SO_4 (1.7%) $FeSO_4$ (0.3%)	—	—	—	—	11.6	18.9	29.2	31.2	9.6	15.2	22.2	23.4
8	$NaHSO_4$ (2.0%)	—	—	—	—	7.0	13.2	22.8	23.4	6.4	12.4	19.0	20.0
9	$Al_2(SO_4)_3.16H_2O$ (1.5%+K_2O_3(1.5%)	—	—	—	—	40.5	10.6	21.2	22.5	30.0	9.2	19.2	21.1

Table 7.6 Compressive Strength of SPIC Phosphoanhydrite Plaster with Different Accelerators

Sl. No.	Accelerators	650°C				750°C				850°C			
		3d	7d	28d	90d	3d	7d	28d	90d	3d	7d	28d	90d
1	K_2SO_4 (2.0%)	—	—	—	—	9.0	15.0	23.0	24.0	8.6	14.8	19.6	22.0
2	K_2SO_4 (2.5%)	6.0	14.0	12.6	19.0	—	—	—	—	—	—	—	—
3	K_2SO_4 (1.0%) + $ZnSO_4$ (0.5%)	—	—	—	—	7.0	12.0	19.6	21.6	—	—	—	—
4	Na_2SO_4 (2.0%)	—	—	—	—	12.6	19.6	28.0	29.3	11.8	18.0	22.8	24.5
5	Na_2SO_4 (2.5%)	8.9	12.0	21.5	24.0	—	—	—	—	—	—	—	—
6	Na_2SO_4 (1.0%) + $FeSO_4$ (0.5%)	—	—	—	—	12.0	19.9	31.6	33.0	9..5	18.0	28.0	30.0
7	Na_2SO_4 (1.7%) $FeSO_4$ (0.5%)	80.0	14.0	24.0	22.0	—	19.9	—	—	—	—	—	—
8	$NaHSO_4$ (2.0%)	—	—	—	—	7.0	14.0	24.0	24.9	—	—	—	—
9	$Al_2(SO_4)_3$.16 H_2O(1.5%) + K_2 CO_3(0.5%)	—	—	—	—	6.0	11.0	21.9	24.0	—	—	—	—

Table 7.7 Compressive Strength of Anhydrite Cement Produced from CFL PG at 1000°C with Different Activators

Sl. No.	Chemical activators by mass (%)	Compressive strength (MPa)			
		1 day	3 days	7 days	28 days
1	$(NH_4)_2SO_4$				
	1.0	1.4	5.50	5.80	24.30
	2.0	1.3	5.40	8.60	32.23
	3.0	5.2	9.62	15.20	36.0
2	$Ca(OH)_2$(3%) + $CaCl_2$(0.5%) + Na_2SO_4(2%)	13.2	18.4	28.70	37.20
3	$Ca(OH)_2$(3%) + $CaCl_2$(0.5%) + Na_2SO_4(2%)	12.30	20.70	23.70	36.0
4	$K_2Cr_2O_7$(1%) + K_2SO_4(1%)	2.10	13.0	3.30	32.90
5	Na_2SO_4(0.5%) + K_2SO_4(0.5%)	7.50	15.50	25.30	38.90

Table 7.8 Compressive Strength of Anhydrite Cement produced from SPIC PG at 1000°C with different Activators

Sl. No.	Chemical activators by mass (%)	Compressive strength (MPa)			
		1 day	3 days	7 days	28 days
1	$(NH_4)_2SO_4$				
	1.0	1.45	6.80	7.90	28.00
	2.0	1.80	6.40	9.80	34.30
	3.0	6.80	10.80	18.20	38.0
2	$Ca(OH)_2$(3%) + $CaCl_2$(0.5%) + Na_2SO_4(2%)	14.20	19.40	39.70	39.40
3	$Ca(OH)_2$(3%) + $CaCl_2$(0.5%) + Na_2SO_4(2%)	11.90	22.60	25.70	37.80
4	$K_2Cr_2O_7$(1%) + K_2SO_4(1%)	4.00	15.0	14.80	34.60
5	Na_2SO_4(0.5%) + K_2SO_4(0.5%)	7.80	15.60	26.80	39.80

and exotherms confirm the formation of gypsum from anhydrite. Thus, DTA reaffirmed the results of X-ray diffraction data.

The chemically combined water of phosphoanhydrite cement plaster hydrated for different periods is reported in Fig. 7.5. Data show that chemically combined water increased with the development of hydration of anhydrite cement. The increase in chemically combined water with the progress of hydration confirmed increased inversion of anhydrite into gypsum which can be correlated with the periodical increase in the strength of anhydrite cement on curing.

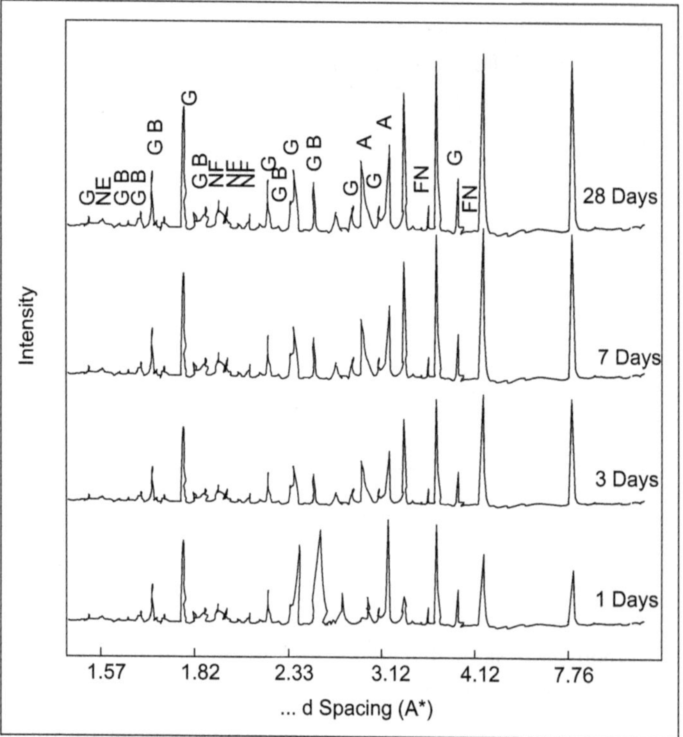

Fig. 7.3 X-ray Diffractograms of Anhydrous Plaster/Cement Hydrated for Different Periods in presence of Mixture of $Na_2SO_4.10H_2O$ + $FeSO_4.7H_2O$ Activators (G-gypsum, NF- $Na_2SO_4.Fe_2(SO_4)_3$, FN-ferrinatrite [$Na_3.FeSO_4.3H_2O$)], GB-Glauberite ($Na_2SO_4.CaSO_4$))

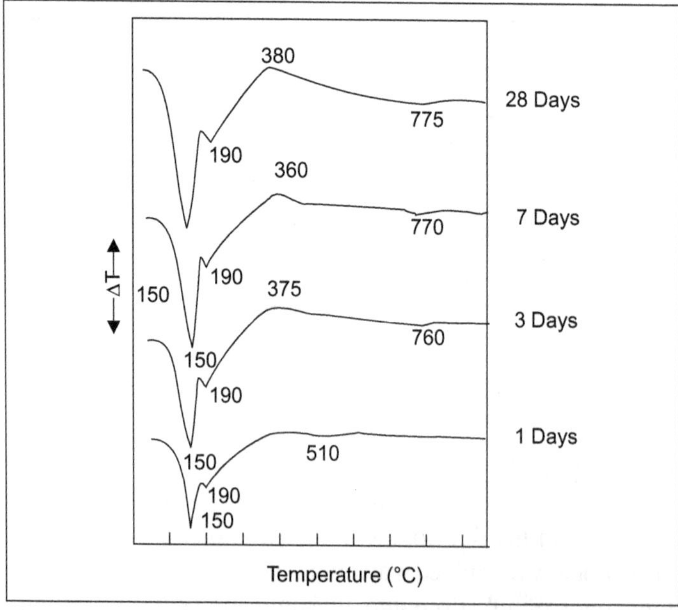

Fig. 7.4 DTA of Anhydrite Cement Hydrated for Different Periods

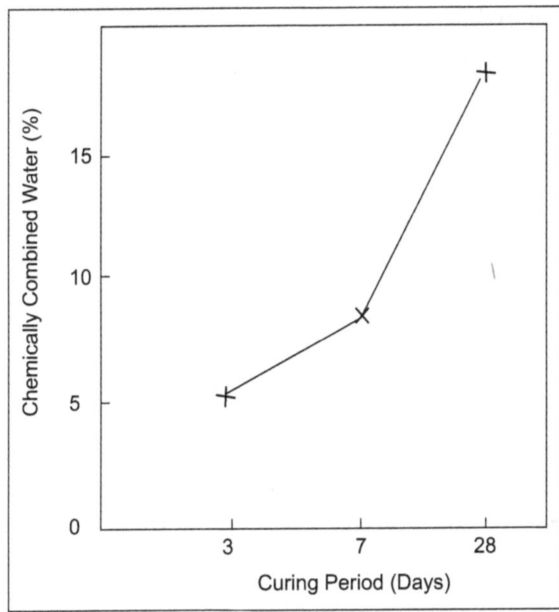

Fig. 7.5 Correlation between chemically Combined Water and Curing Period of Anhdrite Cement

7.1.4 Hydration Mechanism of Anhydrite

Some of the observations recorded by different researchers regarding hydration of anhydrite plaster/binder are described. The hydration of anhydrite binders where gypsum is the reaction product is based on the intergrowth and matting of the crystals which are formed in the shape of rods and needles[12–13]. However, there is still some uncertainty about what causes the different strengths of the anhydrite activated in different ways. During the investigation of relationship between hydration and strength of anhydrite binders. Ottemann[14] discovered that in addition to the degree of drying, the degree of hydration of the binder is also a critical factor. On the other hand, Hajjouli and Murat[15] deny any direct relationship between strength development and gypsum content in the anhydrite binder. They consider that the pore size depending upon the resulting gypsum crystals, is responsible for the strength of the hardened material. According to their observations the crystal size of gypsum which is formed is in turn strongly dependent on the nature of the activator cations used. Older and Robler[16] investigated the effect of different activators on the morphology and the microstructure of setting of gypsum pastes. They found that a needle habit of the gypsum crystals is of paramount significance for high strength of set material due to good matting and interlocking of the gypsum crystals. Whereas Riedel *et al.*[17] were unable to establish any differences in the microstructure which equated with the strength of the set and hardened gypsum and clarified that strength is governed by the quantity of anhydrite which is converted to gypsum.

Further, it is not the gypsum content which is responsible for the strength of the set plaster but also the extent of the over all hydration process which, in addition to the conversion of anhydrite to gypsum, is also characterized by a change in the Crystalline habit. There is striking lack of typical gypsum needles, the interlocking of which is generally held responsible for gypsum hardening. The habit of hardened anhydrite is platy and thus the strength is caused by the bonding of the large, unconverted anhydrite particles by the gypsum plates. In contrary to this, Hunger, et. al[18]. and Budnikov et. al[19]. have shown that preferred growth of gypsum crystal was found in the fibre direction at the super saturation ratios for gypsum of 1.5 to 4.

Infact, accelerators added to the anhydrite plaster act in the following ways to bring about hydration.

The salt content considerably retards the evaporation of water due to the reduction in vapour pressure and therefore the (*i*) gauge water is available for longer time to facilitate hydration for prolonged period (*ii*) the solubility of anhydrite is increased in sulphate solution, thus hydration is augmented with gypsum crystallizing rapidly out of the calcium sulphate solution. The production of gypsum by hydration of anhydrite with alkali sulphate is believed to proceed through the processes and reconstruction of displacement on crystallized scale. The double salt decomposes in solution into calcium ions, activator cations, and sulphate ions and consequently supersaturate the solution with respect to the gypsum. The activator may be considered to increase the dissolution rate of anhydrite .

$$\text{CaSO}_4 \text{ (anhydrite)} \xrightarrow{H_2O} \text{CaSO}_4\text{. Activator ions } Ca^{2+} + SO_4^{2-} +$$

$$\text{Activator ions} \xrightarrow{H_2O} \text{CaSO}_4.2\text{H}_2\text{O (Gypsum)}$$

The plausible mechanism of transformation of anhydrite into gypsum in presence of chemical activators ($Na_2SO_4.10H_2O + FeSO_4.7H_2O$) may be explained as follows.

The attainment of high strength in the gypsum anhydrite cement probably takes place by production of gypsum ($CaSO_4.2H_2O$) through the formation of intermediate unstable salts such as ferrinatrite ($Na^+{}_3FeSO_4.3H_2O$), $Na_2SO_4.Fe_2(SO_4)_3$ and glauberite ($Na_2SO_4.CaSO_4$). These salts are formed by the high concentration of colloidal particles of activators on the surface of anhydrite ($CaSO_4$) in an unsaturated solution. Consequently, active nucleating centers are created around which crystallization sets when the solution becomes super-saturated. The unstable intermediate salts decompose into Ca^{2+}, Na^+, Fe^{3+} and SO_4^{2-} ions which then bind water to anhydrite to form prismatic, tabular and rhombic shaped crystals that in turn, cement the unconverted anhydrite grains to give high mechanical properties. It is, therefore, deduced that activators tend to increase the rate of dissolution of anhydrite through formation of intermediate compounds. The transformation of anhydrite into gypsum, thus involves a dissolution – nucleation – growth process. Several workers have reported the use of chemical activators that increase the hydration rate of the solid anhydrite[20–24].

7.1.5 Microstructure of Hydrated Anhydrite

The setting and hardening process in anhydrite is related to its hydration and crystallization. The accelerator apparently separates out of the solution in the form of minute crystals, thereby establishing potential centers for crystallization of hydrated plaster. Each accelerator is characterized by its own form of gypsum crystals ranging from the acicular to radiating crystals that are fairly large[25]. The hydrated anhydrite plasters in presence of $Na_2SO_4.2H_2O$ or $Na_2SO_4.2H_2O$ + $FeSO_4$ accelerators have shown the formation of prismatic shaped crystals with abundance of radiating of anhydrite with platy habit. The formation of prismatic crystals is due to the effect of $Na_2SO_4.2H_2O$ while the radiating crystals are considered to be due to combined effect of $Na_2SO_4.2H_2O$ + $FeSO_4$. With other accelerators, similar type of crystals of variable sizes and stacking are obtained giving different level of strengths.

7.1.6 Effect of Temperature

A. The effect of temperature on hydration and strength development in anhydrite was studied. The results are depicted in Table 7.9. It can be seen from the Table that compressive strength is adversely affected by the rise of temperature of curing. Serious reduction of strength takes place at 40°C. Theoretical considerations indicate that at temperature above 42°C, water does not combine at all with anhydrite. The reason for this is apparent from the solubility curves of gypsum dihydrate and anhydrite (Referred Fig. 1.4, Chapter-1). Up to a certain temperature solubility of anhydrite is higher than that of gypsum, but above this temperature, the solubility of gypsum is higher. Therefore, at temperature above the limiting temperature, all the dissolved anhydrite will settle out again as anhydrite and not as gypsum. In other words, anhydrite plaster will not set at temperatures in excess of about 42°C. This behaviour obviously imposes limitation on the use of this material in tropical climates. It is evident that the temperature of curing should not exceed 35°C in any case.

The adverse effect of temperature on hydration was pointed out earlier by Andrew[26], who showed experimentally that 20°C is the optimum curing

Table 7.9 Effect of Temperature on the Development of Compressive Strength

S. No.	Temperature (°C)	Compressive Strength (MPa)							
		CFL (1000°C)				SPIC (1000°C)			
		Na_2SO_4 (1.7%) + $FeSO_4$ (0.3%)				Na_2SO_4 (1.0%) + $FeSO_4$ (0.5%)			
		3d	7d	28d	90d	3d	7d	28d	90d
1	27.0	7.50	15.50	25.30	38.90	7.80	15.60	26.80	39.80
2	35.0	3.40	7.90	21.40	34.40	4.30	7.80	17.50	35.20
3	40.0	3.00	5.20	19.20	33.10	3.80	5.40	16.80	29.40

temperature for getting maximum hydration. However, a summer temperature as high as 42°C is not significant after the material has set and hardened. Similar results were reported by Tenny and BenYair[27] from their field experiments in Israel.

B. Tests carried out to study the effect of temperature on durability showed that the anhydrite plaster did possess excellent durability. The cubes tested for compressive strength after 15 and 30 cycles of heating and cooling showed no loss in weight and compressive strength.

From this study it can be concluded that anhydrite plaster can be manufactured from the phosphogypsum with the help of suitable accelerators. Its production will bring in the market a new type of plastering material of high strength. High humidity is a favourable factor for hydration and development of strength. The set and hardened plaster is not affected by summer temperature. This plaster gives a hard smooth finish. Its use can be safely recommended in situations where it is not exposed to continuing dampness.

7.1.7 Uses of Phosphogypsum Anhydrite in Construction Works

The suitability of phosphogypsum anhydrite plaster was evaluated for use in (*i*) masonry mortars, (*ii*) internal plastering, (*iii*) light weight blocks and (*iv*) acoustic tiles[28]. The results are discussed hereafter.

1. Masonry Mortars

Mortars used in construction work should have adequate strength, good durability and high water retention against suction. Data on compressive strength and water retentivity of different mortar mixes prepared with the anhydrite plaster and two sands of fineness modules 1.75 and 1.22 is shown in Table 7.10.

It can be seen that all the mortars prepared with anhydrite plaster have much higher water retentivity than the commonly used 1:6 cement-sand mortar. Considering that the mortar prepared with 1 part of plaster and 1 part of sand (F.M. 1.75 and 1.22) has the maximum water retentitivity and develops considerable higher strength than the 1:6 cement-sand mortars, its use can be made in place of 1:6 cement-sand mortars for internal masonry work.

2. Internal Plastering

To find out suitability of phosphogypsum anhydrite plaster for use in plastering, mortars of mix proportion 1:1 and 1:2 plaster : sand were prepared and used in the plastering over burnt brick wall. The fineness modulus of sand was kept 1.22. It was found that a plaster-sand mix of 1:2 proportion by weight (1:2:8 by volume) is suitable for internal plastering. The plaster developed good strength after 24 hours of application having smooth finish, the plaster was hard and showed good bond with bricks. Cost-wise mortar mix 1:2 plaster-sand is cheaper than 1:6, cement-sand mortar.

Table 7.10 Properties of Masonry Mortars prepared with Phosphogypsum Anhydrite Plaster and Sand

Sl. No.	Mortar Mix proportion (by wt.)	Compressive Strength (kg/cm^2)			Water Retention (per cent)
		3d	7d	28d	
1.00	Plaster : Sand				
	1:1	20.64	48.00	99.40	95.22
	(FM +1.75)	—	—	—	—
	1:1	18.13	37.55	96.96	88.39
	(FM 1.22)	—	—	—	—
2.00	Plaster : Sand				
	1:2	13.20	22.40	38.85	87.61
	(FM +1.75)	—	—	—	—
	1:1	10.36	16.69	34.40	83.80
	(FM 1.22)	—	—	—	—
3.00	Cement : Sand				
	1:6	9.20	21.00	34.60	9.5
	(FM 1.75)	—	—	—	—
	1:6	4.40	6.50	15.60	6.0
	(FM 1.22)	—	—	—	—

* FM : Fineness Modules

As a Finish and Base Coat Plaster

Suitability of phosphogypsum anhydrite plaster was checked for use as finishing and base coat plastering material. In the first instance, 9 mm thick base coat consisting of 1:2 (1:2:8 by volume) anhydrite-sand and 1:6 cement-sand mixes were applied over unplastered burnt brick wall separately. Immediately 3 mm thick neat anhydrite plaster paste was applied over the base coat. The total thickness was kept at 12 mm. Finished coat developed good strength in 24 hours, developed a hard surface with glossy pinkish appearance. Thus, anhydrite plaster can be used as a finish as well as base coat plastering material.

3. Light Weight Blocks

The phoshogypsum anhydrite plaster was tested for its suitability for making light weight blocks as described below:

Anhydrite-Saw Dust Blocks

Effect of addition of washed saw dust was studied on the bulk density and compressive strength of plaster. Based on 5 cm cube data, 40 × 20 × 10 cm blocks

Table 7.11 Strength of Blocks Produced Using Phosphogypsum Anhydrite Plaster and Saw Dust

S. No.	Saw Dust Per cent	Bulk Density (kg/m^3)	28 days Compressive strength at 28 days (MPa)
1	5.0	1620	1.25
2	10.0	1510	0.80

were cast using anhydrite plaster and saw dust. The blocks were demoulded and cured in high humidity for 28 days at 27 ± 2°C and then dried in the sun to constant weight. The blocks were tested for bulk density and compressive strength. The results are given in Table 7.11.

It can be seen that on using 10 per cent saw dust, the density is lowered and the strength data comply with the minimum strength of 2.0 MPa specified in IS:2849–1983, a specification for non-load bearing gypsum partition blocks (solid and hollow type). These blocks are mainly suitable for internal non-load bearing partition walls or for inner leaf of cavity wall construction. The blocks should not be used under damp conditions as they are liable to suffer deterioration as their strength is seriously reduced.

Anhydrite-Foamed Slag Blocks

Effect of addition of foamed slag of size 6 mm and down having density 303 kg/m^3 was studied on the bulk density and compressive strength of anhydrite plaster to develop insulative blocks. Based on 5 cm × 5 cm × 5 cm cube data, a mix proportion of 60:40 anhydrite plaster : foamed slag was chosen for casting large size blocks. The blocks of size 40 × 20 × 10 cm were cast, demoulded, cured in high humidity at 27 ± 2°C for 28 days, dried at 42°C and tested for their bulk density and compressive strength. The results are reported in Table 7.12. It can be seen that the blocks develop strength-conforming to IS : 2849–1983, Indian specification for gypsum blocks. These blocks can thus be used for non-load bearing partition wall for internal purpose only.

The vitreous materials known to have far lower thermal conductivity than crystalline material. Since foamed slag used in making of blocks is a vitreous material, therefore 300 × 300 × 30 mm size boards were cast using the above composition. The boards after curing for 28 days were dried and subjected to the thermal conductivity test as per heat flow method specified in ASTM (518–1971), specification for method of thermal conductivity of building insulating materials. The results are given in Table 7.13.

Table 7.12 Strength of Blocks Produced Using Phosphogypsum Anhydrite Plaster and Foamed Slag

Sl. No.	Mix Proportion (By Wt.) Plaster	Foamed Slag	Bulk Density (kg/m^3)	28 days compressive strength (MPa)
1	60	40	1070	1.40

Table 7.13 Thermal Conductivity of Plaster: Slag Mix at 28 Days of Curing

Sl. No.	Mix Proportion (By Wt.) Plaster : Foamed slag	Bulk density (kg/m³)	Thermal conductivity K.cal/m/h/°C
1	60:40	1070	0.28
2	Light weight concrete	1200	0.60
3	Common burnt clay bricks	1700	0.70

Table 7.14 28 Days Compressive Strength of Neat Anhydrite Plaster Block

Sl. No.	Bulk density (kg/cm³)	Compressive strength (MPa)
1	1900	17.7

The data shows that the blocks produced using foamed slag have lower thermal conductivity than the conventional building materials like light weight concrete and common burnt clay bricks.

Neat Plaster Blocks

To make high density and high strength blocks, anhydrite plaster (without aggregate) was gauged with water required for normal consistency. The kneaded mass was put into the moulds of size 40 × 20 × 10 cm. After 24 hours of hydration, hardened blocks were demoulded, cured for 28 days in high humidity and then dried at 42°C before testing for compressive strength and bulk density. The results are given in Table 7.14. The strength of blocks is much higher than the minimum specified value of 50 kg/cm^2 given in IS : 3590–1976, specification for load-bearing light weight concrete blocks. Hence, the blocks produced with neat plaster at normal consistency are suitable for load-bearing internal partition walls.

The typical photographs of blocks made using (A) 10% saw dust, (B) 40% foamed slag and (C) neat anhydrite plaster are shown in Fig. 7.6.

Fig. 7.6 Building Blocks Made with Phosphogypsum Anhydrite Plaster using (A) 10 % Saw Dust (B) 40% Foamed Slag and (C) Neat Anhydrite Plaster

4. Acoustic Tiles

Acoustic tiles are used in large offices and auditoria to mitigate the reflected sound by absorbing the sound energy. For making acoustic tiles, the phosphogypsum anhydrite plaster and washed saw dust in proportion 90:10 parts by weight were mixed with water at workable consistency. The kneaded mass was then put into the moulds of size 300 × 300 × 300 mm uniformly. After one hour of hardening, the mix was pressed from top to bottom with a wooden plate containing several projected pins to form slots in the tiles. The tiles were demoulded after 36 hours of hardening, cured in high humidity for 28 days at 27 ± 2°C and subjected to sound absorption coefficient test. The results are reported in Table 7.15. The data shows that sound absorption coefficient of slotted phosphogypsum anhydrite tiles compares fairly well with the imported slotted gypsum tiles and indigenous sitatex fibre board tiles. Manufacture of these tiles can be taken up as a cottage industry. Figure 7.7 shows a typical photograph of phosphogypsum-anhydrite-saw dust tiles.

Table 7.15 Sound Absorption Coefficients of Tiles

Sl. No.	Characteristics	Frequency (c.p.s.)					
		125	250	500	1000	2000	4000
1	Slotted phosphogypsum anhydrite tiles	0.15	0.16	0.28	0.30	0.31	0.35
2	Slotted gypsum tiles (UK)	0.5	0.10	0.25	0.30	0.15	0.20
3	Sita tex plain fibre board tile (India)	0.13	0.18	0.21	0.18	0.17	0.30

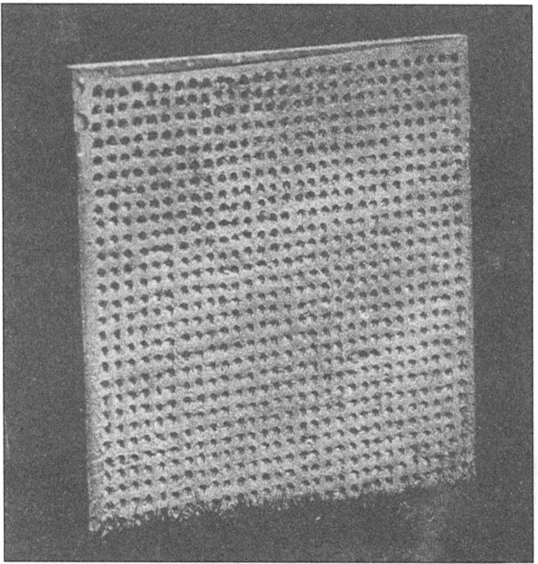

Fig. 7.7 Phosphogypsum Anhydrite Plaster-Saw Dust Acoustic Slotted Tile

It can be seen that all the mortars prepared with anhydrite plaster have much higher water retentivity than the commonly used 1:6 cement-sand mortar. Considering that the mortar prepared with 1 part of plaster and 1 part of sand (F.M. 1.75 and 1.22) has the maximum water retentivity and develops considerable higher strength than the 1:6 cement-sand mortars, its use can be made in place of 1:6 cement-sand mortars for internal masonry work.

5. Internal Plastering

To find out suitability of phosphogypsum anhydrite plaster for use in plastering, mortars of mix proportion 1:1 and 1:2 plaster : sand were prepared and used in the plastering over burnt brick wall. The fineness modulus of sand was kept 1.22. It was found that a plaster-sand mix of 1:2 proportion by weight (1:2:8 by volume) is suitable for internal plastering.

7.2 Polymerised Anhydrite Plaster for Use in Flooring Composition

There are two major objectives of the study reported under this head, to develop high strength anhydrite plaster from phosphogypsum using suitable chemical activators and to produce flooring tiles from the activated anhydrite plaster admixed with monomer, glass fibre, pigments, etc.

To produce high strength gypsum plaster, phosphogypsum was calcined to form β-anhydrite and then blended with suitable chemical activators followed by fine grinding. The anhydrite plaster thus, produced was mixed with optimum quantities of predetermined monomer methyl methacrylate properly catalyzed, different metallic oxide pigments, fly ash and red mud, cut glass fibres (E-type) and the quartz sand to cast flooring tiles by vibration moulding process followed by moist curing at ambient temperature, grinding and polishing. The properties of tiles such as density, water absorption, abrasion resistance, compressive strength and modulus of rupture were determined as per Indian Standards. The hydration and the durability characteristics of the polymerised anhydrite as checked by alternate wetting and drying and heating and cooling cycles is reported in the this Chapter. The applications of the polymerised anhydrite plaster other than producing flooring tiles has been discussed.

7.2.1 Methodology

Phosphogypsum, fly ash and the red mud samples collected from various sources were analysed for their chemical composition (Table 7.16). These materials were used along with methyl methacrylate monomer and pigments (iron oxide, chromium oxide, Prussian blue, TiO_2, etc.) as per relevant Indian Standards, laboratory grade chemical activators and the glass fibre (E-type, 12 mm long) as the raw material for moulding the flooring tiles.

Table 7.16 Chemical Composition of Raw Materials

Constituents	Phosphogypsum (%)	Fly ash (%)	Red mud (%)
P_2O_5	0.47	—	—
F	0.86	—	—
Organic matter	0.59	—	—
SiO_2 + insoluble in HCl	0.29	70.60	13.50
$Al_2O_3 + Fe_2O_3$	0.54	24.40	64.50
CaO	31.09	2.60	Tr.
MgO	1.31	0.73	Tr.
SO_3	43.21	—	Tr.
TiO_2	—	—	5.0
Na_2O	0.29	0.20	9.0

7.2.2 Production of Anhydrite from Phosphogypsum

Anhydrite was produced by heating phosphogypsum at an optimum temperature of 1000°C in the furnace for a period of 4–5 hours, cooled and ground to pass 75 micron IS sieve (450–500 m^2/kg, Blaine). The mechanism of hydration and strength development has been explained in 7.1.4 in detail. Effect of three different type of activators mainly sulphates, chlorides and hydroxides of alkali and alkaline earth materials independently or in the mixture forms was studied on the properties of anhydrite as per ASTM designation C 61–50 (1981) and IS : 2542 (Part I)–1978. The strength development of anhydrite was monitored with the help of differential thermal analysis (Stanton Red Croft, UK) and scanning electron microscopy (LEO 438 VP, UK).

7.2.3 Preparation of Flooring Tiles

Flooring is an essential component of all building and construction activities. Flooring tiles are used in large quantities in domestic, commercial and industrial buildings in various colours, styles, sizes, textures and configurations. These tiles are laid over a strong substrate of concrete or wood, forms ultimate wearing and decorative surfaces. Flooring tiles of various thicknesses having variable strength characteristics are produced[28].

Tiles provide ecstasy, life and colour to the offices, showrooms and residential buildings and involves low maintenance cost to the reception rooms, lobbies and heavy duty areas.

Several materials, natural and synthetic, are used for moulding flooring tiles. Choice of tiles may vary according to their use and the factors like their sound absorption, colour, finish, resilience, chemical and abrasion resistance. Different materials such as plastic, rubber, cork, asphalt, vinyl and wood veneer constitute thin floor coverings and tiles.

The inorganic tiles are mostly cast from cement, either in plain, plain coloured or terrazzo tiles. Terrazzo tiles consist of marble or stone chippings embedded in the binder matrix generally coloured cementitious material. Sometimes oxychloride cement and sand mix with colouring agents are employed to mould flooring tiles. These tiles are produced by pressing the mixture of cement and aggregate into moulds in two layers – backing layer and top or wearing layer. Inorganic tiles have also been made by sintering the ceramic materials like clay in different configurations and colours.

The flooring tiles of sizes 200 mm × 200 mm × 200 mm and 300 mm × 300 mm × 200 mm were cast by vibration moulding of the mix containing phosphoanhydrite powder with different pigments, catalyzed monomers (methyl methacrylate (2–4%) with benzoyl peroxide (0.5–1.5%)), a small quantity of glass fibre and coloured stone chips at normal consistency. The fly ash and red mud industrial solid wastes were also added to the anhydrite mixes for moulding these tiles. These tiles after demoulding were cured in high humidity (over 90%), dried at 42 ± 2°C and tested for properties such as flexural strength, compressive strength, water absorption, wear resistance and porosity as per IS : 1237–1980[29].

The durability of gypsum flooring tiles was studied by wetting and drying and heating and cooling cycles at 27°, 42° and 55°C respectively. One cycle of wetting and drying comprised of heating the 28 days hardened cubes (2.5 cm) of the anhydrite plaster at above temperatures for a period of 16.0 hours followed by cooling to ambient temperature for 1 hr and followed by immersion in water for 7.0 hours. The heating and cooling cycles comprises heating the cubes at above temperatures for 6.0 hours followed by cooling the cubes to ambient temperature for a period of 18.0 hours.

7.2.4 Properties of Phosphoanhydrite and Its Hydration

The physical properties of phosphoanhydrite plaster are shown in Table 7.17.

It can be seen that the setting time and strength requirements of anhydrite plaster complies with the requirements as laid down in ASTM and the Indian

Table 7.17 Physical Properties of Phosphoanhydrite Plaster

Activator	Consistency (%)	Setting time (Min.)	Compressive strength (MPa)			Expansion (%)
Desgn..			3d	7d	28 d	
A (3%)	22.14	180	9.6	15.2	36.0	0.05
B (2%)	23.57	228	18.4	28.7	37.2	0.06
C (2%)	23.0	200	15.8	22.6	35.1	0.06
ASTM:C 61—50(1981) Limits	—	20–360	—	—	Min. 17.5	—
IS : 2547–1976 Limits	—	—	—	—	—	Max. 0.5 at 96 hours

standards. All the chemical activators (A, B and C) show development of high compressive strength at 28 days period in the anhydrite plasters. In case of ordinary plaster of Paris, maximum of 12–15 MPa, compressive strength is attained.

7.2.5 Hydration of Anhydrite in Presence of Polymer

The polymerized product (Polymerized Acrylate Molecules) formed as per following chemical reaction fills up voids and pores in the gypsum matrix during hydration process and thus improves density, strength and durability of the anhydrite towards water. The strength development in anhydrite takes place just like hydration of cement. It continuously increases with the increase in curing period and reaches maximum at 28 days. The mixing of fly ash or red mud wastes with the anhydrite works not only as filler but add colour to the anhydrite matrix or tile. The addition of small quantity of glass fibre improves strength and texture of the tiles.

$$(ArCO_2)_2 \longrightarrow 2ArCO_2^* \longrightarrow CO_2 + Ar^*$$

$$Ar^* + CH_2{=}C(CH_3){-}C({=}O){-}O{-}CH_3$$

Aryl radical + methyl methacrylate (M)

Ar* + H—CH2 — C(COOCH3)=H2C   CH3—C(COOCH3)=H2C   CH3—C(COOCH3)=H2C

Ar—H

$$H_2C{=}C(COOCH_3){-}\left[CH_2{-}C(CH_3)(COOCH_3){-}CH_2{-}C(CH_3)(COOCH_3)\right]_n{-}H_2C^*$$ continued linkage this side

Polmerized Acrylate Molecule, where M is the monomer and n is the number of repeated units put in bracket, that is length of chain of polymer

7.2.6 Properties of Flooring Tiles

The properties of flooring tiles moulded from polymerized and activated phosphoanhydrite are listed in Table 7.18. It can be seen that the phosphogypsum flooring tiles complied with the requirements of flexural strength, water absorption and wear resistance as given in IS : 1237–1980.

The tiles produced using fly ash or red mud in place of pigments also complied with the standard requirements. These tiles are suitable for use in flooring for general purposes for places where light loads are taken up by the floor such as office buildings, schools, colleges, hospitals and residential buildings.

Table 7.18 Properties of Polymerised Phosphogypsum Tiles

Property	Value	IS : 1237–1980 Limits
Perpendicularity, %	0.7	Max. 2.0
Straightness, %	0.6	Max. 1.0
Flatness, mm	0.4	Max. 1.0
Flexural Strength, N/mm^2	11–15	Min. 3.0
Compressive strength, N/mm^2	30–40	N.S.
Water Absorption, %	2.0–3.5	Max. 10.0
Wear resistance, mm	2.0–4.0	Av.3.5,Individual 4 mm, Gen. purpose
Porosity	5.0–8.0	N.S.

7.2.7 Durability of Phosphogypsum Tiles

The effect of wetting and drying cycles and heating and cooling cycles on the compressive strength of polymerised phosphogypsum cubes was studied (Figs.7.8 and 7.9). Data showed a little fall in strength of the cubes in the cyclic investigations. The heating and cooling cycles showed less fall in strength than the wetting and drying cycles. It was also noted that fall in strength increased

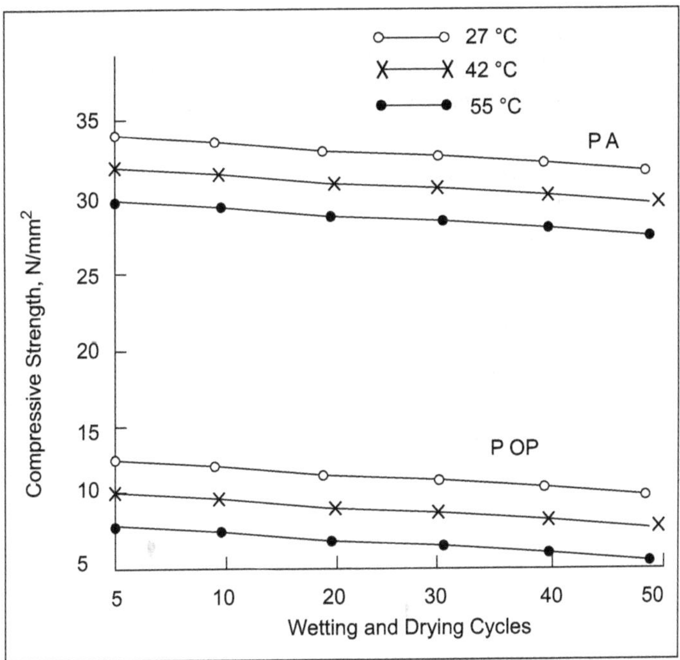

Fig. 7.8 Effect of Wetting and Drying Cycles on the Compressive Strength of Phosphoanhydrite (PA: Phosphoanhydrite, POP: Plaster of Paris)

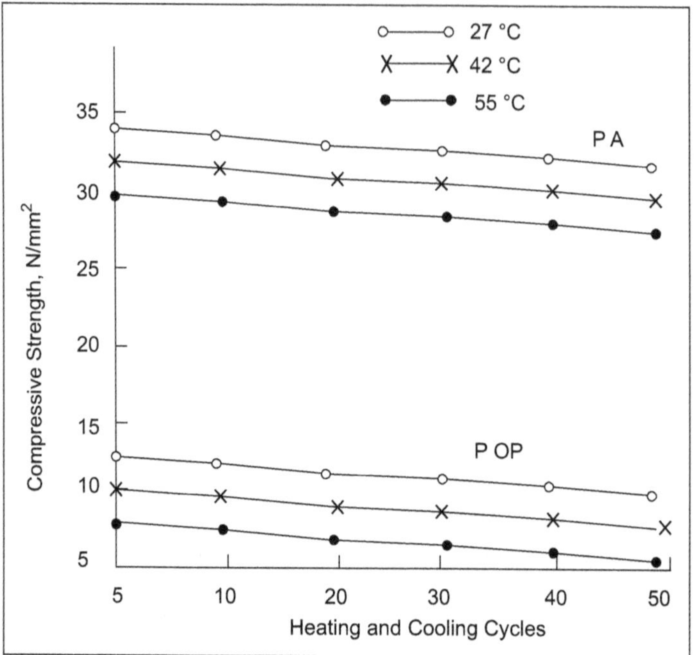

Fig. 7.9 Effect of Heating and Cooling Cycles on the Compressive Strength of Phosphoanhydrite

with the temperature from 27°C to 55°C where as the plaster of Paris (POP) cubes could not be tested for the strength during wetting and drying cycles beyond 40 cycles at 42°C and 55°C indicating better resistance of the polymerised phosphogypsum towards water. The weight loss was 3 to 5% for the polymerised phosphogypsum cubes verses 18 to 20% found for the POP cubes over 50 wetting and drying cycles. No weight loss was, however, found during heating and cooling cyclic study.

7.2.8 Techno-economic Feasibility

A plant of capacity 12,000 m^2 per year has been envisaged to produce flooring tiles from the phosphogypsum plaster and other ingredients. The major equipment are moulds, mixer, vibrating tables, curing chamber, drying chamber, rotary kiln, ball mill, grinding and polishing machine and demoulding plates. The production cycle of these tiles involves mixing of phosphogypsum with polymer, chemical activators, glass fibre, etc. Moist mass is then put into moulds and the coloured stone chips are mixed with the top wearing surface and vibrated to remove excess of water to form cohesive tiles. The tiles are cured for a period of 28 days, dried, ground and polished.

The cost of phosphogypsum tiles has been estimated at ₹300/m^2 which are much cheaper than the natural marble tiles being sold at ₹450 to 600/m^2. The decorative and protective values of phosphogypsum tiles are comparable to those of natural marble or dolomite. The cost of phosphogypsum tiles can be

Fig. 7.10 Flooring Tiles Produced from Phosphogypsum

further reduced to ₹250/m^2 if the synthetic pigments are replaced with the fly ash or red mud. The fly ash and red mud impart grey and redish colours to the tiles. The production and use of phosphogypsum tiles is recommended in the mega cities as most of the phosphatic fertilizer plants are located close to these cities. Moreover, the transporation of natural marble from Rajashtan to mega cities can be reduced. The phosphogypsum tiles may be produced in different colours and desirable designs and may replace cement and ceramic tiles too. Details of making flooring tiles out of phosphogypsum has been covered by the Indian Patent Application No. 696/DEL/2000 filed by CBRI Manjit Singh, Mridul Garg. A typical photograph of flooring tiles produced out of phosphogypsum is shown in Fig. 7.10.

7.2.9 Newer Applications of Polymerised Gypsum

Polymers such as styrene actyl acrylate, styrene malic anhydrite copolymer, poly acrylic acid hydrozide, urea formaldehyde, polybutadiene, polyurathane etc., may be blended with the gypsum plasters along with other ingredients are preferably used in building materials specially gypsum based building products to form value added building materials and components. The building materials such as high strength water resistant mortars (for rehabilitation of old structures, monuments) and for casting fibre reinforced building boards, blocks, panels, claddings etc. (for use in door shutters) with improved strength and weather-resistance may be produced. Waste materials such as phosphogypsum, fly ash, slag, lime sludge, mine tailings etc., can be blended with these polymeric materials which can be shaped into variety of building materials for use in ceiling, walling and flooring materials.

The crux of the whole investigations reported in this chapter showed that high strength anhydrite plaster can be produced by heating the waste phosphogypsum at 1000°C whereby the impurity of P_2O_5 is converted into an inert insoluble calcium pyrophosphate and the fluoride partially get voltalized. Hence, no wet treatment of the phosphogypsum is required. The phosphogypsum anhydrite is suitable for casting flooring tiles when blended with polymer, glass fibres, pigments, fly ash/red mud at normal consistency. The euhedral prismatic platy nature of phosphoanhydrite plaster can bind various constituents together to form high strength durable flooring tiles for use in flooring and walling construction works.

7.3 Phosphogypsum Based Water-Resistant Anhydrite Binder

A useful anhydrite binder has been developed from waste phosphogypsum by heating the waste at 800–1000°C for 4 hours in an oven or in a continuous rotary kiln. The phosphoanhydrite obtained contained P_2O_5 0.69%, F 0.97%, SiO_2 + insoluble 1.26%, R_2O_3 0.50%, CaO 38.60%, MgO 0.06%, SO_3 54.20%, Loss on ignition 0.32%. The phosphogypsum anhydrite was made water-resistant by the addition of granulated blast furnace slag, alkali, and alkaline earth hydroxide and sulphates. The physical properties of the binder as tested as per IS : 4031–1968 are reported in Table 7.19.

Data showed continuous increase in compressive strength of the binder with curing period up to 28 days. The plain phosphoplaster attained constant strength from 1day to 28 days. The binders were sound as it did not cross maximum limit of 5 mm specified in IS : 6909–1980, specification for supersulphated cement. DTA and XRD studies have shown formation of ettingite, C_4AH_{13}, CSH and $Ca(OH)_2$ as the hydraulic phases in the binder to contribute strength.

7.3.1 Durability of Phospho Anhydrite Binder

Durability of anhydrite binder was studied by its long term performance in water and by alternate wetting and drying cycles and by the alternate heating

Table 7.19 Properties of Water-resistant Anhydrite Gypsum Binder

Properties	Anhydrite binder Values	Plain Phosphoplaster
Fineness, m^2/g	405	300
Setting time, hr.		
Initial	3.5	0.30
Final	6.5	
Bulk density, g/cm^3	1.35	1.2 -1.28
Compressive strength, MPa		
1 day	12.36	13.03
3 days	23.56	13.18
7 days	28.80	13.20
28 days	56.00	13.20
Soundness, mm	1.5	0.50

and cooling cycles at 27 to 60°C as per reference[30] respectively. One cycle of wetting and drying comprised of heating the 2.5 cm cubes of the 28 days cured binder for 16 hours at different temperatures followed by cooling for one hour and immersion in water for 7.0 hours. After certain number of cycles, the weight loss and compressive strength of cubes were determined. One cycle of heating & cooling consisted of heating the cubes at temperatures from 27°C to 60°C separately for 6.0 hours at room temperature for 18 hours.

1. Performance of Binder in Water

The effect of the total time of immersion of hardened binder in water is shown in Table 7.20.

Data show no leaching of binder takes place with immersion period of 2 to 8 hours. However, in plain plaster, after 3 days of immersion, plaster dissolves in water. The better performance of the binder in water may be ascribed to the presence of CSH gel which fills the voids and pores in the binder and improves its stability towards water.

2. Wetting and Drying Cycles

The effect of wetting and drying cycles on the compressive strength of the binder kept at 27–60°C temperatures is shown in Fig. 7.11.

It can be seen that as compared to plain plaster, the anhydrite binder showed fall in strength with increase in temperature. The fall in strength was maximum

Table 7.20 Performance of Anhydrite Binder in Water

Curing Period (Days)	Immersion Period (Hours)	Water Absorption (%)	
		Anhydrite Binder	Phospho Plaster
1	2.0	13.51	26.92
	8.0	13.56	30.62
	24.0	15.85	32.10
3	2.0	13.21	33.70
	8.0	13.30	—
	24.0	15.70	—
7	2.0	11.49	Leaching
	8.0	12.72	—
	24.0	12.72	—
28	2.0	10.24	Leaching
	8.0	10.43	—
	24.0	10.43	—

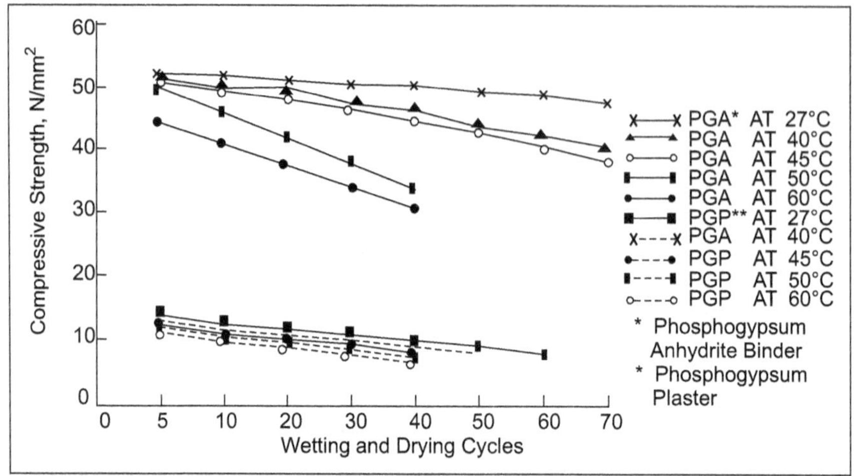

Fig. 7.11 Effect of wetting and drying cycles on the compressive strength of anhydrite binder at different temperatures

at 60°C. At 60°C, the strength could be measured only up to 40 cycles only. The compressive strength at 40 cycles was only 59% of the original strength.

It can be seen that in PGA binder, weight loss was below 2% after 20 cycles (Fig. 7.12). At higher temperature, the weight loss increased and was maximum at 60°C. As compared to PGA, the weight loss in PGP, was much higher at all temperatures.

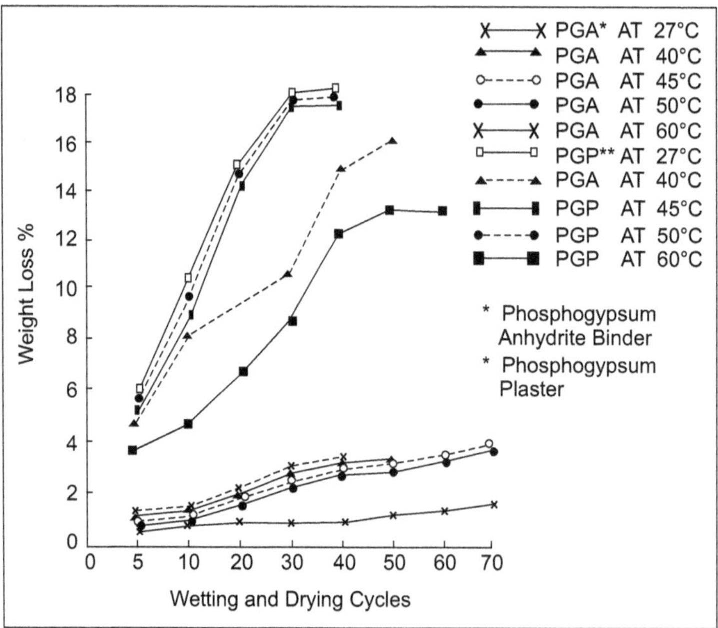

Fig. 7.12 Effect of wetting and drying cycles on weight loss of anhydrite binder at different temperatures

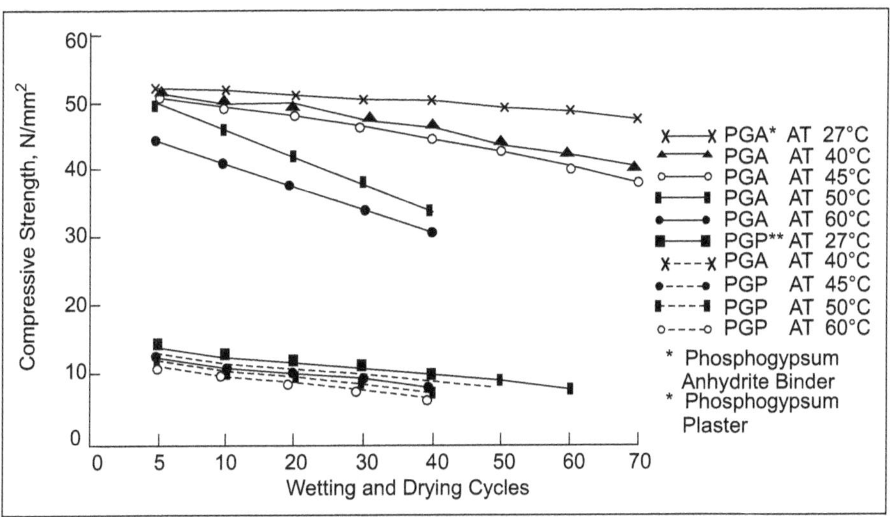

Fig. 7.13 Effect of heating and cooling cycles on Compressive strength of anhydrite binder at different temperatures

3. Heating and Cooling Cycles

The effect of alternate heating and cooling cycles on the compressive strength of the anhydrite plaster/binder is plotted in Fig. 7.13. The results showed negligible fall in strength at 27°C. With increase in temperature, strength decreased but maximum fall was noticeable at 50° and 60°C. There was no loss of weight. The above studies conclude that binder is suitable for use in external construction with ambient temperature under 45°C.

7.4 Utilization of By-product Fluorogypsum

The use of fluorogypsum which is available to the extent of 1.0 million tonne per annum in India is important. Fluorogypsum contains impurities of fluoride and free acidity. These impurities particularly free acidity may interfere with the setting and strength development of plaster/building components. Since the utilization of fluorogypsum is limited, thus, there is a disposal threat of the waste and it has become essential to suitably utilize the material.

The fluorogypsum has been studied for use in supersulphated cement[31–33]. Researches carried out have shown that water-resistant binding materials can be produced from phosphogypsum, slag, cement, fly ash and lime sludge and have been dealt in detail in previous chapters. In this chapter, investigations have been undertaken to characterize and to beneficiate fluorogypsum to make high strength binding plaster. Effect of various chemical activators/additives on setting, strength, water absorption, porosity, etc. of the fluoro plaster has been studied. The hydration and microstructure properties of the binder are reported. The suitability of fluorogypsum plaster for making building bricks containing various admixtures and its use in plastering works has been discussed.

The sample of fluorogypsum was collected from M/s Navin Fluorine International, Bhestan, Gujarat. It was analyzed for chemical composition as per IS : 1288–1983, methods of analysis for mineral gypsum. The fluorogypsum contained Fluoride 1.32%, SiO_2 + insoluble in HCl 0.65%, $Al_2O_3 + Fe_2O_3$ 0.65%, CaO 41.19%, MgO Tr., SO_3 56.10 % and Loss on ignition 0.61% and pH 5.0. Data showed that fluorogypsum possess high purity *i.e.,* $CaSO_4.2H_2O$ besides fluoride as the major impurity. The low pH value shows presence of free acidity. Minor earthly impurities of SiO_2 and $Al_2O_3 + Fe_2O_3$ have also been identified.

The chemical activators of laboratory grades ranging from sulphates to chloride of alkali and alkaline earth hydroxides were used to activate hydration of fluoroanhydrite. Differential thermal analysis (DTA) (Stanton Red croft, UK) and Scanning electron microscopy (SEM LEO 438VP) of the raw gypsum and hydrated plaster were studied. DTA (Fig. 7.14) shows endotherm and exotherm at 140°C and 250°C due to the inversion of poorly weathered anhydrite into hemihydrate ($CaSO_4.1/2\ H_2O$) and anhydrite [$CaSO_4$ (II)] peaks.

The SEM of fluorogypsum sample (Fig.7.15) shows majority of crystals in fluorogypsum are anhedral. to subhedral prismatic interspersed with lath in the agglomerated form. Twinning of some crystals may also be noted.

7.4.1 Beneficiation of Fluorogypsum

As the fluorogypsum contains free acidity which may rust the grinding media (balls etc.) and the lining of the ball mill and sometimes the plaster/ binder may become hygroscopic, it is therefore, essential to neutralize the acidity to get the suitable product. With this objective, effect of addition of dry hydrated lime [Ca $(OH)_2$] was studied on the pH of fluorogypsum. It was found that at

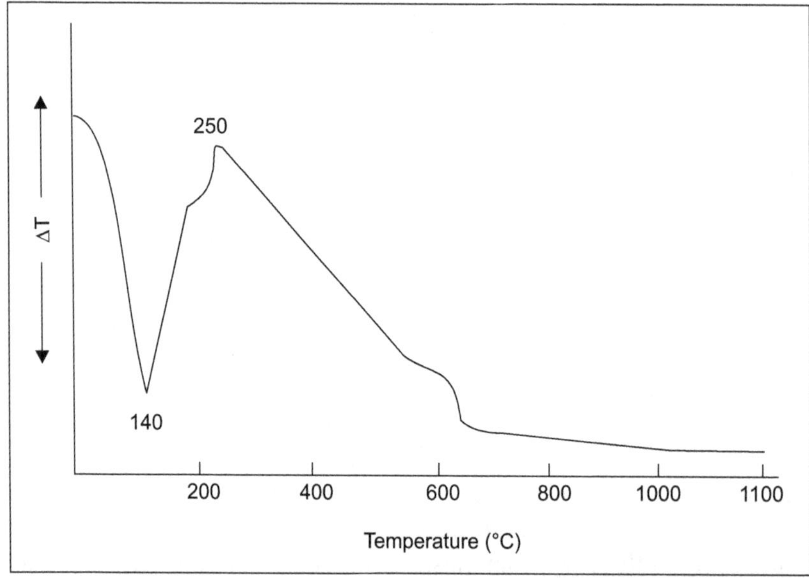

Fig. 7.14 DTA of Fluorogypsum Anhydrite

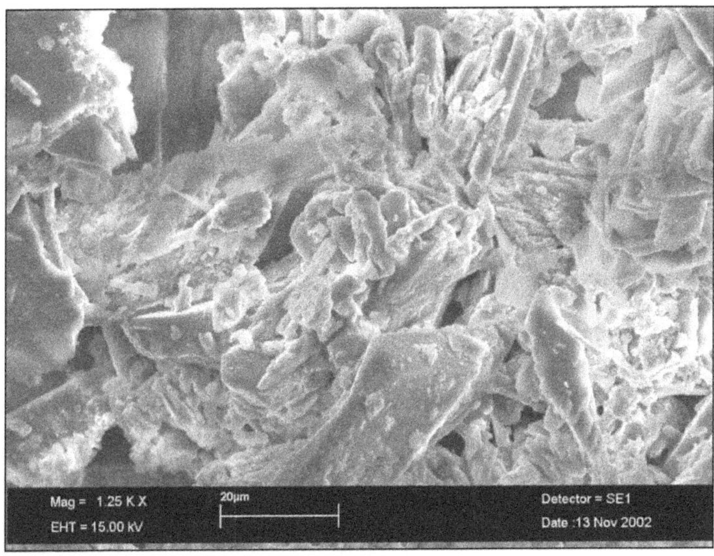

Fig. 7.15 SEM of Unbeneficiated Fluorogypsum

1.0% addition of [$Ca(OH)_2$] to the fluorogypsum, a neutral pH value of 7.0 was obtained. The SEM of beneficiated fluorogypsum is shown in Fig. 7.16. It can be seen that gypsum crystals are euhedral platy prismatic and lath shaped in nature with out agglomeration showing absence of any exotic impurity.

7.4.2 Preparation of Gypsum Binder/Plaster

The fluorogypsum dried at 42 ± 2°C, was ground in the ball mill to a fineness of 99% passing through 90 micron (Indian Standard 9 sieve) sieve. The

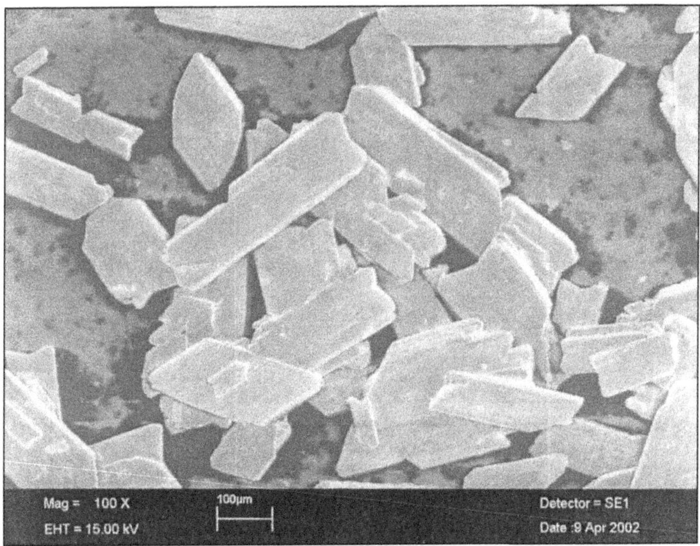

Fig. 7.16 SEM of Beneficiated Fluorogypsum

ground material was then blended with different chemical activators for 1 hour in the blender/powder mixer to get a uniform product. The binder was tested and evaluated as per IS : 2542 (Part 1) – 1978. The water absorption and porosity of fluorogypsum plaster was examined by immersing the 2.5 cm × 2.5 cm × 2.5 cm cubes of the plaster (28 days cured) in water for a period of 2 hours, 8 hours and 24 hours. The porosity of cubes was evaluated by multiplying the water absorption with bulk density of the hydrated plaster.

7.4.3 Properties of Fluorogypsum Plaster/Binder

The physical properties of fluorogypsum plaster activated by the chemical activators are reported in Table 7.21.

Data show that with the use of $(NH_4)_2\ SO_4$ activator, the setting time of fluorogypsum plaster was beyond the maximum specified limit of 6.0 hrs as per ASTM C 61–50. At the same time, the compressive strength was also much less. However, with the use of combined chemical activators *i.e.,* $Ca(OH)_2 - CaCl_2 - Na_2\ SO_4$, the setting time was accelerated and was much with in the limit and the rate of strength development was even quite high at 3 days of hydration. The hydration of fluorogypsum plaster/binder was supplemented by DTA. The therograms are shown in Fig. 7.17. It can be seen that intensity of endotherms at 140–150°C,190–200°C and exotherms at 360–370°C were increased due to

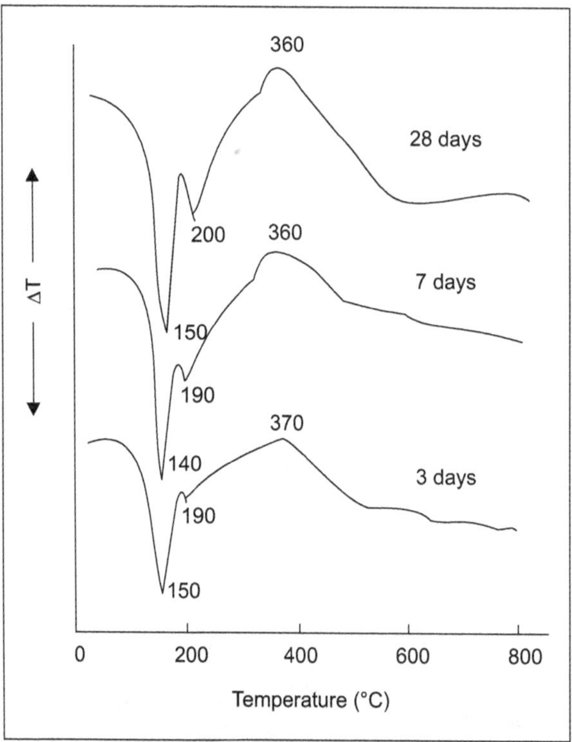

Fig. 7.17 DTA of Hardened Fluorogypsum Plaster

Table 7.21 Effect of Chemical Activators on the Properties of Fluorogypsum

Sl. No.	Chemical Activators (By wt.%)	Properties								
		Consistency (%)	Setting Time (Hour)	Compressive Strength(MPa)			Bulk density (kg/m^3)			Expn.(%)
				3d	7d	28d	3d	7d	28d	
1	$(NH_4)_2$ SO_4									
	1.0	24.5	7.0	4.2	9.32	27.7	1890	1900	1920	0.080
	2.0	25.7	8.0	4.9	10.3	12.2	1660	1950	1950	0.088
	3.0	25.0	8.5	6.0	11.8	12.5	1881	1880	1875	0.070
2	$Ca(OH)_2 - CaCl_2 - Na_2SO_4$									
3	3–0.5–1.5	29.0	1.58	41.9	44.3	45.2	1930	1970	2200	0.060
3	3–0.5–1.0	32.0	2.20	42.9	44.0	49.9	1830	1970	2080	0.065
3	3–0.5–0.5	26.0	2.13	36.2	47.8	50.3	1970	2080	2110	0.060
ASTM 61–50 (1981) limits IS : 2547 (Part-1) 1976 limits		—	20–360	—	—	Min.17.5	—	—	—	Max 0.5 at 96 hrs.

dehydration of gypsum and inversion of $CaSO_4$(III) in to β-$CaSO_4$. The intensity of endotherms and exotherms found to be increased with curing period indicating increase in gypsum formation. The fluorogypsum has been found sound in nature.

7.4.4 Water Absorption and Porosity of Fluorogypsum Binder/Plaster

The results of water absorption and the porosity of fluorogypsum binder is shown in Table 7.22.

Data show that fluorogypsum binders produced with the chemical activators $Ca(OH)_2 - CaCl_2 - Na_2SO_4$ possess lower water absorption and the porosity values than the $(NH_4)_2SO_4$ activator. On the basis of strength development, water absorption and porosity properties of the fluorogypsum binder, the addition of chemical activators *i.e.,* $Ca(OH)_2 - CaCl_2 - Na_2\ SO_4$ (3.0 – 0.5 – 0.5 by wt.%) were selected for further studies.

7.4.5 Effect of Lime Sludge on the Fluorogypsum Binder Produced by Blending Activators

The addition of lime sludge from the paper industry on the properties of flourogypsum binder containing activators ($Ca(OH)_2:CaCl_2:Na_2SO_4$) is shown in Table 7.23.

The trend of results show an increase in consistency and decrease in strength *[$Ca(OH)_2:CaCl_2:Na_2SO_4$::3:0.5:0.5] values with the addition of lime sludge. However, the attainment of strength is quite high. Data showed that with increase in lime sludge addition, the water absorption and the porosity values were

Table 7.22 Water Absorption and Porosity of Fluorogypsum Binders in Presence of Different Chemical Activators

Chemical Activators (%)			Water Absorption (%)			Porosity		
			2 hr	8 hr	24 hr	2 hr	8 hr	24 hr
1. $(NH_4)_2\ SO_4$								
1.0	—	—	8.69	8.69	8.7	16.24	16.3	16.4
2.0	—	—	18.72	18.80	18.8	34.2	34.2	34.3
3.0	—	—	13.70	13.70	13.75	26.5	26.5	26.6
2. $Ca(OH)_2 - CaCl_2 - Na_2SO_4$								
3.0	0.5	1.5	4.48	4.48	4.90	9.38	9.38	9.38
3.0	0.5	1.0	5.29	5.29	5.30	10.19	10.19	10.20
3.0	0.5	0.5	3.12	3.12	4.68	6.25	6.25	9.35

Table 7.23 Effect of Addition of Lime Sludge on the Properties of Fluorogypsum Binder Containing $Ca(OH)_2$:$CaCl_2$:Na_2SO_4*

Lime Sludge (%)	Consistency (%)	Compressive strength (MPa)			Bulk density (kg/m^3)		
		3d	7d	28d	3d	7d	28d
0.0	26.0	36.1	47.8	50.4	1970	2080	2110
5.0	33.7	36.5	40.8	40.0	1960	2060	2006
10.0	34.3	32.6	39.0	41.8	1930	1960	2000
15.0	37.1	30.4	38.0	40.2	1810	1850	1852
20.0	38.5	24.9	32.5	33.5	1730	1750	1870

increased with the enhancement of immersion period. At 20.0% addition of the lime sludge, maximum water absorption (21.90%, 23.60%, 26.40% at 2hr, 8hr, 24 hr) and the porosity values (40.60, 43.70, 48.90 at 2hr, 8hr, 24 hr) were attained. These studies suggest that fluorogypsum plaster may be partly replaced with the lime sludge to economize the use of such binders. Lime sludge available from sugar, acetylene (carbide) and fertilizer industries may also be used in addition to paper sludge.

7.4.6 Development of Bricks from Fluorogypsum Binders

Burnt clay bricks are essential ingredients for providing shelter to the millions, and is most popular because it can be adopted for any size or shape of construction. Brick constitutes about 13% of the total cost of the building materials required for construction of moderate house. The burnt clay brick industry as it exists today, is not capable to meet the demand of the modern construction agencies which require bricks of higher strength, better shape and of lower water absorption. The escalating cost of energy and unprecedented pressure on activities have undoubtedly caused great set-back to the production and quality of the bricks. It is, therefore, imperative that an alternative material is required to bridge the gap which is environment friendly and is acceptable to the construction agencies. It would therefore be useful to consider certain feature of technological development in this country with particular reference to the current thinking on the need to use waste materials. Thus, utilization of fluorogypsum binder for making building bricks can be considered a new and useful preposition in building sector.

The efforts were therefore, made to use fluorogypsum waste for making building bricks. The effect of various materials such as saw dust, rice husk, exfoliated vermiculite etc. on the properties of fluorogypsum binder was studied to arrive at optimum mix composition for casting bricks. 5 cm × 5 cm × 5 cm cubes were cast at the workable consistency for the compressive strength, bulk density, water absorption and the porosity properties.

Table 7.24 Effect of Saw Dust and Rice Husk on the Properties of Fluorogypsum Binder Containing $Ca(OH)_2$:$CaCl_2$:Na_2SO_4 Activators

Materials (By Wt. %)	Compressive strength (MPa)			Bulk density (kg/m³)		
	3d	7d	28d	3d	7d	28d
Saw Dust						
5.0	8.0	19.1	24.5	1810	1820	1820
10.0	6.26	11.2	16.4	1600	1650	1660
Rice Husk						
5 .0	8.42	10.6	11.0	1770	1860	1870
7.5	4.30	4.91	7.6	1640	1710	1750
10.0	1.83	1.92	2.40	1520	1600	1520

1. Effect of Saw Dust and Rice Husk

The effect of saw dust and rice husk on the compressive strength and bulk density of the fluorogypsum binder are listed in Table 7.24. It can be seen that with the addition of saw dust and the rice husk to the fluorogypsum binder, the compressive strength and the bulk density values were reduced. In case of saw dust, the fall in strength was comparatively less than the addition of rice husk. However, the decrease in the bulk density was much more with the addition of rice husk than the saw dust. In view of reduction in the density values, the manufacture of lightweight bricks can be contemplated.

The effect of saw dust and rice husk on the water absorption and the porosity of the fluorogypsum binder was studied. The results showed that the strength values are higher in case of addition of rice husk than the addition of saw dust. This may be attributed to the organic impurities particularly sugars present in the saw dust.

However, there is decrease in bulk density of the plaster with increase in saw dust and rice husk. With the addition of 10% saw dust to the plaster, the water absorption was found to be 8.75%, 9.42% and 9.98%, while the porosity was 15.58, 16.78 and 17.78 at 3, 7 and 28 days of curing. In case of addition of 5% rice husk, the water absorption of the plaster was 9.38%, 10.35% and 11.46% and the porosity was 17.40, 19.20 and 19.40 at 3, 7 and 28 days respectively.

2. Effect of Addition of Exfoliated Vermiculite on the Properties of Fluorogypsum Binder

The effect of addition of exfoliated vermiculite on the compressive strength and the bulk density of fluorogypsum are reported in Table 7.25.

Data show that with the increase in vermiculite content, the compressive strength and the bulk density are reduced. However, there is an increase in the strength and density values with the increase in curing period. It can be noted that

Table 7.25 Effect of Exfoliated Vermiculite on the Properties of Fluorogypsum Binder Containing $Ca(OH)_2$:$CaCl_2$:Na_2SO_4 Activators

Vermiculite (By Wt. %)	Compressive strength (MPa)			Bulk density (kg/m^3)		
	3d	7d	28d	3d	7d	28d
5 .0	17.98	23.11	29.28	1950	2080	2120
7.5	17.86	22.53	26.07	1850	1920	1950
10.0	17.67	19.30	20.00	1825	1850	1875
15.0	16.36	18.13	18.63	1750	1760	1850

the density can be further reduced by increasing the vermiculite content but the cost of the composition may also be increased. At 10% addition of vermiculite, the water absorption was 17.64%, 18.18% and 20.45% while the porosity was 30.51, 32.54 and 36.61. However, at 10.0% addition of vermiculite, adequate strength and the density values are achieved.

3. Preparation of Full Size Bricks

On the basis of properties obtained by the addition of an optimum quantities of saw dust (10%), rice husk (5%) and the vermiculite addition (10%) to the fluorogypsum binder, the full size bricks (19 × 9 × 9 cm) were cast at normal consistency. These bricks were tested for physical appearance, compressive strength, water absorption and efflorescence as per IS : 3495 (Part 1)–1976[34]. The properties of bricks are listed in Table 7.26. The strength and water absorption values complied

Table 7.26 Properties of Bricks Produced by Admixing Saw Dust, Rice Husk and Vermiculite with Fluorogypsum Binder

Property	Bricks Based on		
Physical state	Saw dust Sharp edges, corners	Rice husk Sharp edges, corners	Vermiculite Sharp edges, corners
Compressive strength, MPa			
7 days	4.27	6.36	11.5
28 days	10.84	6.60	12.0
Bulk density, kg/m^3			
7 days	1165	1171	1815
28 days	1166	1165	1875
Water absorption, %			
7 days	16.0	19.2	18.2
28 days	18.0	20.0	19.8
Efflorescence	Slight	Slight	Slight

Fig. 7.18 Building Bricks Cast from Fluorogypsum Binder/Plaster using Saw Dust, Rice Husk and Exfoliated Vermiculite

with the properties of IS : 12894– 1990[35] except those bricks prepared with rice husk. Typical photograph of binder-saw dust, binder-rice husk, binder-exfoliated vermiculite and binder-lime sludge bricks are shown in Fig. 7.18.

4. Suitability of Fluorogypsum Binder in Plastering

To find out suitability of fluorogypsum binder for use in internal plastering works, mortars of mix proportions 1:1, 1:2 and 1:3, by volume were prepared at mason consistency to smear plaster on the burnt brick wall. Mortar mixes 1:1, 1:2 and 1:3, binder-sand in 12 mm thickness were applied over the internal brick wall. The fineness modulus of the sand was kept at 1.91 (50:50 Badarpur and Ranipur river sand). The finish coat of 3 mm of neat binder was applied over 9 mm of 1:2, binder-sand under coat. Before applying binder-sand plaster, the brick wall was well watered so that mortar water may not be evaporated before the mortar was set. The plastered patches were examined for their various characteristics after 24 hours and onward. It was found that plaster patches developed adequate strength and hardness after 24 hours of application. The texture of the plaster was smooth and hard and showed good adhesion with the bricks. The cost of fluorogypsum binder/plaster of a plant of capacity 1000 tonnes per day in three shifts has been estimated to be ₹1800/-per tonne taking ₹500/tonnes, the cost of fluorogypsum.

7.5 Fluorogypsum Based Cementitious Binders

The fluorogypsum waste can be used in the production of cementitious binders. The binder is produced by blending fluorogypsum with 60–70% fly ash, hydrated lime sludge with and without Portland cement and chemical activator in different proportions.. The details of manufacture, properties, and durability of the binder are discussed.

7.5.1 Preparation of Cementitious Binders

The cementitious binders were prepared by intimately blending the ground fly ash, fluorogypsum, hydrated lime sludge, Portland cement of chemical composition (Table 7.27) with suitable chemical activator, in different proportions (Table 7.28) followed by inter- grinding in a ball mill to the specific surface area of 410–450 m^2/kg, (Blaine). The cementitious binders were tested and evaluated for their physical properties as per methods specified in IS : 4031–1976 and IS:2542 (Part I)–1981 The hydration of cementitious binder was studied by differential thermal analysis (Stanton Red Croft, U.K.) and SEM (Model LEO 438 UK). The cementitious binders were cast into 25 mm × 25 mm × 25 mm cubes at normal consistency for compressive strength and durability test. The cubes were cured under high humidity (>90%) at 27 ± 2°C for a period of 28 days and tested for compressive strength.

The durability of cementitious binders was studied by examining their behaviour in (*i*) water (*ii*) by wetting and drying and (*iii*) by heating and cooling cycles. One cycles of wetting and drying comprised heating the cubes for 16 hours at different temperatures (27° to 50°C) followed by cooling for one hour and immersion in water for 7 hours, where as one heating and cooling cycle consisted of heating the cubes at temperatures from 27° to 50°C separately for 6 hours and cooling at room temperature for 18 hours. After a certain

Table 7.27 Chemical Composition of Fly Ash, Fluorogypssm, Lime Sludge and Portland cement

Constituents	Fly Ash	Fluorogypsum	Lime sludge	Portland cement
P_2O_5	—	—	3.60	—
F	—	1.20	1.00	—
Organic matter	—	—	—	—
Cl	—	—	0.10	—
Na_2O+K_2O	0.76	—	—	—
SiO_2	62.90	0.67	3.10	22.50
R_2O_3 (Al_2O_3+Fe_2O_3)	28.35	0.61	0.50	9.60
CaO	1.50	40.44	52.00	61.50
MgO	0.80	Tr.	0.31	2.65
SO_3	0.20	56.00	0.16	1.75
LOI	1.50	0.62	41.00	2.00

Table 7. 28 Mix Composition of Cementitious Binder based on Fly Ash, Fluorogypsum, Lime Sludge and Portland Cement

Binder Activator Design.	Fly Ash	Fluorogypsum	Portland Cement	Lime Sludge	Chemical Activator
	(1)	(2)	(3)	(4)	(5)
F1	70	15	—	15	—
F2	65	15	—	20	—
F3	60	20	—	20	—
F4	50	30	—	20	—
F5	70	15	5	10	1.0
F6	65	10	10	15	1.0
F7	60	10	15	15	1.0
F8	50	20	15	15	1.5

number of cycles, the compressive strength of dry cubes was measured as per method laid down in IS : 4031–1976.

1. Properties of Cementitious Binders

The properties of cementitious binders containing fly ash, fluorogypsum and lime sludge with and without Portland cement are reported in Table 7.29.

Table 7.29 Physical Properties of Cementitious Binders with and without Portland Cement

Sl. No.	Binder Desgn.	Physical Properties						
		Consistency (%)	Setting Time (Hours)		Compressive Strength (MPa)			
			Initial	Final	3d	7d	28d	90d
1	F1	33.0	4.25	7.25	4.33	4. 80	13.88	18.32
2	F2	32.7	4.00	5.20	3.51	5.65	13.40	34.22
3	F3	32.3	4.50	7.60	4.43	5.92	18.35	24.14
4	F4	30.6	4.25	7.55	2.91	4.89	14.25	25.23
5	F5	32.3	4.10	7.80	4.93	10.91	15.67	17.12
6	F6	32.0	4.50	7.25	7.31	11.33	20.60	28.86
7	F7	30.0	4.80	7.50	13.72	16.80	30.50	39.28
8	F8	29.0	4.50	7.10	14.12	16.43	23.41	34.13

It can be seen that in all the compositions, the compressive strength increased with increase in curing period. Data show that mix composition F2 developed adequate strength at 90 days of curing. On addition of Portland cement the compressive strength of all mixes (F5, F6, F7 and F8) increased with much faster rate with curing period as compared to F1, F2, F3 and F4. The maximum strength was achieved in mix F7 at 90 days. The setting time of cementitious binders is retarded but they are within the specified value of 30 to 600 minutes as given in IS : 269–1989, specification for Ordinary Portland cement, 33 Grade. The hydration products of the cementitious binders (F2 and F7) has been shown by DTA (Fig. 7.19) as the formation of ettringite, gypsum, wollastonite, C-S-H gel and calcium carbonate. The intensity of ettringite, C–S–H and wollastonite phases is more pronounced in binder F7 than the binder F2, confirming development of higher strength in former than later case.

7.5.2 Durability of Cementitious Binder

1. Performance in Water

The cementitious binder cubes hardened for 28 days were dried and then immersed in water to measure their water absorption and porosity after different periods. The effect of immersion period on water absorption and porosity of cementitious binders is shown in Tables 7.30 and 7.31. It can be seen that porosity and water absorption of the cementitious binders increased with increase in immersion period. These results clearly showed that the cementitious binders

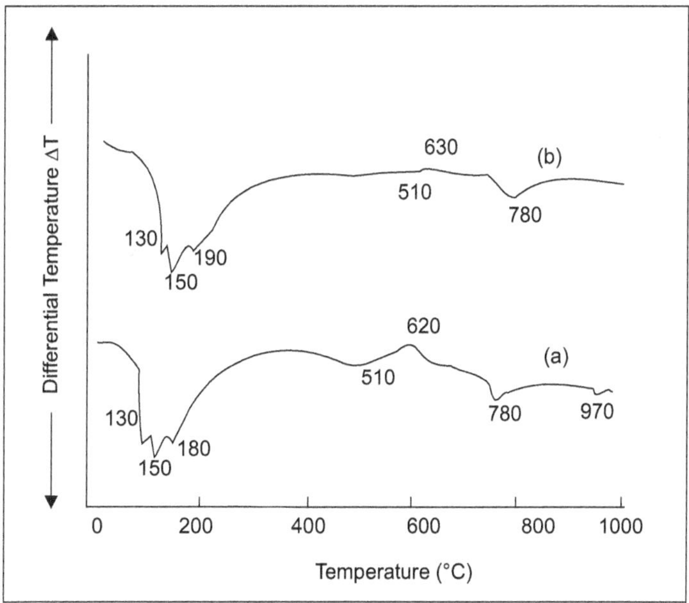

Fig. 7.19 Differential thermograms of cementitious binder (*a*) binder F2 (*b*) binder F7 hydrated for 90 days

Table 7.30 Water Absorption of Cementitious Binders

Binder designation	Water absorption (%)			
	1d	3d	7d	28d
F1	16.30	16.39	16.80	16.97
F2	15.70	16.24	16.24	16.60
F3	15.69	15.69	16.06	16.06
F4	17.33	17.40	17.38	17.40
F5	12.80	12.99	12.99	12.99
F6	8.22	8.88	9.37	9.70
F7	5.64	6.67	6.67	7.50
F8	4.82	5.78	5.78	6.10

Table 7.31 Porosity of Cementitious Binder

Binder designation	Porosity			
	1d	3d	7d	28d
F1	24.20	24.59	25.18	25.18
F2	27.17	28.10	28.10	28.73
F3	26.20	26.20	26.83	26.83
F4	30.85	30.85	30.90	31.00
F5	21.13	21.44	21.50	21.50
F6	15.62	16.87	17.81	18.43
F7	10.32	12.21	12.21	13.76
F8	9.35	11.22	11.22	11.85

containing Portland cement have better water resistance and low porosity than the cementittious binders without Portland cement. The significant reduction in porosity and water absorption of the cementitious binders containing Portland cement may be due to filling of voids, and pores in the hardened matrix by the enhanced quantities of ettringite and tobermorite. Moreover, these observations clearly exhibit that binders have no erosion effect and are durable.

2. Wetting and Drying Cycles

The effect of alternate wetting and drying cycles on the compressive strength of cementitious binders F2 and F7 studied up to 50 cycles at temperatures from 27° to 50°C are shown in Figs. 7.20 and 7.21.

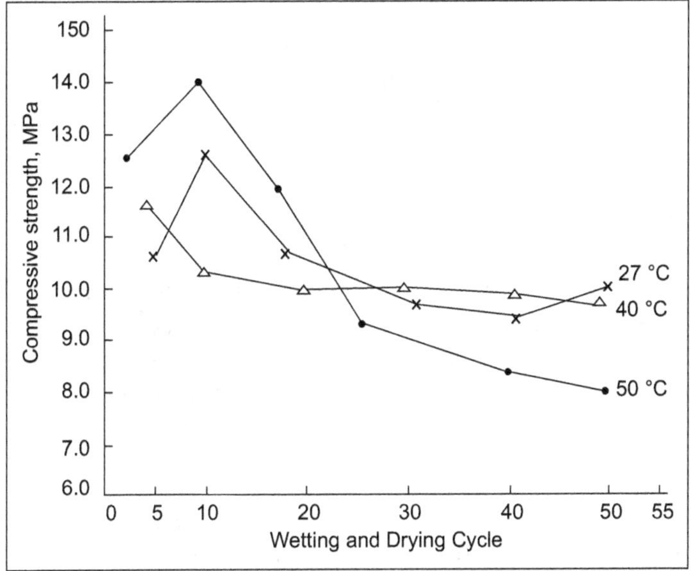

Fig. 7.20 Effect of Wetting and Drying Cycles on Compressive Strength of Binder F2

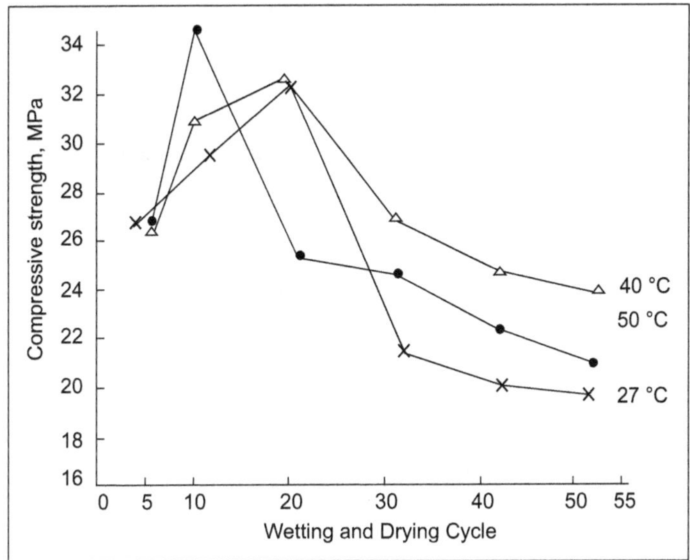

Fig. 7.21 Effect of Wetting and Drying Cycles on Compressive Strength of Binder F7

Figure 7.20 shows that on increasing temperature from 27° to 50°C, the maximum attainment of strength was found at 10 cycles and afterwards strength dropped gradually with the increase in cycle and temperature. The increase in strength up to 10 cycles may be attributed to the increase in pozzolanacity of unreacted fly ash with increase in temperature. The loss in strength after 50 cycles at 27°, 40° and 50°C was found to be 8.50%, 9.32% and 30.5% respectively against the pristine strength values. In Fig. 7.21 (Binder F7), the maximum strength was obtained at 20 cycles at temperatures 27°, 40° and at 10 cycles at 50°C respectively and then fall in strength was observed. However, the strength measured at 50 cycles at 27°, 40° and 50°C was 18.0%, 14.20% and 16.15% respectively of the pristine strength values.

3. Heating and Cooling Cycles

The effect of alternate heating and cooling cycles at temperatures from 27°C to 50°C on the compressive strength of cementitious binders is shown in Figs. 7.22 and 7.23. Data show that the compressive strength of cementitious binders F2 and F7 decreased with the increase in heating and cooling cycles. The compressive strength of binder reduced to 36.15%, 43.78%, and 33.0% at 27°, 40° and 50°C respectively of the original strength value after 50 cycles.

4. SEM Studies

Scanning Electron Microscope (SEM) was used to study the morphology of the hydrating binders. The identification of shape and growth of crystals assists in evaluating the pattern and stacking of the crystals. The development of crystals with sharp edges and proper stacking is important to get desired hydraulic products in such type of cementitious binders.

Wetting and Drying Cycles

The SEM micrographs of cementitious binders cured at 27°C (28 days) and subjected to 40° and 50°C showed formation of subhedral to anhedral prismatic crystals

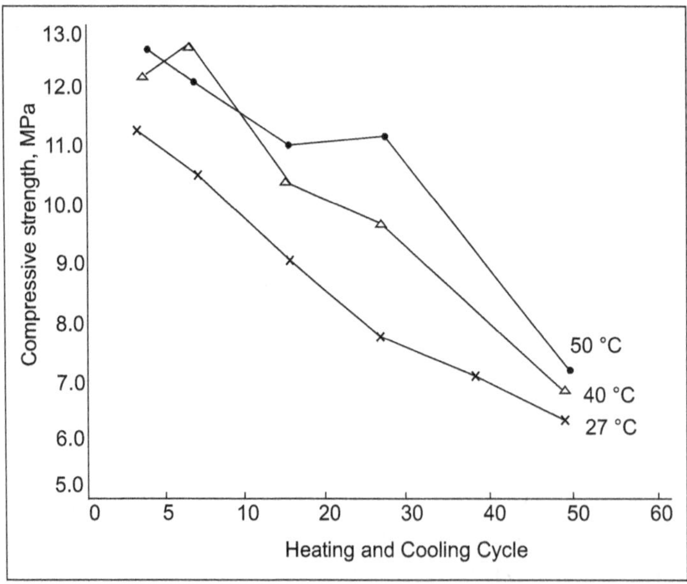

Fig. 7.22 Effect of Heating and Cooling Cycles on Compressive Strength of Binder F7

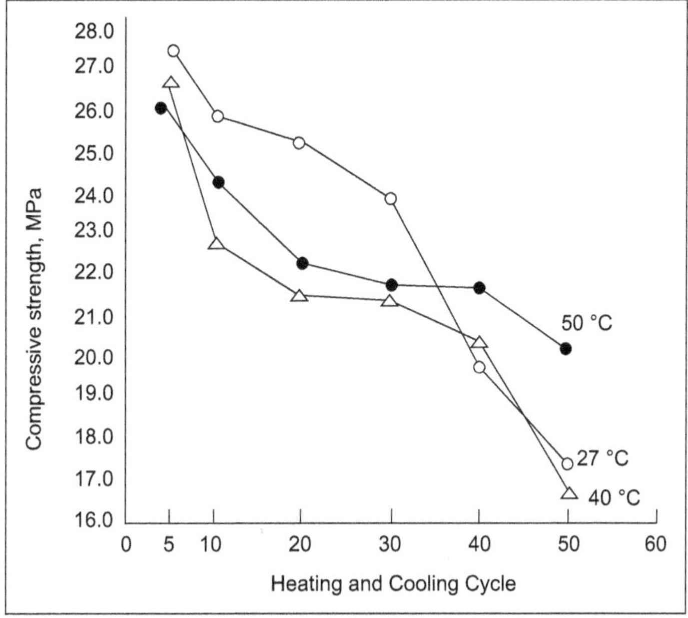

Fig. 7.23 Effect of Heating and Cooling cycle on Compressive Strength of binder F7

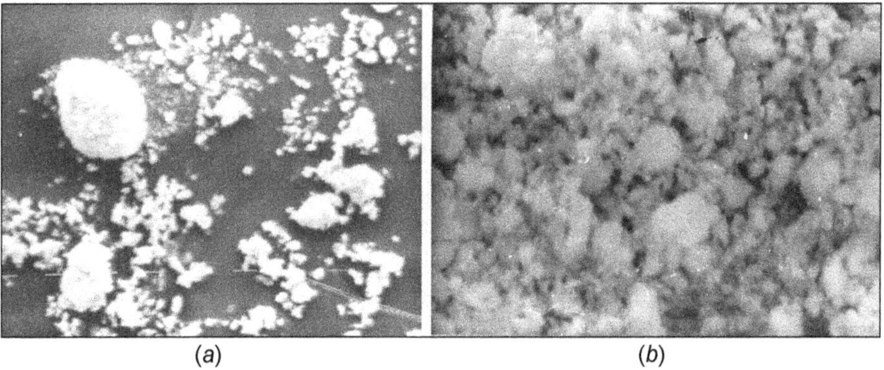

(*a*) (*b*)

Fig. 7.24 SEM of Cementitious Binders F2 (*a*) and F7 (*b*) at 50°C after 50 Wetting and Drying Cycles

of gypsum and hydrated plates of tobermorite with irregular stacking interspersed with partially hydrated fly ash spheres. The micrographs of cementitious binders F2 and F7 cured at 50°C after 50 cycles of wetting and drying cycles are shown in Figure 7.24 (*a*) and (*b*).

The micrograph of binder F2 (*a*) illustrates appearance of subhedral to anhedral rounded bodies of unhydrated fly ash spheres in association with prismatic and lath shaped crystals of gypsum and unreacted fluorogypsum particles. While the micrograph of cementitious binder F7 (*b*) showed formation of euhedral, hydrated fly ash sphere with microcrystalites of gypsum and stellated fluorogypsum bodies.

(a) (b)

Fig. 7.25 SEM of Cementitious Binders F2 and F7 at 50°C after 50 Heating and Cooling Cycles

Heating and Cooling Cycles

The role of heating and cooling cycles on the compressive strength of cementitious binders F2 and F7 at temperatures 50°C at 50 cycles is shown in Fig. 7.25. It can be seen from binder F2 (*a*) there is formation of euhedral to anhedral agglomerated crystals of partially hydrated and unhydrated fly ash spheres wherein binder F7 (*b*), euhedral to anhedral prismatic and lath shaped crystals of ettringite and gypsum are formed in association with hydrated plates of tobermorite embedded in binder matrix.

The cementitious binder can be produced by blending industrial wastes like fly ash, fluorogypsum, lime sludge with and without OPC in suitable proportions. Ettringite, CSH and Wollastonite have been identified as the major hydraulic products responsible for strength development of the cementitious binders. Water absorption and porosity of cementitious binders increased with increase in curing period, indicating, thereby absence of leaching and increased level of stability towards water. When subjected to alternate wetting and drying cycles at 27°, 40° and 50°C, variation in strength development was observed due to variable hydration of the binders. When subjected to alternate heating and cooling cycles at 27°, 40° and 50°C, the strength is reduced with an increase in temperature. The maximum fall in strength occurs at 50°C. The cementitious binder can be used for masonry work and preformed products under 50°C.

The chemically activated fluorogypsum can also be used for producing flooring tiles as made from phosphogypsum anhydrite[36–37].

References

1. Mac Taggart, E.F., 'Gypsum Anhydrite as a Base for Plaster Product (Anhydrite Plasters)', Rock Prod., 1937, Vol. 40, No. 7, pp. 46–48.
2. Moyer, F.T., 'Gypsum and Anhydrite', Inform. Circ., 1939, No. 7049, U.S. Bur. Min. p. 45.
3. Pippard, W.R., 'Calcium Sulphate Plasters', Bull. No. 3, Build. Res., London., Suppl., 1940, Pt. 2, pp. 41–61, (Revised Edn.).

4. Andrews, H., Gypsum and Anhydrite Plasters', Nat. Build. Stnd. Bull. 1948, No. 6, p. 16.
5. Gypsum and Anhydrite Building Plasters, Brit. Stand. No. 1191, 1944, p. 22.
6. Grine, H., Zement-Kalk-Gips, 1962, Vol. 5, pp. 285–298,
7. Eipertaur, E., Utilization of phosphoric acid gypsum sludge, Tonind Ztg. 1973, Vol. 97, pp. 4–8.
8. Mooney, R.W., Tuma, S.Z., Gold Smith, R.L., J. Inorg. Nucl. Chem., 1968, Vol. 30, pp. 1669–1675.
9. IS-2547(Par-1)–1976, Specifications for Gypsum Building Plaster, Part-1, Excluding Premixed Light Weight Plasters, Bureau of Indian Standards, New Delhi, 1976, p. 5.
10. Berezovskii, V.I., Phase Transformation in the Annealing of Phosphogypsum and strength of Phosphoanhydrite cement, Zh. Prikl. Khim., 1965, Vol. 38, pp. 1653–1657.
11. IS-2542 (Part-1)–1978, Materials of Tests for Gypsum Plaster,concrete and Products Part-1, Bureau of Indian Standards, New Delhi, 1978, p. 31
12. Simanovskaya R.E., Transations of the Ya.S., Samoilov, Scientific Research Institute of Fertilizers and Insectfungicides, 1940, p. 153.
13. Knauf, A.N. Kronert, W und Haubert, P., Die Rasterelektronemikroskopie, eine erganzende Methode zur Unsuchung von Gipsen, ZKG INTERNATIONAL, 1972, Vol. 25, No. 11, pp. 546–552.
14. Otterman, J., Beiziechungen zwiscen Hydratation und Festigkeit.Mitterilungen aus den Laboratorien des Gelogoischen Dienstes Berlin, Neue Folge, No. 2, 1951.
15. Hajjouli E., A. und Murat, M, Strenth Development and Hydrate Formation Rate, Investigation on Anhydrite Binders, Cem.Concr, RES., 1987, Vol. 17, No. 5, pp. 814–820.
16. Older, I., Robler, M., Zusammenhange zwischen Porengefuge und Festigkrit abgebundener Gipspasten. Teil 2, EinFlB Chemischer Zusatze, ZKG INTERNATIONAL, 1989, Vol. 42, No. 8, pp. 419–424.
17. Riedel, W, Bimberg,R und Gohring Ch.,: RinfluB von Anregren auf die Eigenschaften eines synthetischen Anhydritbinders aus Flouranhydrit. Bauustoffind. 1989, No. 2, pp. 62–65.
18. Hunger, K.J. und Henning, Zur,Bildung von Gip: Faserkistallen aus wa Brigen Losungen. Silikattechnik, 1988, Vol. 39, No. 1, pp. 21–24.
19. Budnikov, P, and Morova A.A. (USSR), 'Hydration Condition for Natural Anhydrite', Cem-Wapno-Gips, 1969, 24/36(4), pp. 106–8, (Polish).
20. Shabanova, E.E. Segalova and P.A. Rehbinder, Dokl, Akad. Nauk, USSR, 1963, Vol. 161, P. 403.
21. Rehbinder, A., Pure & Appl, Chem., 1965, Vol. 10, p. 337.
22. Ridge, M, J., Aust. J. Appl. Sci., 1959, Vol. 10, p. 218.
23. Anhydrite Building Materials, Verlag Technik, Berlin, 1942.
24. Hamri, G. Murat, M. and Foucault, M (Eds.), Proc. Int. RI:FM Symposium on Calcium Sulphate and Derived Materials, Saint-Remy Les Chevreuse 1977.
25. Singh Manjit & Garg Mridul, Making of Anhydrite Cement from Waste Gypsum, Cement and Concrete Research (U.S.A.), Vol. 30, 2000, pp. 571–577.
26. Andrew, H., 'The Production, Properties and Use of Calcium Sulphate Plasters', Building Research Congress, Division 2, (London), 1951, pp. 138–40.
27. Tenny, M. and Ben-Yair, Cement Lime Mfr., May 1965, pp. 49–53.
28. Singh Manjit, Use of Phosphogypsum Anhydrite in Construction works, Civil Engineering & Construction Review, May, 1994, Vol. 7, No. 5, pp. 5–8.
29. IS:1237–1980, Specifications for Cement & Concrete Flooring Tiles, Bureau of Indian Standards, New Delhi, 1980, pp. 1–19.
30. Singh Manjit, Garg Mridul & Rehsi, S.S. Durability of phosphogypsum based water resistant anhydrte binder, Cem. Concr. Res., 1990, Vol. 20, No. 1, pp. 271–276.

31. Singh Manjit, Garg Mridul, Microstructure of Glass Fibre Reinforced Water Resistant Gypsum Binder Composites, Cem. Concr. Res., 1993, Vol. 23, pp. 213–220.
32. Singh, M., From Waste to Wealth-Developing Potential Construction Materials from Industrial Wastes, (Edited by Prof. P.C. Trivedi), Avishkar Publishers, Distributers, Jaipur, (2006) 36–62.
33. Taneja, C. A. et.al., Supersulphated Cement from Waste Anhydrite, Res & Ind., 1974, Vol. 19, pp. 51–52.
34. IS : 3495 (Part 1)–1976. Methods of Test for Burnt Clay Building Bricks: Part 1- compressive strength and Water absorption, 1976 .
35. IS : 12894–1990, Specifications for Fly ash-Lime Bricks, Bureau of Indian Standards, New Delhi, 1990.
36. Singh Manjit, Recent advances in the utilization of waste fluorogypsum, Global Gypsum Magazine, (U.K.), No. ember 2011, pp. 18–29.
37. Singh Manjit, Durable cost effective building materials from waste fluorogypsum, New Building Materials and Construction World, Vol. 17, No. 9, March, 2012, pp. 201–217.

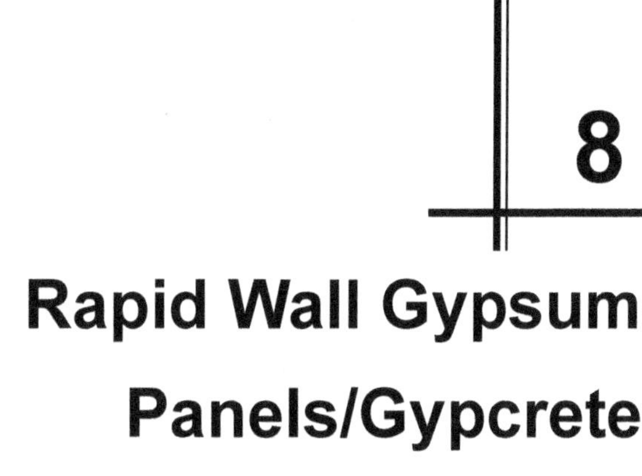

8 Rapid Wall Gypsum Panels/Gypcrete

8.1 Rapid Wall Panels for Buildings

Steel, Cement, Sand, Concrete and blocks, RCC, Bricks etc. are the building materials currently used for building and construction. Extensive use of these materials adversely affect the environment. Occurrence of natural disasters like earthquakes, cyclones, high tides, fire etc. have become more frequent than ever before. Design and construction of disaster resistant houses and buildings are very costly and unaffordable to the common people/masses who in most such disasters bear the brunt. Their unsafe houses and structures can not withstand such fury of the nature.

There is absolute need and necessity for an efficient and alternative building material which is available at affordable cost to build safer houses and buildings without causing adverse impact on the environment. The dwellings should be able to withstand such disasters and protect lives and property. Such building material is ever more relevant now after the unprecedented occurrence of tsunami in the recent past. It is essential that we have access to technological advancement in this field so that the society/people will be benefited at large.

8.2 Rapid Wall Panels

Now there is an efficient, eco-friendly alternative building material available at competitive cost which has many advantages and benefits. This technology was developed in the 1990's in Australia. The technology provides for manufacture of large load bearing wall panels which can be used for walling and roofing.

World's largest load bearing gypsum building panels (12 × 2.85 mtrs. = 34 sq. mtrs.) called Rapid wall can be manufactured using advance Australian building technology developed by Rapid Wall Building Systems (RBS). One such wall panel can replace one lorry load of conventional bricks. No plastering of the walls is required. The smooth and superior finish is ready for a primer coat and painting.

Numerous buildings including large multistoried apartments with more than 3000 units and buildings up to 17 storey have been built in Sydney, Adelaide, Brisbane, Melbourne, Perth etc., in Australia using this material. The technology has been given to Malaysia and China since 1999. In China, multistoried buildings up to 7 storey have been built using the material as load bearing wall. In China internal walls of 62 storey have built using the Rapid wall panels.

8.3 Residential Buildings Using Rapid Wall Panels

Houses and Buildings designed and built with the panels can resist natural disasters like earthquakes, cyclones, tidal waves etc. It is resistant to fire, water, corrosion and free from rot and termite. The product has high compressive strength, tensile strength, flexural strength and ductility. One of the greatest advantage is the modular cavities of the large panel which enables to use it in combination with concrete and RCC. It forms composite material by which the strength of the material can be increased many folds. By filling the cavities of the roofing panel with RCC and linking down to the foundation through the wall panel, the structure can resist cyclones and disasters like sunami. It is most ideal for coastal regions as the panel can resist corrosion.

8.3.1 Low Cost Mass Housing

A typical low cost mass housing schemes with carpet area of 25 sq. ft. will require 101 sq.mtrs. of Rapid wall wall Panels. Construction corner joints can be made using infill of RCC. These panels can be used as load bearing wall panel up to two story's without concrete infill. For multistoried buildings more than two stories, concrete can be filled in the cavities by providing steel reinforcement. By filling up the cavities of the wall panel with nominal steel reinforcement superstructure can with stand very heavy lateral load.

The weight of one 12 mtrs. × 3 mtrs. Panel is only 1.5 tons. Compared to this an equivalent 23 cm. thick brick wall weight is about 19 tons. Weight of equal thick RCC wall will be 9 tons. Whereas it's strength will be many folds than the RCC slab. The light weighted material with high compressive, tensile, flexural strength and high ductility helps to withstand earth quakes. It is most ideal for tropical climate like India as its high thermal insulation and making interiors cooler during summer and warmer during winter. It does not cause CO_2 emission and does not require cement plaster.

Use of eco-friendly Rapid wall Panels shall save precious soil, sand and water. This can provide low cost housing for slum clearance and help Poverty Alleviation Programme.

This material is cost effective and provides fast track method of construction which can save more than 50% times. It saves about 25% cost of construction compared to conventional building construction. The basic raw material is phosphogypsum or mineral gypsum which is calcined to form gypsum plaster. The gypsum plaster is reinforced by micro-strand glass roving's. Special emulsions and chemicals are added. The new product is called Rapid wall panels.

The Rapid wall panel material has been subjected to extensive testing by SERC/CSIR Chennai and IIT Chennai. SERC carried out earth quake testing of six structures built with this material and all the six structures withstood earthquakes up to Richter Scale 8. Considering the above test reports with assessment and analysis by SERC and IIT Madras the product has been approved by Building Material Technology Promotion Council (BMTPC), a National Body to approve new building materials in the country as mandated by Govt. of India Gazette Notification No. 1–16011/5/99 H-11 published in Gazette of India No. 49 dated 4th Dec. 1999. BMTPC has evaluated the product for construction buildings up to 10 stories. The Rapid wall panels in stacking position are shown in Fig. 8.1.

8.4 Manufacture of Wall Panels in India

Rashtriya Chemicals and Fertilizers Ltd., (RCF) a Goverment[1–2] of India enterprise engaged in the manufacture of chemical fertilizers and industrial chemicals, produces phospho-gypsum as byproduct from their Phosphoric Acid plant. This phosphogypsum can be used to manufacture the above load bearing wall panels and other plaster products using Australian technology. The RCF phosphogypsum was extensively tested in Australia and suitability of the same for these end products has been established. RCF plans to manufacture wall panels and wall plaster and Putty. The rapid wall plant has an annual production capacity of 14.2 lakh sq meters. Similarly Gypsum Building India Ltd. (GBIL) are also

Fig. 8.1 Gypsum Rapid Wall Panels

Table 8.1 Properties of Rapid wall Building Panel at a Glance

Compressive Strength empty panel: 73.10 kg/cm^2 concrete in fill: 180.70 kg/cm^2	**Flexural Strength Empty panel: 21.25g/cm^2 Concrete in fill: 20.80 kg/cm^2**	**Tensile Strength 28 kN/m**	**Ductility 4.11**
Water absorption <2%(24 hrs)	Fire resistance 4 hrs rating 700 °C to 1000 °C.	Thermal resistance 0.36 R	Sound Transmission Class s-STC 40
Free from Corrosion	Axial vertical load on empty panel of 2.85 m high × 1 m. 14.04 tons	Axial vertical load on Concrete in filled panel of 2.85 n high × 1 m^2.	Rot and termite proof
Light weight 40 kg/m^2 (1/8th weight of concrete)	Net Density 1140 kg/m^3	Earthquake resistant: Test withstood earthquakes up to Richter Scale 8	Eco-friendly (green material) (raw material recycled waste and no emission of CO_2)
Saving • Water • River sand • Agri soil • Cement • Steel	Saving • Time of construction	Very low maintenance Saving in recurring energy cost by cooling (air condition) [88%]	Cost-effective (affordable cost)

making these panels in Koyambedu, Chennai, Tamilnadu[3–6]. The properties of these panels are shown in Table 8.1.

8.4.1 Production of RWP

A Rapid wall plant can be designed to include up to six casting tables. Installation of peripheral equipment may include a Rapid cure dryer, for curing the panels and the Rapid saw, for processing whole panels into specific applications.

Rapid Building Systems of Australia sells the plant and equipment and has licensed the rights to manufacture Rapid wall and Rapid flow plaster for certain geographical areas around the globe.

The key to manufacturing Rapid wall in any country is having a readily available and reliable source gypsum or high quality gypsum plaster. If high quality gypsum plaster is not available then, Rapid flow calciner is needed to produce plaster.

The Rapid flow calciner can produce high quality plaster from either natural gypsum or chemicalgypsum obtained as a by-product of either the fertiliser industry (phospho gypsum) or from the desulphurization of coal during the production of electricity, from a coal fired power station (flue-gas gypsum).

With a source of gypsum or plaster available, manufacturing Rapid wall or Rapid flow plasters can bring both social and financial rewards. At the same time as providing high quality, affordable housing and the untold benefits this will bring a fifty per cent return on the investment in one year after production.

Fig. 8.2 Rapid Flow Gypsum Calciner

8.4.2 Manufacture of Gypsum Plaster

1. Rapid Flow Gypsum Calciner

The Rapid Flow Gypsum Calciner[7] (Fig. 8.2) is based on the fluidised bed concept that differs significantly from other fluidised bed calciner designs. It uses a unique process that allows the introduction of hot gases into the fluidised bed in a completely controlled manner.

While other calciners produce over and under burnt plaster because of different expansion and contraction rates, Rapid flow plaster particles are calcined at exactly the right temperature.

The Rapid Flow Gypsum Calciner is designed to produce consistently high quality Rapid flow plaster that is suitable for use in the production of the highest quality compounds, other high grade plaster products and in the manufacture of Rapid wall, plaster board and other materials for the building industry.

The most important aspect of the novel design features incorporated into the Rapid flow Calciner is the ability to produce a very high strength beta plaster. Other aspects of the design provide enhanced versatility and economy of operation particularly in terms of the types of gypsum that can be processed.

2. Manufacture of RWP

Casting Table

The RWP are produced by reinforcing E-type glass fibre cut into desired size to yield optimum strength in composite on a specially designed casting tables. This plaster layer is lightly screed after which the traveling crab assembly automatically chops and dispenses a predetermined quantity of glass-fibre rovings over the entire liquid-plaster surface. This layer of glass-fibre is then rolled into the plaster to position it centrally within the 15 mm thick skin to provide reinforcement to the plaster. The glass fibre roll is shown in Fig. 8.3.

The casting-table (Fig. 8.4), in each computer-controlled plant, comprises a flat steel epoxy-coated surface with sides that are raised to contain the plaster when in the fluid state[8].

Fig. 8.3 Micro-strand glass roving role

Prior to the commencement of the manufacturing process the casting table is first lightly oiled /greased. Commencing from the start position the crab assembly moves over the Casting table accurately dispensing the special plaster mix comprising water, Rapid flow gypsum-plaster, water repellents and additives, over the entire table to a depth of 15 mm.

In every 250 mm length of Rapid wall a 230 mm by 94 mm cell is formed using teflon coated removable plugs that are laid at right angles to the 12-metre panel. The core-table mechanism positions all 48 plugs over the plaster and glass-fibre layer on the casting table. The final quantity of plaster is dispensed onto the casting table filling between the plugs and forming the top skin of the panel.

The travelling crab then dispenses a further layer of chopped glass-fibre over these cores and the tamping process is undertaken. A final quantity of chopped glass-fibre is again automatically and uniformly dispensed over the entire panel surface by the traveling crab. The crab assembly then automatically returns to its cleaning and filling station to be prepared for the production of the next Rapid wall panel.

Fig. 8.4 Gypcrete/Rapid wall Mfg Plant in China – Jinan City Showing Casting Table (Internal View)

Using a mesh roller the surface of the Rapid wall panel is then rolled to position this final layer of glass-fibre centrally within the plaster top skin. Final screeding and smoothing of the cast is completed manually by two operators. To this point the process has taken only 20 minutes. After this, the panel is left to cure until the temperature and the consistency of the plaster allows final screeding. Once the plaster has completed its initial set, a further 20 minutes, the core-table mechanism advances and locks onto the core formers and slowly withdraws them from the set panel.

To remove the panel from the casting-table three perimeter edges of the casting table are opened and two panel-supports are extended. The table is then automatically tilted to approximately 88 degrees off vertical. The entire two-tonnes weight of the wet panel is taken by the bottom supports.

To this point, the entire process has taken 45 minutes. Finally a multi-directional truck, fitted with a transfer frame, removes the Rapid wall panel from the tilted casting-table and places it either in air drying racks or in the Rapid cure drying oven for final curing prior to it being stored and ultimately cut to dimensions for installation on a specific building project. The elapsed time of the complete manufacturing process, including full curing in the Rapid cure dryer, is less than two hours for each panel. The cut RWP and their cross section are shown in Figs. 8.5 and 8.6 respectively.

Fig. 8.5 Rapid Wall Panels (RWP)

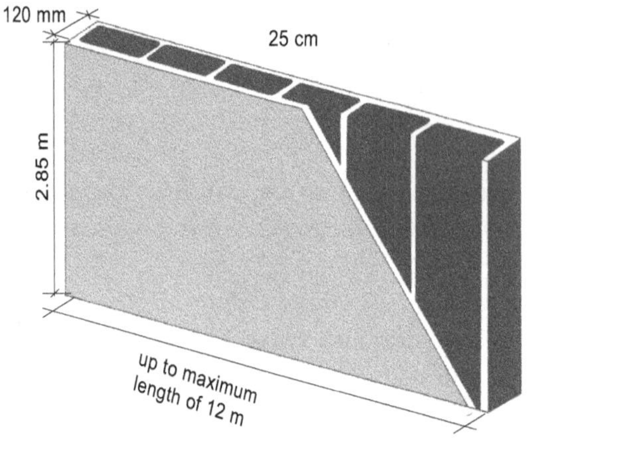

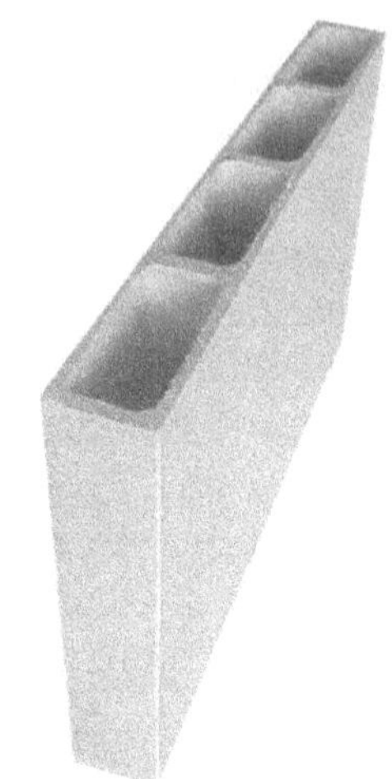

Fig. 8.6 Cross-section of Rapid Wall Panel

8.5 Characteristics of RWP

Panel is reinforced by micro-strand glass roving (13 microns) by both ways on all four sides - both flanges and webs. This glass roving itself is a hi-tech product manufactured in India.

Reinforcement is provided during the automated hi-tech production process itself. This reinforcement provide load bearing capability & "confinement property" when cavities filled up with concrete. Tensile strength of glass roving (single filament) is 3100–3800 MPa (tensile strength of glass roving is much higher than even that of steel use for concrete reinforcement which is maximum @ 500 MPa the automated hi-tech product manufactured in India. Elastic Modulus of glass roving is 76–78 GPa. This provides very high load bearing capability to the panel making into a composite material. The properties of Rapid Wall Panels with infill are shown in Table 8.2. The reduction in mass on using RWP is shown in Table 8.3.

8.5.1 How Gypcrete buildings will be able to resist earthquakes better compared to conventional load bearing brick/ concrete block construction?

- Gypcrete panel is very large load bearing panel without joints.
- Gypcrete panel is light weight with modular cavities.
- Gypcrete panel has the uniqueness of the combination of high compressive strength, flexural-tensile strength and ductility.
- Gypcrete panel has very high confinement property, by which when cavities in filled with concrete increases load bearing capability of the panel to 11 folds.

*Buildings with Gypcrete panel will be more resistant to lateral loads due to earthquakes or wind, provided both vertical joints between the walls and horizontal joints between the walls and slabs are properly designed and detailed to ensure structural stability.

So, small buildings can be built with Gypcrete panels resistant to earthquakes at lesser cost.

Table 8.2 Testing Report of RWP Tested by Structural Engineering Research Centre (SERC)/CSIR, Chennai

1.	Compressive Strength without infill	7.31 N/mm^2
2.	Compressive Strength with concrete (M20) infill	18.07 N/mm^2
3.	Flexural-tensile strength by single point load without infill	2.13 N/mm^2
4.	Flexural-tensile Strength by single point load with concrete M20 infill	2.08 N/mm^2
5.	Flexural-tensile Strength by double point load without infill	3.80 N/mm^2
6.	Flexural-tensile Strength by double point load with concrete M20 infill	4.59 N/mm^2
7.	Axial Compression of wall panel 2.85 m height × 2 m width × 120 mm thick without infill	281.70 kN (140.85 kN/m)
8.	Axial Compression of wall panel 2.85 m height × 0.775 m width × 120 mm thick with concrete (M20) infill	1200 kN (1548.3 kN/m)
9.	Ductility Factor of Gyp crete empty panel (Ductility of Concrete is 1.75)	4.11
10.	Earthquake Resistance	—
All the structures/buildings/houses tested withstood the artificial earthquake with a maximum peak ground acceleration of 0.36g satisfying the requirement of IS : 1893 (Part – I)–2002 for Zone – V. This is equivalent to earthquake intensity of Richter Scale 8.		

Table 8.3 Reduction of Mass (weight) Using Gypcrete Building Panels

Sl. No.	Panels	Weight,	kg.
1.	Weight of 1 sq.m brick wall 9" thick with 2 sides cement plastering	400	—
2.	Weight of 1 sq.m Gypcrete panel	40	90.%
3.	Weight of 1 sq.m panel with concrete infill	250	37.5%
4.	Load bearing walls + non-load bearing internal walls using filled and unfilled panel	200 (avg)	50.0%

8.5.2 Protection against Corrosion

- Corrosion of reinforcement occurs not only in marine environments, but also in urban environment due to automobile exhaust and industrial environments.
- To protect high rise apartments/buildings, protective measures with recurring additional cost is necessary.
- Reinforcement in Gypcrete buildings encased/embedded within the cavities of the panel and Gypcrete panel is free from corrosion.

For Rapid wall buildings/ Housing a conventional foundation like spread footing, RCC column footing, raft or pile foundation is used as per the soil condition and load factors. All around the building RCC plinth beam is provided at basement plinth level. For reaction of panel as wall, 12 mm dia. vertical reinforcement of 0.75 m long of which 0.45 m protrudes up and remaining portion with 0.15 m angle is placed into the RCC plinth beams before casting. Start up rods are at 1 m centre to centre.

8.6 Rapid Wall Panels for Use in Rapid Construction

Rapid wall enables fast track method of construction[9–11]. Conventional building construction involves various cumbersome and time consuming processes, like (*i*) masonry wall construction (*ii*) cement plastering requiring curing, (*iii*) casting of RCC slabs requiring centering and scaffolding and curing (*iv*) removal of centering and scaffolding and (*v*) plastering of ceilings and so on. It also contributes to pollution and environmental degradation due to debris left on the site.

In contrast, Rapid wall construction is much faster and easier. There will be no debris left at site. Construction time is minimized to 15–20%. Instead of brick by brick construction, Rapid wall enables wall by wall construction. Rapid wall also does not require cement plastering as both surfaces are smooth and even and ready for application of special primer and finishing coat of paint.

8.6.1 Rapid Construction Method

As per the building plan[12], each wall panel will be cut at the factory with millimeter precision using an automated cutting saw. Door/window/ventilator, openings for AC unit, h etc. will also be cut and panels for every floor is marked relating to building drawing. Panels are vertically loaded at the factory on spillages for transport to the construction sites on trucks. Each still age holds 5 or 8 pre-cut panels. The still ages are placed at the construction site close to the foundation for erection using vehicle mounted crane or other type of crane with required boom length for construction of low, medium and high rise buildings. Special lifting jaws suitable to lift the panels are used by inserting into the cavities and pierced into webs, so that lifting/handling of panels will be safe. Panels are erected over the RCC plinth beam and concrete is in filled from top. Protruded start up rods go inside cavities as can be seen from Fig. 8.7.

8.6.2 Joints

Wall to wall 'L', 'T', '+' angle joints and horizontal wall joints are made by cutting of inner or outer flanges or web appropriately and infill of concrete with vertical reinforcement with stirrups for anchorage. Various construction joints are illustrated (Figs. 8.8 to 8.9).

All the panels are erected as per the building plan by following the notation. Each panel is erected level and plumb and will be supported by lateral props to

Fig. 8.7 Foundation

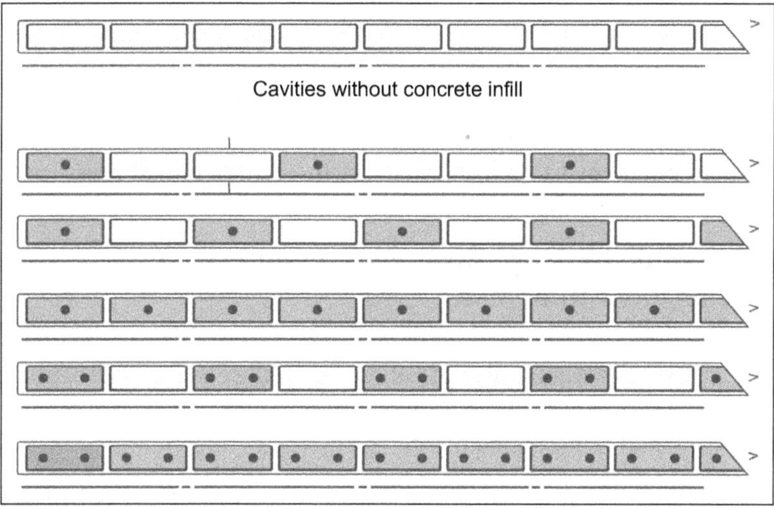

Fig. 8.8 RCC Infill to increase load capabilities

keep the panel in level, plumb and secure in position. Once wall panels erected, door and window frames are fixed in position using conventional clamps with concrete infill of cavities on either side. Embedded RCC lintels are to be provided wherever required by cutting open external flange. Reinforcement for lintels and RCC sunshades can be provided with required shuttering and support.

8.6.3 Concrete Infill

After inserting vertical reinforcement rods as per the structural design and clamps for wall corners are in place to keep the wall panels in perfect position, concrete of 12 mm size aggregate will be poured from top into the cavities using a small

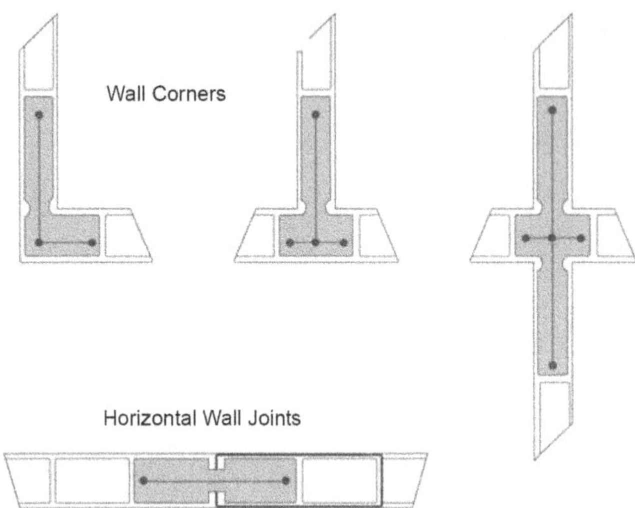

Fig. 8.9 Various Construction Joints

hose to go down at least 1.5 to 2 m into the cavities for directly pumping the concrete from ready mix concrete truck. For small building construction, concrete can be poured manually using a funnel. Filling the panels with concrete is to be done in three layers of 1 m height with an interval of 1 hr between each layer. There is no need to use vibrator because gravitational pressure acts to self compact the concrete inside the water tight cavities.

8.6.4 Embedded RCC tie Beam All Around at Each Level Floor/Roof Slab

An embedded RCC tie beam to floor slab is to be provided at each floor slab level, as an essential requirement of national building code against earthquakes. For this, web portion to required beam depth at top is to be cut and removed for placing horizontal reinforcement with stirrups and concreted[13].

8.6.5 Rapid Wall for Floor/Roof Slab in Combination with RCC

Rapid wall for floor/roof slab will also be cut to required size and marked with notation.

First the wall joints and other cavities and horizontal RCC tie beams are in-filled with concrete; then wooden plank of 0.3 to 0.45 m wide is provided to room span between the walls with support wherever embedded micro beams are there; finally roof panels will be lifted by crane using strong sling tied at mid-diagonal point, so that panel will float perfectly horizontal (*See* Fig. 8.10).

Each roof panel is placed over the wall in such a way that there will be at least a gap of 40 mm. This is to enable vertical rods to be placed continuously from floor to floor and provide monolithic RCC frame within Rapid wall. Wherever embedded micro-beams are there, top flanges of roof panel are cut leaving at

Fig. 8.10 Floor/Roof Panels

least 25 mm reinforcement rods are provided, door/window frames fixed and RCC lintel cast. Then concrete is filled where required and joints are filled. Then RCC tie beam all around are concreted. Roof panel for upper floor is repeated same as ground floor. For every upper floor the same method is repeated.

8.6.6 Finishing Work

Once concreting of ground floor roof slab is completed, on the 4th day, wooden planks with support props in ground floor can be removed. Finishing of internal wall corners and ceiling corners, etc. can be done using wall putty or special plaster by experienced POP plasterers, simultaneously, electrical work, water supply and sanitary work, floor tiling, mosaic or marble works, staircase work, etc. can also be carried out. Every upper floor can be finished in the same way[14].

Some of the photographs of Rapid Wall Panels with cut sections and used in the buildings in Australia and China are shown in Figs. 8.11 to 8.17.

Fig. 8.11 Panel for Pitched Roof

Fig. 8.12 With Computerized Cutting Saw Panels can be cut to required sizes with Door Window Openings etc., in Factory

8.7 Concluding Remarks

Rapid wall Panel provides a new method of building construction in fast track, fully utilising the benefits of prefabricated, light weight large panels with modular cavities and time tested, conventional cast-*in-situ* constructional use of concrete and steel reinforcement. By this process, man power, cost

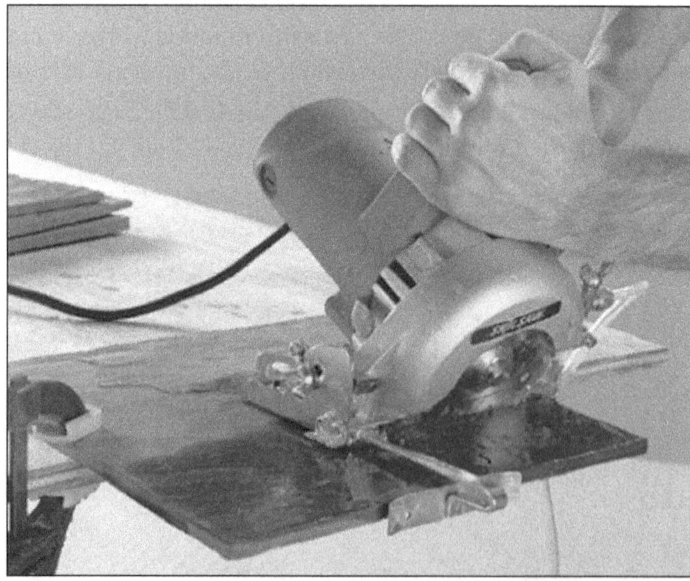

Fig. 8.13 Panels can be cut to required size with Door/Window opening etc at Construction Sites also using Electric Hacksaw or Marble Cutting Small Machines/Motors.

Fig. 8.14 Start up Rods are provided in Conventional Foundation Plinth Beam for Erecting Wall Panels

and time of construction is reduced. The use of scarce natural resources like river sand, water and agricultural land is significantly reduced. Rapid wall panels have reduced embodied energy and require less energy for thermo-regulation of interiors.

Fig. 8.15 Construction Methods Using RWP (Door/window openings are made in the factory itself)

Multistorey Construction

Use in External and Internal Walls

Multistoried Gypcrete building "CROWN OF THE HILL" at Pennant Hill, Sydney.

18 Storyes Marbury Apartments in Sydney

Fig. 8.16 Multistoried Construction Using RWP

Fig. 8.17 Use of RWP in Compound Walls

Rapid wall buildings thereby reduce burdening of the environment and help to reduce global warming. Rapid wall use also protect the lives and properties of people as these buildings will be resistant to natural disasters like earthquakes, cyclone, fire etc. This will also contribute to achieve the goal of much needed social inclusive development due to its various benefits and advantages with affordability for low income segments also. Fast delivery of mass dwelling/ housing is very critical for reducing huge urban housing shortage in India. Rapid wall panels will help to achieve the above multiple goals.

References

1. Rapid wall/Gypcrete Building Material Innovative, Energy efficient eco-friendly disaster resistant (RCF-Mumbai Report) 2010.
2. Prefabricated Walls-Rapid wall, RCF in Mumbai is about to complete a plant and produce Rapid wall for the Indian market and there is also a plant in Chennai currently www.indiarealestateboard.com/forums/www.rapid wall.com.au (Source: Internet).
3. Gypcrete Building India Limited (GBIL), Tamilnad, (Source: Internet) gypcreteindia.com/aboutus.html, gypcreteindia.co.in/project-gallery, highbeam.com/doc/1P3–1135050791, jejuqohuwicuxu.idoo.com/gypcrete (Source: Internet).

4. Gypcrete Building Panel/Rapid Wall Panel, Gypcrete Building India (P) Ltd. 134, Arihant MajesticTowers, 216, Jawaharlal Nehru Road,. Chennai – 600 107. Performance Appraisal. Certificate No. ... www.bmtpc.org/pacs/certs/cpac503.pdf (Source: Internet).
5. Gypcrete Building India Ltd., 1–3–4, AM Towers, 216, J,N,Road, Gypcrete Building India Ltd. 1–3–4, AM Towers, 216, JN Road, Koyambedu Chennai. No. th Tamil Nadu, Building Material Supplier in Chennai No. th, ... www.realestateindia.com/profile/gypcr... (Source: Internet).
6. Gypcrete Building India. (P) Limited. Chennai 600017. 044 2475 5272, 5219 3 /2475.5272(f)gbil@gypcreteindia.com; www.gypcreteindia.com (Source: Internet).
7. The Rapidflow Gypsum Calciner (Internet) gypcreteindia.com/rapidflow.asp. 12th Dec.2006... and building industry by setting up rapid flow gypsum calciner plant Kamayani has reported a net profit of ₹ 90 lakh at a turnover .www.business-standard.com/india/news (Source: Internet).
8. The Rapid wall Plant (Internet)...Horizon Industrial Development Company, MuscatSultanate of Oman)The Rapidwall plant was belt in Sohar Industrial area north ofSaltentofOman and the ..... More>>. HIDC is bringing this technology in to the region to ...www.horizons-hidc.com/ - (Source: Internet).
9. The Rapid Wall System is made up of a variety of Wall Types (exterior wall ... to offer a modular in-plant wall system and it included this unique design. ... n-p.com/products/rapidwall/index.aspx –supplier database. (Source: Internet).
10. Gypcretein GypcreteWall Panel at Gypcrete Building India Ltd., Gypcrete Wall Panels GYPCRETE WALL PANEL is a revolutionary, low-cost, prefabricated walling product with broad construction applications from load-bearing ...gypcreteindia.co.in/features/gypcrete. (Source: Internet).
11. India. Gypcrete Underlayment/OnlineTips.org, The Maxxon company, which specializes in flooring materials, has developed a product called Gypcrete underlayment as an answer to this problem. ...www.onlietips.org/gypcrete-underlayment co.in/features/gypcrete...(Source: Internet).
12. Features of Gypcrete Wal Panels – Gypcrete Building India Ltd. The finish of gypcrete wall panel is smoother and flatter than an equivalent precast concrete, in-situ-concrete or rendered masonry walls. ...www.gypcreteindia.com/gypFeatures.asp (Source: Internet).
13. www.Place-crete-com=Approved Gypcrete Applicators Located in Place-Crete Systems Ltd. is a full service specialty contracting firm that focuses on concrete restoration work, traffic deck membranes, cementious and ...www.place-crete.com (Source: Internet).
14. Gypcrete Wall Panel Products, buy GYPCRETE Wall Panel, GYPCRETE WALL PANEL, Find complete details about WALL PANEL from Gypcrete Building India Limited. Other WALL PANEL products www.alibaba.com/product-free/25225605... (Source: Internet).

9

Radioactivity in Gypsum

INTRODUCTION

Building materials contain different types of natural radioactive nuclides in various concentrations. For example, rock based materials derived and soil contain mainly natural radio nuclides of the uranium (^{238}U) and thorium (^{232}Th) series and the radioactive isotope of potassium (^{40}K). In the uranium series, the decay chain segment starting from radium (^{226}Ra) is radio logically most significant and, therefore, reference is often made to radium instead of uranium. The world-wide average concentrations of radium, thorium and potassium in the earth's crust are about 40 Bq/kg, 40 Bq/kg and 400 Bq/kg, respectively[1]. Increased interest in measuring radio nuclides and radon concentration in building materials is due to the health hazards and environmental pollution.

It is interesting to note that timber-based building materials have low concentration of natural radioactive substances. In stone-based building materials the concentration depends on the constituents. Materials used by the building industry that may be of radiological significance include marl, blast furnace slag, fly ash, phosphogypsum, Portland cement clinker, anhydrite, clay, radium-rich and thorium-rich granites (used as aggregates in concrete or in stone products).

Most of people spend more than 80% of their time indoors and natural radioactivity in building materials is a source of indoor radiation exposure[2]. Indoors-elevated dose rates may arise from high activities of radio nuclides in building materials. Chronic exposure of human beings to low doses of ionizing radiation can cause health damages which may appear 5–30 years after the exposure[3]. The most critical damage which can result from such exposure is an increase in the probability of contracting malignant diseases by the person who was exposed and by his offspring. The risk increases with the dose, and the probability of the appearance of the damage is greater when the exposure starts at a younger age. It appears that the large scale use of by-products with enhanced levels of radioactivity as a raw material in building products can increase considerably with the exposure of the population and therefore, pose a real potential risk.

Radiation exposure due to building materials can be divided into external and internal exposure. The external exposure is caused by direct gamma radiation. The internal exposure is caused mainly by the inhalation of radon (^{222}Rn) and its short lived decay products. Radon is part of the radioactive decay series of uranium, which is present in building materials. Because radon is an inert gas, it can move rather freely through porous media such as building materials, although usually only a fraction of that produced in the material reaches the surface and enters the indoor air. This fraction is determined by so called emanation ratio (or emanation coefficient) of the building product.

As the presence of radon gas in the environment (indoor and outdoor), soil, ground water, oil and gas deposits contributes the largest fraction of the natural radiation dose to populations, tracking radon concentration is thus of paramount importance for radiological protection.

The most important source of indoor radon is the underlying soil. In most cases the main part of indoor radon on the upper floors of a building originates from building materials. Typical excess indoor radon concentration due to building materials is low about 10–20 Bq/m^3, which is only 5%–10% of the design value introduced in the European Commission Recommendation (200 Bq/m^3) [4]. However, in some cases the building materials may be an important source also. For example, in Sweden, the radon emanating from building materials is a major problem. There are about 300,000 dwellings with walls made of lightweight concrete based on alum shale.

9.1 European and National Regulations of Natural Radioactivity of Building Materials

The European Basic Safety Standards Directive (BSSD) sets down a framework for controlling exposures to natural radiation sources arising from work activities[5]. Title VII of the directive applies to work activities within which the presence of natural radiation sources leads to a significant increase in the exposure of workers or of members of the public. Amongst the activities identified in the BSS as potentially of concern are those "which lead to the production of residues which contain naturally occurring radio nuclides causing significant increase in the exposure of members of the public". Such materials may include coal ash from power stations, by-product gypsum and certain slags which are produced in large volumes and which may potentially be used as building materials. The purpose of setting controls on the radioactivity of building materials is to limit the radiation exposure due to materials with enhanced or elevated levels of natural radio nuclides. The recently published EC document[6] provides guidance for setting controls on the radioactivity of building materials in European countries. This guidance is relevant for newly produced building materials and not intended to be applied to existing buildings.

The guidelines of the European Commission[4] are the first comprehensive document issued by the EC, which sets the principles of radiological protection principles concerning the natural radioactivity (both external and internal) of building materials. RP-112 states that restricting the use of certain building materials might have significant economical, environmental or social consequences locally and nationally. Such consequences, together with the national levels of radioactivity in building materials, should be assessed and considered when establishing binding regulations.

Gamma doses due to building materials exceeding 1 mSv/year are very exceptional and can hardly be disregarded from the radiation protection point of view. Therefore, recommends that national controls should be based on a dose in the range 0.3 – 1 mSv/year. This is the excess gamma dose to that received outdoors. This criterion is aimed to restrict exceptionally high individual doses. (1 m Sv = 0.001 Sv, 1Sv = 1J/ kg = 1Gy) Milli Sievert named after Rolf Maximilian Sievert, a radiation SI Unit.

When gamma doses are limited to levels below 1 mSv/year, the ^{226}Ra concentrations in the materials are limited, in practice, to levels which are unlikely to cause indoor radon concentrations exceeding the design level of 200 Bq/m^3. At the same time, some countries apply separate regulation for ^{226}Ra content, which requires that the amount of radium in building materials should be restricted to a level where it is unlikely that it could be a major cause for exceeding the design level for indoor radon introduced in the Commission Recommendation (200 Bq/m^3). For example, Nordic countries (like Denmark, Finland, Iceland, Norway and Sweden) recommend 100 and 200 Bq/kg, respectively, as exemption and upper levels for the activity concentration of ^{226}Ra in building materials for new constructions as a source of indoor radon[7] [NOR 2000].

The activity index in the EC document RP-112 and in the other national standards regulating radioactivity of building materials is calculated on the basis of the activity concentrations of radium (^{226}Ra) in the uranium (^{238}U) decay series, thorium (^{232}Th) in the thorium (^{232}Th) decay series, and potassium (^{40}K). Other nuclides are sometimes taken into consideration as well; for example, the activity concentration of cesium (^{137}Cs) from fallout is regulated in the Finnish guidelines[8].

If the activity index exceeds 1, the responsible party is required to show specifically that the relevant action level is not exceeded. If the activity index does not exceed 1, the material can be used, so far as the radioactivity is concerned, without restriction. The criterion of meeting the standard is the non-dimensional value of so called activity concentration index taking into account the total effect of three main natural radio nuclides, which can present in building materials (I = ^{226}Ra/300 + ^{232}Th/200 + ^{40}K/3000, concentrations are given in Bq/kg). According to RP-112, the activity concentration index I shall not exceed the following values depending on the dose criterion and the way and the amount the material is used in a building (Table 9.1).

The EC guidelines allow for controls to be based on a lower dose criterion, if it is judged that this is desirable and will not lead to impractical controls. It is recommended to exempt building materials from all restrictions concerning their radioactivity, if the excess gamma radiation originating from them increases the annual effective dose of a member of the public by 0.3 mSv at the most.

Table 9.1 Dose Criterion Recommended by EC [RP-112 1999]

Dose criterion	0.3 mSv/year	1.0 mSv/year
Materials used in bulk amounts, *e.g.*, concrete	$I \le 0.5$	$I \le 1$
Superficial and other materials with restricted use: tiles, boards, etc.	$I \le 2$	$I \le 6$

Most of the European countries apply their controls based on the upper end of the dose scale (1.0 mSv/year), however, the recent Danish regulations [9] apply the strictest criterion based on the lower end of the dose scale (0.3 mSv/year). Among non-EU countries only Israel applies the strict regulations based on the maximum allowable dose excess of 0.3 mSv/year [SI : 5098–2007]; the rest of the countries, which have similar regulations, apply more liberal dose criteria. The decision to apply a strict criterion of 0.3 mSv/year in these two countries can be explained by relatively low radioactive background resulting from the local geological conditions, because the majority of mineral resources in both Denmark and Israel are of sedimentary origin[10].

As we can see from Table 9.1, the EC regulations are different for building products having different thickness, products used in "bulk amounts" and relatively thin products such as superficial materials. The separation between these two groups is not defined precisely, however this approach, even in such a simplified form, is encouraging, because it reflects an attempt of the legislator to take into account the overall mass of radio nuclides in dwellings, which is indeed a very important forming the radiation dose. The consideration of the product geometry in the norms is a big step forward in comparison with the most of existing national standards, which still do not address the effect of the product thickness.

At the same time, the EC guidelines and most of the existing national standards still do not consider the density of building products, which is another important component primarily influencing the overall radiation dose of the inhabitants. The first Israeli standard SI : 5098 published recently tries to overcome this shortage and provides the information on the maximum allowable activity concentrations of all the three main radio nuclides (^{226}Ra, ^{232}Th and ^{40}K) depending on the mass per unit of surface (kg/m^2) of the building products in walls, ceilings, floors, coatings etc.[7]. In addition to testing the activity concentrations of these radio nuclides, this standard requires testing radon emanation of building products isolated from the sides (such a test arrangement simulates conditions of the each unit in the wall of the given thickness with two other dimensions infinite, where the number of radon atoms entering each building unit is equal to that exhaling toward its "neighbour"), and thus the contribution of radon gas into the internal radiation exposure (which is also dependent on the specific surface mass of the product) is taken into account.

According to SI : 5098, the activity concentration index $I = {}^{226}\text{Ra}/A({}^{226}\text{Ra}) + {}^{232}\text{Th}/A({}^{232}\text{Th}) + {}^{40}\text{K}/A({}^{40}\text{K}) + e^{226}\text{Ra}/A({}^{222}\text{Rn})$ should not exceed 1 for building products used in bulk amounts and 0.8 for superficial (thin) products, respectively. Coefficients A(^{226}Ra), A(^{232}Th), A(^{4o}K) and A(^{222}Rn) are determined where e is emanation ratio.

9.2 Industrial By-products Incorporated in Building Materials

The building industry uses large amounts of by-products from other industries. In recent years there is a growing tendency in European and other countries to use new recycled materials with technologically enhanced levels of radioactivity. The most known examples are coal fly ash and phosphogypsum (which is a by-product from phosphorous fertilizers production[11–12].

As can be seen from Table 9.2 radioactivity concentrations found in fly ash, phosphogypsum and in some other industrial by-products can be significantly higher in comparison with most common building materials.

Large quantities of coal fly ash are expelled from coal-fired thermal power plants and these may contain enhanced levels of radio nuclides along with other toxic elements. More than 180 Million Tonnes of coal ash (fly ash and bottom ash combined) are produced annually in India. About 40 Million Tonnes of these are used in the production of bricks road and cement. Since most of the process residues further processed into building materials do not meet the required

Table 9.2 Typical and maximum activity concentrations in common building materials and industrial by-products used for building materials in Europe [RP-112 1999].

Material	**Typical activity concentration (Bq/kg)**			**Maximum activity concentration (Bq/kg)**		
	^{226}Ra	**^{232}Th**	**^{40}K**	**^{226}Ra**	**^{232}Th**	**^{40}K**
Most common building materials (may include by-products)						
Concrete	—	30	400	240	190	1600
Aerated and light-weight concrete	60	40	430	2600	190	1600
Clay (red) bricks	50	50	670	200	200	2000
Sand-lime bricks	10	10	330	25	30	700
Natural building stones	60	60	640	500	310	4000
Natural gypsum	10	10	80	70	100	200
Most common industrial by-products used in building materials						
Phosphogypsum	390	20	60	1100	160	300
Blast furnace slag	270	70	240	2100	340	1000
Coal fly ash	180	100	650	1100	300	1500

technical specifications, they are typically mixed with pristine raw materials. The net effect is a dilution of the NORM (Naturally Occurring Radioactive Material) content relative to the process residues . In 1996, it has been estimated that up to 15% of phosphogypsum was recycled and that within the European Union some 2 Mt were recycled annually[13]. Its activity concentration depends on the origin and the chemical treatment of the raw material: for example, phosphogypsum from phosphate rocks generally contains considerably higher concentrations of ^{226}Ra than gypsum from carbonate rocks. In any case, not only the ^{226}Ra concentration, but also the radon exhalation from it can be higher than normal.

Blast furnace slag is used mainly as crushed aggregate in concrete as well as a finely ground mineral additive in cement. The activity concentration in slag depends on the ore type, the origin of the raw material and the metallurgic processes. The use of coal fly ash and slag in concrete is a well-recognized source of gamma exposure that is due to the presence of activity concentrations of ^{226}Ra, ^{232}Th and, to a lesser extent, ^{40}K, while its effect via radon exhalation is controversial, due to the low emanation coefficient from the ash [14]. Phosphogypsum used, for example, in the production of plasterboard may give rise to a concern about extremely high concentrations of ^{226}Ra, and high radon exhalation.

Decommissioning or rebuilding the structures made of fly ash, slag, phosphogypsum and the ensuing dust generation or land filling of secondary wastes may lead to exposure. These waste materials are used to make a variety of mainly lightweight construction materials. Lightweight building blocks and plasterboard are typical examples with a potential to result in external exposures.

Some regulations address the radioactivity of the waste materials and industrial by-products specifically, but others do not distinguish between building products containing the waste materials and regular building products.

The Finnish Guide[8] can serve as an example of the first group of regulations. According to this document, when there are plans to incorporate industrial by-products or wastes in building materials and it is discovered or there is reason to suspect that these contain radioactive nuclides in greater amounts than normal, the activity concentrations of these radioactive nuclides in the final product shall be measured. Where necessary, also other nuclides than ^{226}Ra, ^{232}Th and ^{40}K shall be taken into consideration. If a by-product or wastes containing radioactive nuclides are incorporated in building materials, it must be confirmed that the action level of 1 mSv/year is not exceeded.

The Guide ST 12.2 regulates radioactivity of both building materials and fly ash (including its handling and uses in construction) in one document, which seems to be a useful approach. For example, when ash is added into a material that will be used in building, ST 12.2 sets that the gamma radiation from the contained cesium (^{137}Cs) shall not contribute more than 0.1 mSv per year to the total effective dose of the population due to the material. The action level of 0.1 mSv/year is not exceeded, if the activity concentration of ^{137}Cs in the ash is less than 1000 Bq/kg and the maximum amount of ash incorporated in the concrete is 120 kg/m^3. If the amount of ash incorporated in the concrete is less than 120 kg/m^3, the activity concentration of the ash may be correspondingly higher.

It has to be noticed that the level of 120 kg/m^3 is usually not exceeded in concrete mixes applied in dwelling construction. The high-volume fly ash (HVFA) concrete mixes containing 125–225 kg/m^3 of fly ash [15] can be cast successfully in pavements and road construction applications, and thus these mixes are not of concern from radiological point of view.

9.3 Radon Emanation from Building Materials and Regulations

Most of the existing regulations do not address radon emanation from building materials. At the same time, the inclusion of radon emanation test in the standards regulating radioactivity of building materials has both pros and contras, and sometimes is even desirable. In particular, RP-112 recommends considering separate limitations for radon isotopes (radon ^{222}Rn and thoron ^{220}Rn) exhaling from building materials, where previous evaluations show that building materials may be a significant source of indoor radon or thoron and restrictions put on this source are found to be an efficient and a cost-effective way to limit internal radiation exposures.

It is known that the macrostructure, specific surface area of the solid phase and porosity play an important role in radon emanation behaviour. Let us consider two extreme cases of radon emanation from building products made of coal fly ash (for example, concrete or masonry blocks) and from phosphogypsum (for example, gypsum wallboard).

For example, the emanation ratio for gypsum was found as 30%–50%, and for fly ash as less than 1% [16], Stoulos *et al.*[17]. In other words, the emanation abilities of gypsum and fly ash differ by two orders of magnitude, representing opposite ends of the emanation scale. The emanation power of radon is thus strongly dependent on the microstructure and morphology of the solid particles. It is well known, for instance, that fly ash particle has a dense glassy structure with most of the mass concentrated in the particle shell, preventing radon atoms from escaping the material. In addition, fly ash particles are known for their ideally spherical shape having the minimum surface to volume ratio among all possible particles geometries.

In contrast, gypsum crystals usually are of longitudinal (fibroid) shape with well-developed surface area and have lower density. The typical "layered" structure of gypsum crystal is relatively weak (for example, it easily disintegrates under heating, resulting in the formation of calcium sulfate hemihydrate crystals of high specific surface area, up to 10 m^2/g). All these features make the process of radon release from gypsum relatively easy.

As was mentioned before, radon emanation test is required by the Israeli Standard SI 5098, which includes the dose from radon inhalation in the total dose excess for the inhabitants (0.3 m Sv/year). Exposure to radon gas is also addressed in the Austrian Standard ÖENORM S 5200 Steger[18]. However, the radon emanation test in this standard is not mandatory, in contrast to the Israeli Standard SI : 5098. For the calculation of the activity concentration index I ÖENORM S 5200 allows to use the precondition value e = 10%, if the emanation factor is not known. The real emanation factor can be also determined in the direct experiment, but its value should not be higher than the precondition value. The coefficients A(^{226}Ra), A(^{232}Th) and A(^{40}K), which consider external radiation exposure, and the coefficient A(^{222}Rn), which is responsible for radon inhalation in the final dose criterion, are 1000, 600, 10000 and $(0.15 + e\,\rho\,d)/1000$, respectively, where ρ is the density and *d* is the wall thickness. The precondition values for the wall thickness are d = 0.3 m, for the density ρ = 2000 kg/m^3 and for the emanation factor e = 10%. It can be seen that the part of this combined dose criterion responsible for gamma exposure does not depend on the density and geometry of the building element, however the part responsible for radon inhalation does.

The comparison between the two standards addressing radon emanation properties, SI : 5098 and ÖENORM S 5200, shows that the Austrian Standard, which guarantees that inhabitants of dwellings do not receive a higher dose from natural radioactivity as 2.5 mSv/year, is several times more liberal, than RP-112 and other national standards.

On one hand, knowing the real radon emanation properties of the product tested in the laboratory makes the decision by legislators more accurate. On the other hand, there are still difficulties with recommending an optimum standard test procedure for getting the results reliable and reproducible. For example, it is well-known that moisture of the porous building products significantly influences the radon exhalation rate. That is why the climatic conditions at the sample preparation/curing and during the radon test itself should be chosen as stable as possible. There are also other factors, which can influence the emanation test result, for example, the dimensions of the sample and radon chamber, temperature, test duration, method of the approximation of "exhalation rate – time" dependence needed to calculate the emanation ratio, age of the materials, changing their properties in time (cementitious materials, for example). As a consequence, the radon-dependent part of the dose criterion is more uncertain, which should be compensated by more liberal criterion (similar to "safety factors" accepted in the structural design). This correction is a part of a non-radiological justification of the regulations, which should be based on technological, social, environmental and economical considerations. At the same time, such non-radiological criteria are seldom applied in the legislation practice, because of the difficulties related to the methodology of cost-benefit analysis.

9.4 Radiation in Waste Gypsum

Gypsum has been used in several forms since the beginning of civilization. At the present time, natural gypsum is utilized as building material (gypsum board, plaster ingredient, component of Portland cement, binder, in medicine (surgical splints), as fertilizer and soil conditioner, in hygienic products (foot creams, shampoos and hair products). The shortage of this product, its price or the non-existence of gypsum phosphogypsum which would have the additional advantage of resolving partially the environmental problems created by fertilizer industries. Phosphogypsum can replace some of the natural components of building materials. However, it contains a higher radioactivity concentration than the natural products (Table 9.1) and its use in houses may lead to increase radiation doses to the inhabitants.

Simple models can be applied to the concentration results to calculate the external and internal irradiation for people living in a common house constructed with this Gypsum. For practical monitoring purposes, investigation levels can be presented in the form of an activity concentration index (Ra equivalent) to ensure that the gamma dose rate inside a room due to the building materials does not exceed 1 mSv. In spite of these radioactivity levels, the phosphogypsum would not represent a high risk in the building industry. However, as only 4% of the phosphogypsum world production is used in agriculture, in gypsum board and cement industries, some 120 million are accumulated annually, most of which are stock piled close to the fertilizer factories, therefore creating a complex environmental problem. On the one hand, large extensions of land are occupied by stacks containing the phosphogypsum, zones that could be available for other uses (recreation, parking, building, agriculture, roads).

On the other hand, the emanation of Radon from the gypsum stacks produces its incorporation in the atmosphere increasing the content of its non-volatile daughters (Pb^{210}, 210 Bi, 210 P) in the surroundings areas. The contamination of aquatic ecosystem could be originated for engineering barriers that prevent its filtration to ground water, or the overflowing of liquids when decanting the waste phosphogypsum from the fertilizer factories. Expensive countermeasures for protecting the surroundings ecosystems and avoiding not only the visual and aesthetic impact but also the radiological problem are usually taken by the other radio nuclides migrations to the stack deepest layers. The recycling of phosphogypsum (PG) in the building industry could be a solution for the environmental problem created in the vicinity of fertilizer factories located in Huelva (Spain) as PG is used extensively in cement, wallboard, and other building materials in Europe, Japan, and Australia.

Due to the presence of uranium daughters and other contaminants, the conversion of phosphate rock to fertilizer may be of significant contaminants that may preclude a future use of the tailings and the surrounding area after decommissioning are identified. A field study on the phosphogypsum tailings at the Western Co-operative Fertilizer Ltd. plant in Calgary, Alberta investigated one measuring the radon flux using passive activated charcoal collectors. The radon flux was then used as the governing basis for the design of a proposed cover to isolate the contaminants. The study reveals that radon, dust, particulates and gamma radiation would preclude a future residential use of the tailings area and the adjacent land within 800 m of the tailings. Earth or liquid covers could reduce contamination leaving an inactive tailings area enough to technically allow some reduction to the 800 m setback distance. However, the inability to guarantee the chosen cover's integrity over the very long term would preclude a residential end use for the area.

The study also reveals that discharges of radon, dust and particulates from uncovered tailings will increase when the tailings are not in active use. The instability of the tailings will forestall the placement of inflexible covers designed to reduce radon flux, and thus dust drying of the tailings occurs. Drying times exceed two years.

This study was supported by Alberta Environment and Western Co-operative Fertilizers Limited. It compliments measurements made by Sene's Consultants Limited on behalf of the Atomic Energy of Canada, Low Level Waste Management Office, in the national study into the decommissioning of phosphogypsum piles.

The use of phosphogypsum plaster-board and plaster cement in buildings as a substitute for natural gypsum may constitute an additional source of radiation exposure to both workers and members of the public, both from inhalation of radon progeny produced from radon which is exhaled from the plaster-board and from beta and gamma radiation produced by radioactive decay in the plaster-board. The calculations presented in this paper indicate that if phospho-gypsum sheets 1 cm thick containing a ^{226}Ra concentration of 400 Bq kg^{-1} are used to line the walls and ceiling of a room of dimensions up to 5 m × 5 m × 3 m, the annual effective dose from gamma radiation for a person continually occupying the room should not exceed approximately 0.13 mSv. This compares with a measured annual average effective dose from gamma radiation in Australian homes of 0.9 mSv. The annual effective dose from such thin sheets is directly proportional to the concentration in the plaster-board[11].

Radioactivity in Products Derived from Gypsum in Tanzania has been discussed by Masaki and Banzi[19]. Scientific investigations have long concluded that prolonged exposure to low dose radiation can induce deleterious effects in humans. The aim of this research is to investigate the radioactivity of gypsum and gypsum derived products as part of a bigger project aimed at establishing radiation levels in materials or/and products suspected to have natural radioactivity radiation risk in Tanzania. In response to the concern expressed by the users of chalk sticks in some schools in Arusha municipality, it was found necessary to establish levels of radioactivity in this product and associated radiation risk. Natural radioactivity content was determined in chalk dust, natural gypsum and normal background soil using a hyper pure germanium spectrometer (HPGe). The soil measurements were used as control. Results have shown that the concentration of ^{226}Ra and ^{228}Ra nuclides found in chalk dust were 24.25 Bq.kg^{-1} and 22.86 Bq.kg^{-1}, respectively. These levels were lower or comparable to the corresponding 34.2 Bq.kg^{-1} and 21.5 Bq.kg^{-1}, respectively, found in soil. However, the radioactivity levels found in the chalk dust were five times higher than that found in the gypsum 5 Bq.kg^{-1} for ^{226}Ra and 4 Bq.kg^{-1} for ^{228}Ra. These values compare well with the value recorded for natural gypsum in Denmark of 7 $Bq.kg^{-1}$ for ^{226}Ra and 4 $Bq.kg^{-1}$ for ^{28}Ra by UNSCEAR. The calculated external (0.21) and internal (0.31) hazard indices due to radioactivity in chalk dust were respectively, lower than (0.26 and 0.35) hazard indices calculated for normal background soil. In both cases, the hazard indices were lower than the acceptable limits recommended for building materials. This study has shown that natural gypsum and gypsum derived products have traces of radioactivity. However, the associated levels are not detrimental to health.

9.4.1 Radiation Exposures in Phosphogypsum Disposal Environment

As we are aware phosphatic fertilizers are manufactured in India mainly from imported rock phosphate that contain appreciable concentrations of NORM (naturally occurring radioactive material) resulting from ^{238}U and its decay products. Exploitation of this material in large quantities for production of fertilisers redistributes the uranium, radium (^{226}Ra) and other decay products of the U-chain in the environment and results in technologically enhanced natural radiation exposures. In the process, powdered rock is digested with hot sulphuric acid producing phosphoric acid and gypsum ($CaSO_4.2H_2O$). The acid is then used for the production of various fertilizers and the phosphogypsum (PG) is disposed off as land fill in low-lying areas apart from the limited usage in cement industry and soil conditioning. The disposal of huge quantities of gypsum produced has been a concern for such fertilier industries due to its high acidity, fluoride content, trace metals and the presence of ^{226}Ra and progeny.

According to the geological survey report, 147 million tonnes of phosphate rock was mined globally during the year 2007 [20]. The resultant generation of PG can be estimated to be 250 million tonnes approximately. In India, it is estimated that approximately 6.0 million tonnes of PG is generated at nearly 15 sites based on the fertiliser production during the last year. The concentration of ^{226}Ra in phosphogypsum is reported to be of the order of 1Bq g^{-1} (Haridasan and Paul[21], Laiche and Scott[22]). The mobility of radium from phosphogypsum disposal reported that radium is not leached significantly from gypsum stockpiles in Florida. In a extraction study, Rutherford, Dudas, and Arocena[23] reported phosphogypsum as a potential source of elevated levels of radium in groundwater and the subsurface environment around phosphogypsum repositories. Paul and Pillai[24] observed that radium is leach able in water from calcium carbonate fertilizer process waste.

The studies carried out by Haridasan *et.al.*[25] around two fertiliser plants situated at Udyogamandal (Site A) and Ambalamedu (Site B) in Kerala (India) to assess the distribution of uranium and decay products in the raw material, products, wastes and an assessment of environmental radiological impact due to the disposal of PG. The raw material phosphate rock for the plants is being imported mainly from Morocco. These plants together dispo se off nearly 500 000 tonnes per year of

PG on land. Natural radio nuclides in the aquatic environment of the PG disposal site from the study area were reported earlier with an estimate of ingestion dose to public. Elevated levels of ^{226}Ra have been observed in water and sediment from the river Periyar and Chitrapuzha adjoining these plants. Studies are available in the literature on NORM issues in phosphate industries mainly from Florida and Spain among few others[26]. Dissolution characteristics of ^{226}Ra from PG were also reported earlier from the study area. The present study focuses on the evaluation of environmental radiation exposures attributed to PG disposal in a typical tropical site.

1. Materials and Methodology

The study at two sites (Fig. 9.1), Site A and Site B nearly 20 km apart, were chosen. Site A has lower quantity of PG compared with Site B. At Site A, there are two locations, one active site and the other inactive with natural soil toped and vegetation covered. The active area here is approximately of the size 0.5 × 0.5 km^2 with a height of 2 m above ground level. The inactive soil toped area is also of almost the same size and nearly 15 y old. At Site B, PG stack has an approximate dimension of 1 × 0.5 km^2 with a height of 5 m. Of late utilization of PG has been under research and limited quantities are being used in cement industry and soil conditioning purpose from the sites under investigation also. Area around both the sites is heavily populated. The predominant wind direction is from west to east.

2. Sampling and Assessment of Radio Nuclides

Samples of rock phosphate, phosphoric acid and PG were collected from the fertiliser plants and analysed for activity content. Measurements were conducted

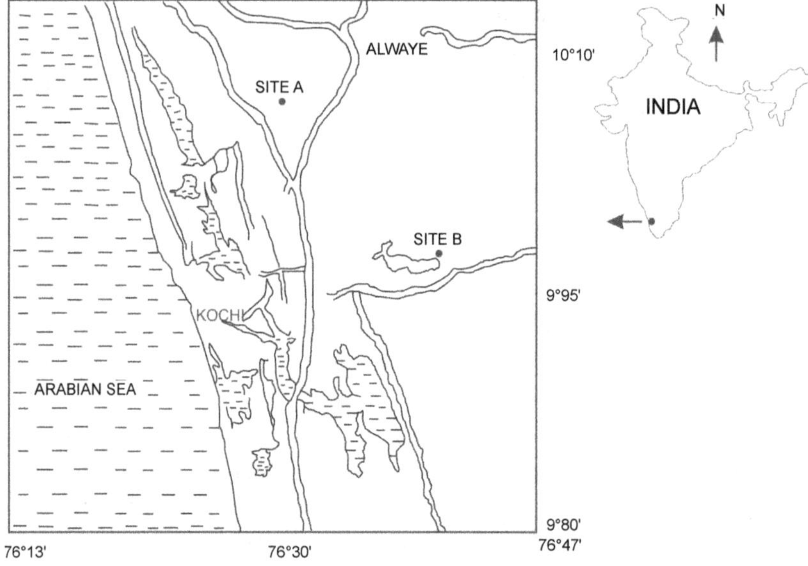

Fig. 9.1 Lotionof Study Sites A and B

using a 4 × 3″ NaI(Tl) scintillation detector coupled gamma spectrometry system. About 300 g of the samples were sealed for equilibrium build-up of radon (^{222}Rn) decay products and counted for 10 000s. From the counts obtained from 1764 and 609 keV peaks of ^{214}Bi, the ^{226}Ra activity was estimated. The minimum detectable activity under the experimental conditions works out to 3 Bq kg^{-1} for 2σ confidence level. Uranium concentration in the samples was estimated by spectrophotometeric method In this method, aliquots of rock phosphate and PG powder were repeatedly treated with concentrated HNO_3 and dissolved in 1:1 HNO_3. The solution was filtered through Millipore filter paper (<0.45 μm) and made up to a known volume. Uranium was extracted into 10% tributyl phosphate in heptane and the organic phase was coloured with ammonium thiocyanate. The absorbance of the organic layer was measured in a spectrophotometer (Schimadzu, Model UV 2100 S) against the reagent blank at 365 nm. A calibration curve showing uranium concentration versus optical density was prepared using standard uranyl acetate solution, and uranium concentrations in the samples were estimated from the graph. In this method, the minimum sensitivity works out to 0.02 mg mL^{-1}. In the case of phosphoric acid, emanometric method was used for the analysis of ^{226}Ra activity. In the emanometric procedure, the minimum activity of ^{226}Ra in phosphoric acid sample was estimated to be 0.1 BqL^{-1} at 95% confidence level.

Gamma exposure rates over the PG disposed at Site A and Site B were measured using a Scintillometer (ECIL Model SM 141D). The exposure rate meter has a sensitivity of 0.05 μ Gyh^{-1}. The radiation field was measured at 1 m above the surface of the piles/ground. Measurements were also carried out in public residential areas near the PG disposal sites. ^{222}Rn gas concentration in the indoor environment of PG disposal area was measured using an Alpha GUARD sampler, Model Genitron, 2000. Round-the-clock sampling was carried out and data were analyzed using the software Alpha EXPERT ver. 3 (Genitron, 2000). Sampling was carried out at seven representative locations at distances varying from 100 m to 1 km away from PG piles at Site A. Locations are identified to assess the impact to representative group of public. In order to estimate the maximum probable ^{222}Rn levels attributable to PG, measurements were carried out at Site B, close to the large PG stack source. Diurnal variation of ^{222}Rn at this location was also measured. The minimum sensitivity of the ^{222}Rn sampler was specified at 2 Bq m^{-3} at 95% confidence level. High volume air samples were collected from eight locations in the PG disposal sites and environs using Envirotech APM 410 Model High Volume Air Sampler having a suction rate of 1 m^3 min^{-1}. Samples were taken, typically for 8 h duration round the clock. What man GFA (Cat. No. 1820–866) filter papers having a collection surface of 23 × 18 cm^2 were used for sampling. The samples were leached with concentrated HNO_3 repeatedly and ^{226}Ra was concentrated by the carrier precipitation method on $BaSO_4$ and activity estimated by the standard alpha counting method. The minimum detectable activity for the measurements was estimated to be 0.02 mBq m^{-3} at 2σ confidence level for 8 h of sampling.

The probable external radiation exposure to representative group of public is estimated as: Dose (mSv) = gamma expo sure rate (mGy h^{-1}) × occupancy factor (8760 h y^{-1}) × 0.7 mSv mGy^{-1}. The effective dose due to the inhalation of ^{222}Rn and progeny is estimated using the dose conversion factors used by the UNSCEAR Report[27]. The dose (nSv) is worked out as: Dose = equilibrium equivalent concentration of ^{222}Rn (Bq m^{-3}) × 7000 h × DCF [9 nSv (Bq h $m^{-3}]^{-1}$).

The committed annual effective dose from the inhaled intake of airborne ^{226}Ra activity is also estimated using the dose conversion factors.

Table 9.3 gives the concentration of radio nuclides, ^{238}U and ^{226}Ra in rock phosphate, PG and phosphoric acid. The mean ^{226}Ra activity in the rock phosphate samples and PG are found to be 1322 ± 90 and 818 ± 100 Bq kg^{-1}, respectively. These activity levels are comparable with the values reported elsewhere. From the concentration data, it is found that on an average 77% of uranium concentrates in phosphoric acid and nearly 83% ^{226}Ra concentrates in PG as expected. The external gamma exposure rates over the PG piles at Site A and Site B including nearby environs are provided in Table 9.4 The radiation field ranged from 0.3 to 0.8 ×Gy h^{-1} at 1 m above the active pile. The maximum field is observed at Site B. Here a large stack of gypsum is heaped up and the quantity disposed is several times higher than that at Site A. Moreover, the concentration of radio nuclides, which depends on the type and origin of the rock phosphate, nature of the process, etc. also contributes to the difference in radiation fields. A significant reduction in the gamma exposure rate is observed at Site A over soil topped areas. At Site A, the general background radiation field is in the range of 0.05–0.1 × Gy h^{-1}. It is evident that soil covers of 1 m thick shields gamma exposure rate to the background levels. The observations at areas frequented by people at about 100 m from the pile edges shows radiation field of 0.15 × Gy h^{-1} and exposure rate at nearby public residential areas ranged from 0.05 to 0.1 × Gy h^{-1}. This suggests an excess incremental radiation field of 0.05 × Gy h^{-1} attributable to the disposal of PG. Hence, an average individual additional exposure to representative person in the local area can be estimated to be 306 × Sy y^{-1} for 8760 h of occupancy near the PG pile. The doses will be much lower

Table 9.3 Distribution of ^{238}U and ^{226}Ra in Rock Phosphate and Products

Sample	^{238}U	^{226}Ra	Per cent distribution	
			^{238}U	^{226}Ra
Rock phosphate	1340 ± 280 (1013–1704)	1322 ± 90 (1285–1370)	100	100
PG	170 ± 30 (140–205)	818 ± 100 (449–939)	17.2 ± 4.7	83.5 ± 12
Phosphoric acid (as P_2O_5)	2442 ± 150 (2280–2578)	6.15 ± 3.50 (3.8–10.7)	77 ± 17	0.2 ± 0.1

Values are expressed as Bq kg^{-1} ± SD and numbers in parenthesis give the range. Number of samples = 20.

Table 9.4 External Radiation Fields over PG Pile.

Location	Radiation field range (×Gy h^{-1})
Gypsum piles at Site A (uncovered)	0.30–0.50
Gypsum piles at Site A (soil topped 0.5 m thick)	0.10–0.15
Gypsum piles at Site A (soil topped 0.5 m thick)	0.10–0.15
Gypsum piles at Site A (soil topped 1 m thick)	0.05–0.10
General background radiation field in the area	0.05–0.10
Gypsum piles at Site B	0.30–0.80
Fresh gypsum (on contact) at Site B	0.60–0.80
Area 100 m away from the PG stockpile at Site B	0.15
Public residential areas around Site B	0.05–0.10
Staff residential colony at Site B	0.05–0.15
General background at locations 2 km away from the Site B	0.05–0.10

as no person is expected to occupy the PG environment throughout the year and in actual situation the exposures will be much lower.

Table 9.5 shows the indoor ^{222}Rn gas concentration in the PG disposal environment. The measurements were carried out using Alpha Guard sampler and the instrument simultaneously records temperature, pressure and relative humidity, the data of which are not shown in the table. Average ^{222}Rn at a distance of 100 m from the active PG pile is observed to be 26 Bq m^{-3}. Measurements at a 200 m distance also indicated levels of the same order. At 1 km distance from the PG pile in different directions, the average ^{222}Rn levels are found to be in the range of 11–19 Bq m^{-3} which are comparable to the local natural background.

Table 9.5 Indoor ^{222}Rn in the PG Disposal Environment

Location	^{222}Rn (Bq m^{-3})		Mean ± error
	Minimum	Maximum	
100 m away from active PG pile (downwind direction)	9 ± 3	45 ± 15	26 ± 8
200 m away from active PG pile (downwind direction)	8 ± 2	54 ± 19	25 ± 9
Near the inactive soil topped PG pile	6 ± 2	41 ± 8	18 ± 6
1 km away from PG pile, south	9 ± 2	42 ± 15	19 ± 7
1 km away from PG pile, east	5 ± 1	30 ± 10	17 ± 5
1 km away from PG pile, north	5 ± 1	20 ± 8	11 ± 4
1 km away from PG pile, west	6 ± 2	23 ± 9	11 ± 5

The regional average values of ^{222}Rn in the indoor environment were reported to be 17 Bq m^{-3}. Comparing to this value, the enhancement in the immediate vicinity of PG pile can be calculated to be 9 Bq m^{-3}. The resultant additional dose to representative person is worked out as 227 × Sy y^{-1} for 7000 h of indoor occupancy during a year assuming an equilibrium factor of 0.4 for indoor concentration. Assuming a similar concentration increase in outdoor environs, the corresponding inhalation dose is worked out to be 85 × Sy y^{-1}. Hence, a total inhalation dose component of 312 × Sy y^{-1} (0.3 mSv y^{-1}) is estimated. Measurements on diurnal variation of ^{222}Rn at a location very close to large PG pile are shown in Fig. 9.2. These measurements are carried out during winter season as higher ground level concentrations are expected during the season. Three sets of samples covering 1 week duration each are collected and the data summarized in the figure. The ^{222}Rn concentration varied between 9 and 80 Bq m^{-3} with an average of 46 Bq m^{-3}. Maximum concentration was observed during the early morning hours and minimum was observed in the afternoon as expected. It is to be noted that International Commission on Radiological Protection (ICRP) recommends ^{222}Rn concentration of 200 Bq m^{-3} in dwellings as the 'action levels' for national regulatory authorities to control the exposures. The present measured levels do not indicate the need of regulation. From the data, it can also be seen that the ^{222}Rn levels near the inactive soil toped PG pile is nearly of the same order of the local background ^{222}Rn levels and it can be recommended that a soil toping of nearly 1 m thick reduces the ^{222}Rn levels in the immediate environs of PG disposal site to natural background levels.

The analysis of airborne activity due to ^{226}Ra in the PG disposal environment is given in Table 9.6. The concentration generally varied in the range of 0.06–0.29 mBq m^{-3} in the immediate vicinity of the pile, whereas the background levels

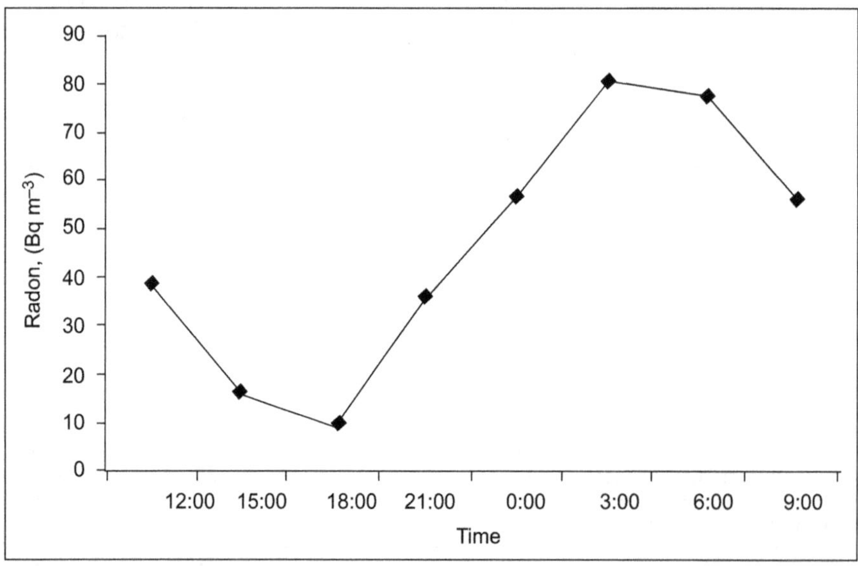

Fig. 9.2 Diurnal Variation of ^{222}Rn at a Location Very Close to Large PG Pile

Table 9.6 Airborne ^{226}Ra Activity in the Environment of PG Stockpiled Area

Location	^{226}Ra activity in air samples (mBq m^{-3})	
	Range	Mean
100 m away from PG stockpile	0.06–0.29	0.15 ± 0.07
500 m western side of the pile	0.07–0.15	0.12 ± 0.03
1 km northern side of the PG pile	0.04–0.26	0.09 ± 0.07
2 km northern side of the PG pile	0.06–0.12	0.09 ± 0.02
2 km southern side of the PG pile	0.04–0.11	0.08 ± 0.03
5 km eastern side of the PG pile	0.03–0.17	0.07 ± 0.05
5 km south-west side of the PG pile	0.04–0.21	0.09 ± 0.06
5 km south-east side of the PG pile	0.04–0.08	0.06 ± 0.02

at far away places ranged between 0.04 and 0.26 mBq m^{-3}. From the average values, it is evident that the ^{226}Ra activity in the airborne dust at PG site is almost double with respect to other locations. Hence, a net excess of 0.06 mBq m^{-3} is attributable to PG disposal and for a full year occupancy the inhalation dose is worked out to be 1.75 × Sy y^{-1} using the dose conversion factor 3.5 × 10^{-6} SvBq^{-1} for type M aerosols as specified by the Inter national Commission Radiological Protection (ICRP), Protection Against Radon at Home and at Work, Pergmon Press, Oxford, 1993, Publication 65. Compared to the external gamma dose and ^{222}Rn inhalation dose this component is insignificant. In an earlier reported study from the site, the probable ingestion dose was estimated as 18 × Sy y^{-1} through the fish and milk intake route. The uptake of radio nuclides in terrestrial foods and resultant dose to humans may be another component of exposure, which needs further study. Hence, the total additional radiation dose to representative person in the immediate vicinity of PG disposal environment at the study area is estimated to be 0.64 mSv y^{-1} above the natural background. This dose can be significantly reduced by natural soil toping on the PG disposal site on level earth.

Average concentration of ^{238}U in rock phosphate is found to be 1.34 Bq g^{-1} and ^{226}Ra selectively concentrates in PG to an extent of 0.82 Bq g^{-1}. About 77% of uranium concentrates in phosphoric acid and 17.2% concentrates in gypsum, during the sulphuric acid digestion of rock phosphate. On an average 83.5% of ^{226}Ra activity concentrates in PG. Study revealed that inhalation of ^{222}Rn and progeny and the external gamma radiation field contribute to radiation dose to public in a PG disposal environment. The additional radiation dose attributable to PG disposal in the study area is estimated to be 0.6 mSv y^{-1} for a representative person who is in close proximity to PG pile. Soil topping over the gypsum pile significantly reduces the external gamma field as well as ^{222}Rn concentration in the area. ^{222}Rn emanation and its diffusion from the gypsum stockpiled area may locally enhance the environmental levels of ^{222}Rn. However, the present levels are below the criteria laid down by the ICRP for categorizing certain areas as 'radon prone' and do not call for immediate control measures.

9.5 Dissolution Characteristics of ^{226}Ra from Phosphogypsum

A study has been reported by Haridasan *et al.*[28] regarding the leach ability of ^{226}Ra in water from phosphogypsum obtained from a phosphate rock processing plant in Kochi, India, which uses imported rock mainly from Morocco as the starting material.

9.5.1 Collection of Phosphogypsum Samples and Methodology

Fresh samples of phosphogypsum were collected in PVC bags from an operating phosphate rock processing plant where the wet process method is used to manufacture phosphoric acid. Leaching experiments were performed in the laboratory using distilled water and rainwater as lea chants, to examine the leach ability of ^{226}Ra from hosphogypsum. Rainwater was collected directly in plastic buckets during the monsoon season, and the pH of the rainwater varied between 5.0 and 5.8. Phosphogypsum samples were leached with water under the following experimental conditions.

(a) Batch wise leaching with distilled water

Ten grams of phosphogypsum sample having a ^{226}Ra activity of 8.5 Bq was placed in a 2 l Marinelli beaker. One litre of distilled water, having a pH of 6.0, was added, stirred for 1 min and kept for 24 h.

The leachate was then collected and filtered through 0.45 m millipore membrane filter paper and analyses for pH and ^{226}Ra activity were carried out. Fresh water was again added, stirred as before, and kept for 24 h. The leachate was collected again and the experiment was repeated up to 10 times by adding fresh distilled water. The experiment was replicated up to three numbers and the experimental error ranged from 15 to 20%.

(b) Batch-wise leaching with rainwater

The leaching study was repeated using rainwater having a pH of 5.8. The experiment was similar to that of experiment (*a*).

(c) Leaching of ^{226}Ra with varying liquid: solid ratio

A leaching study was conducted with different liquid to solid ratios (V/M), using rainwater with a pH of 5.5 as the lea chant. A constant weight of 10 g each of phosphogypsum sample having an initial ^{226}Ra activity of 8.5 Bq was used for the experiment and 100, 200, 300, 400, 500, 1000, 2000, 5000 and 10000 mL of rainwater was added to each sample. The mixture was stirred for

1 min using a glass rod, and was allowed to settle for 24 h. The supernatant was removed, filtered through 0.45 m millipore filter paper and ^{226}Ra analysis was carried out. The experiments were repeated up to five times by adding fresh portions of rainwater into the same phosphogypsum sample. All 45 leachates were subjected to ^{226}Ra analysis and the emulative activity, leached out in five successive batches with different liquid to solid rati os, was calculated.

(d) Leaching of ^{226}Ra with different contact time

Experiments were carried out to examine the leachability of ^{226}Ra from phosphogypsum with respect to change in contact times. Rainwater, having a pH of 5.5, was added batchwise to a phosphogypsum sample at a constant solution to sludge ratio of 50:1. The contact time varied from 10 min to 250 days in separate samples. After the particular contact time had elapsed, each leachate was collected and filtered through millipore filter paper. The leachates were subjected to ^{226}Ra activity analysis. In another experiment, water was added repeatedly to the same phosphogypsum sample and after the particular contact time the supernatant was siphoned out, filtered through 0.45 m millipore filter paper and ^{226}Ra analysis was carried out. The cumulative fraction of activity leached out was estimated. The experiments were repeated and the maximum experimental error was calculated to be 20%.

(e) Leaching with simulating natural conditions

One kilogram of phosphogypsum sample was uniformly spread in a rectangular PVC tray of 50 · 50 cm size having a spout at the middle. Before spreading the phosphogypsum, two Whatman-41 filter papers were placed on the nozzle and sealed. The tray was exposed to rain during monsoon season over a period of one month. The leachates coming through the filter paper were collected every 48 h. Rainfall from the starting time of experiment to each batch collection (every 48 h) was also measured using a standard rain gauge.

Analyses for pH, dissolved solids, and ^{226}Ra activity were carried out in the leachates. Leachates were filtered through 0.45 m millipore membrane filter paper. Calcium phosphate was precipitated in samples by the addition of 10% disodium hydrogen phosphate in ammoniacal medium. $Ba(Ra)SO_4$ was then precipitated after dissolving the calcium phosphate precipitate in 6 N HCl. The precipitate was purified by repeated dissolution in 10% ammoniacal EDTA and reprecipitation using glacial acetic acid. The precipitate was transferred to a 3 cm diameter Al planchet and counted for alpha activity (ZnS(Ag) alpha counter, Model ECIL, RCS4027), after one month when ^{224}Ra completely decayed and the daughter products of ^{226}Ra grew into secular equilibrium. The alpha counting efficiency of the ZnS(Ag) detector was 30% and the standard deviations of the ^{226}Ra activity were restricted to 20% at 3s confidence by selecting appropriate counting intervals. The sample pH was measured using an ELICO Model LI-120 pH meter calibrated to 4, 7 and 9.2 pH levels with standard buffer solutions. Dissolved solids were determined by standard gravimetric procedure.

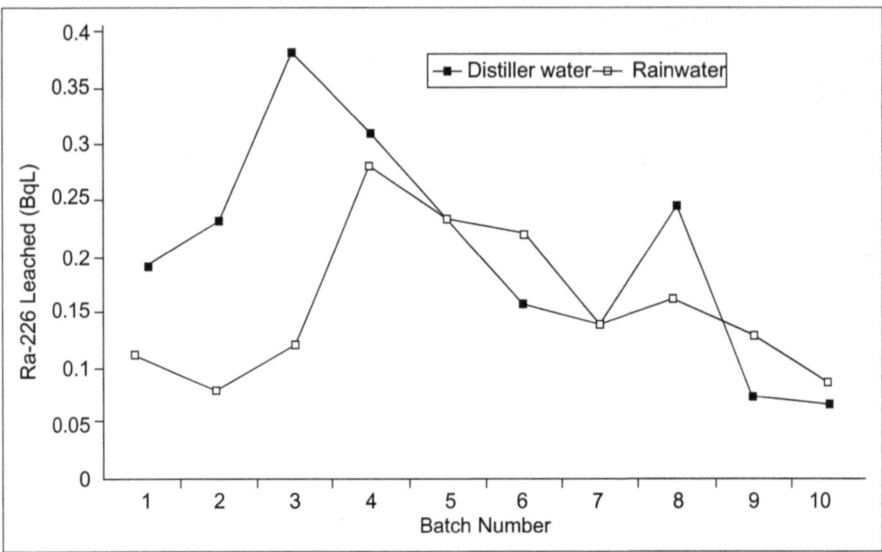

Fig. 9.3. Leaching of ^{226}Ra from phosphogypsum

9.5.2 Leaching Performance of Phosphogypsum

Figure. 9.3 shows the ^{226}Ra activity leached out from phosphogypsum in 10 successive leachings using distilled water and rainwater as leachants, [experiments (*a*) and (*b*)].A solution sludge ratio of 100:1 and a contact time of 24 h were adopted in the experiments. The data points indicate average values obtained from replicate experiments. The concentration of ^{226}Ra in the leachates varied from 0.08 to 0.38 BqL^{-1} in the case of distilled water and from 0.09 to 0.28 BqL^{-1} in the case of rainwater. The initial peaks of the curves in Fig. 9.3 correspond to the loosely bound fraction of ^{226}Ra activity. The pH of the leachates changed in the batches over time from 2.6 to 5.8. The high acidity in the initial batches of the leachates was due to the residual contamination of phosphoric acid and fluro silicic acid in the fresh phosphogypsum samples. The total ^{226}Ra activity leached out was found to be around 24 and 18% in distilled water and rainwater, respectively. Rutherford *et al.* (1995)[23] reported 0.19– 0.65 BqL^{-1} of ^{226}Ra activity in distilled water leachates of phosphogypsum, which is comparable to the values observed in the present study. Rainwater leached less radium compared to distilled water because the former contains dissolved ions, mainly sulphate, from the atmosphere. The results of the leaching study conducted using rainwater having a pH of 5.5 with varying liquid: solid ratios [experiment (*c*)] are presented in Fig. 9.4. Nine different liquid:solid ratios, ranging from 10:1 to 1000:1 were examined. Up to five successive leachings were carried out for each ratio. The cumulative percentage of ^{226}Ra leached out increased from 1 to 25% as the liquid to solid ratios increased. The solution concentration of ^{226}Ra activity was nearly proportional up to a liquid: solid ratio of 200:1, further increase in the ratio did not result in a proportionate increase in the leaching of ^{226}Ra activity. The maximum and minimum concentration of ^{226}Ra in the leachates

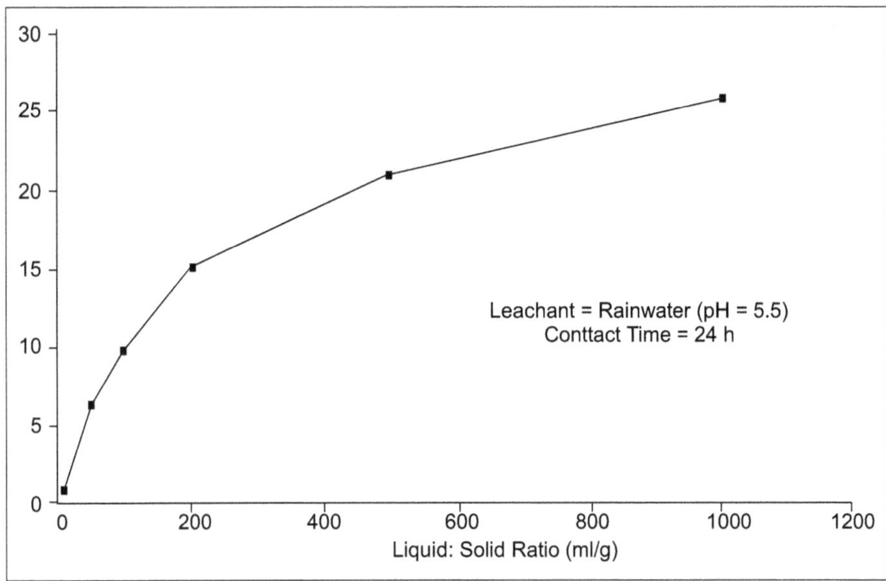

Fig. 9.4. Cumulative Leaching of ^{226}Ra from Phosphogypsum vs Liquid: solid Ratios

were found to be 0.33 and 0.03 BqL^{-1}, respectively, in this experiment. Hence liquid: solid ratio plays an important role in the dissolution characteristics of radium from phosphogypsum.

In order to evaluate the leaching characteristics of ^{226}Ra activity with respect to different contact times corresponding experiments were conducted in the laboratory, (experiment (*d*)). The percentage activity leached out *vs* contact time is plotted in Fig. 9.5. The percentage activity leached out varied in a narrow range of 0.6–1.8% and the maximum value was observed during short contact times. The solution concentration of ^{226}Ra activity showed decreasing trend with increasing contact time. This was possibly due to read sorption of the released ^{226}Ra activity. In another experiment repeated leaching with rainwater in the same

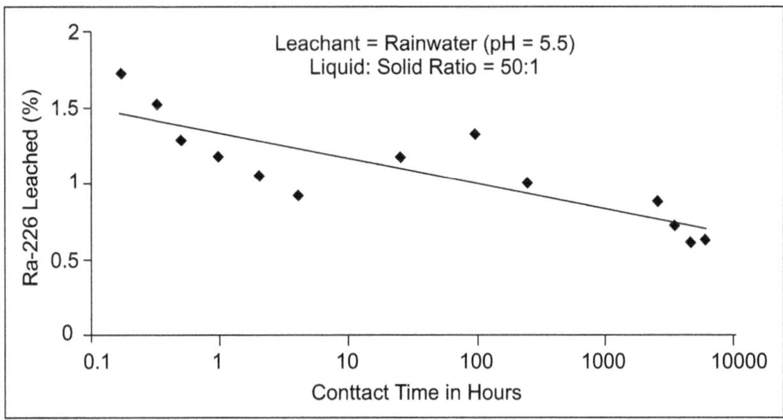

Fig. 9.5 Leaching of ^{226}Ra from phosphogypsum with contact time

phosphogypsum sample at different contact time was examined. The cumulative activity leached out is plotted. Leaching of ^{226}Ra from phosphogypsum with contact time is increased.

The cumulative fraction of ^{226}Ra activity leached out increased with increasing contact time as expected. Results of the experiment that simulated natural conditions by exposing 1 kg of phosphogypsum in a rectangular tray to weather (rains), [experiment (*e*)] are presented in Table 9.7 It was observed that only 0.9% of the ^{226}Ra activity was leached.

Total rainfall during the period: 762 mm (30 days); initial ^{226}Ra activity: 850 Bq; total activity leached out: 7.38 Bq; percentage of ^{226}Ra leached out in 15 batches, over a period of one month: 0.9%. [[28] P.P. Haridasan *et al.* / J. Environ. Radioactivity 62 (2002) 287–294].

From the study reported above, it can be concluded that the dissolution characteristics of ^{226}Ra from phosphogypsum depend largely on the leaching conditions such as solution : sludge ratio, contact time, pH of the solution, etc. Leaching studies indicated that phosphogypsum might enhance ^{226}Ra levels in the phosphogypsum disposal environment. Wide variation in the cumulative leaching of ^{226}Ra activity from phosphogypsum was observed. Leaching of

Table 9.7 Leaching of ^{226}Ra from Phosphogypsum by Simulating Natural Conditions

Leachate No.	Volume of leachate L	pH	Dissolved solids (mg mL^{-1})	226Ra (BqL^{-1})
1	7.60	2.80	2.82	0.28
2	1.63	4.70	2.26	0.53
3	2.55	3.64	2.23	0.25
4	1.37	4.24	2.43	0.11
5	0.50	4.40	2.20	0.24
6	0.48	4.50	2.74	0.25
7	4.20	4.72	2.16	0.11
8	7.00	5.08	2.30	0.09
9	3.10	4.85	2.77	0.09
10	0.70	4.74	2.57	0.24
11	2.15	4.85	2.31	0.25
12	2.15	4.69	2.28	0.16
13	3.90	4.80	2.24	0.10
14	4.70	5.50	2.33	0.07
15	2.70	5.08	2.52	0.08

radium may be slow in field conditions as the experiment simulating natural conditions resulted in the release of only 1% ^{226}Ra of the activity in one month. However, considering the quantity of phosphogypsum disposed at individual sites dispersion of activity in surface water and contamination of ground water is likely at the local environment. Since most of the ^{226}Ra in phosphate rock is entrained in phosphogypsum, and the material produced and stockpiled in large quantities increased, emphasis is required on assessing the environmental impact of phosphogypsum. The results obtained in the study can supplement in evaluating environmental levels of radium attributed to phosphogypsum disposal and hence in evaluating site specific radiation exposures.

Rock phosphate ore processing and disposal of phosphogypsum contribute to enhanced levels of natural radio nuclides in the environment. The studies were carried out[29] in Chitrapuzha River near to Kochi, to assess the impact of disposals on the aquatic environment surrounding a phosphatic fertilizer plant and its phosphogypsum disposal area. Data was evaluated about the distribution of ^{238}U, ^{226}Ra and ^{210}Po in the environment and the associated radiological impact to humans. The concentrations of radio nuclides, especially ^{226}Ra, in the river waters showed enhancement by an order of magnitude relative to the levels in nearby water bodies. It was found that concentrations were influenced by seasonal changes in the river flows during monsoon and summer periods. Ingestion doses via fish and milk have an upper estimate of 18 mSv for the critical population.

References

1. Rutherford P.M., Dudas M.J., Samek R.A., Environmental Impacts of Phosphogypsum. Sci. Total Environ, 1994, Vol. 149, pp. 1–38.
2. US Geological Survey, 2008, Mineral Commodity Summaries (US Geological Survey Publications), http://pubs.er.usgs.gov/usgspubs/recentpubs.jsp.
3. Haridasan P.P., Desai M.V.M., Paul A.C., Natural Radionuclides in the Aquatic Environment of a Phosphogypsum Disposal Area. J. Environ. Radioactivity, 2001, Vol. 53, pp. 155–165.
4. Radiological Protection Principles Concerning the Natural Radioactivity of Building Materials, Radion Protection Report 112, EC, European Commission, Luxembourg, 1999.
5. Natural occurring radioactivity in the No. dic Countries- reccommendations, The Radiation Protection Authorities in Denmark, Finland, Iceland, No. way, and Sweden, Flag-book, Series.
6. Radiological Protection Principles Concerning the Natural Radioactivity of Building Materials, Radion Protection Report 96, EC, European Commission, Luxembourg, 1996.
7. SI 5098, 2007, Content of Natural Radioactivity in Building Products, Standard of Israel No. 5098,, The Standard Institution of Israel, Tel Aviv, Israel.
8. Guide ST 12.2, 2005, The Radioactivity of Building Materials and Ash, STUK, Helsinki, Finland.
9. NIRH 2002, Bkendtgorelse nr. 192 af 2 "Bkendtgorelse om undtagelsesregler fra loy om brug m.v.af radioaktive stoffer", The National Institute of Radiation Hygiene, Herlev, Denmark.

10. Kovler, K., Haquin, G., Manasherov, V., Neeman E and Lavi, N., Limitation of Radionuclides Concentration in Building Materials, Available in Israel, Building and Environment, 2002, Vol. 37, pp. 531–537.
11. O'Brien, R.S., Gamma doses from Phosphogypsum Plaster Board, Health Physics, 1997, Vol. 72, No. 1, pp. 6–96.
12. Beretaka, J., Cioffi, R., Marroccoli, M. & Valenti, G.L., Energy Saving Cements Obtained from Chemical Gypsum and Other Industrial Wastes, Waste Management, 1996, Vol. 16, pp. 231–235.
13. IAEA 2003, Extent of Environmental Contamination by Naturally Occurring Radioactive Material (NORM) and Technological Options, ICRP Publication 60, Annals of the ICRP, Pergmon Press.
14. Kovler, K., Perevalov, A., Steiner, V. & Metzger, L.A., 2005, Radon Exhalation of Cementiotious Materials Made with Coal Fly ash: Part 1- Scientific Backgrounand and Testing of the Cement and Fly ash Emanation, Journal of Environmental Radioactivity, 2005, 82, No. 3, pp. 321–334.
15. Malhotra, V.M., High Performance high Vol. me Concrete, Concrete International, 2002, Vol. 24, No. 7, pp. 30–34.
16. Bassew, P., 2003, The Radon emanation Power of Building Materials, Soils rocks, Applied Radio Isotopes, 59, 389–392.
17. Stouls, S., Manolupoulou, M & Papastefanou, C, 2004, Measurement of Radon Emanation Factor from Granular Samples, Effects of Additives in Cement, Appl. Radiat. Isotopes, 2004, Vol. 60, pp. 49–54.
18. Steger, S., OENORM S 5200, Radioactivity in Building Materials (A Regulation in Austria for Limitation of Natural Radioactivity in Building materials), Radiation Protection Dosumetry, 1992, Vol. 45, No. 7, pp. 21–722.
19. Masaki and Banzi, Radioactivity levels of limestone and gypsum used as building raw materials in Turkey and estimation of exposure doses Radiat Prot Dosimetry 2010 Vol. 140, No. 4, pp. 402–407.
20. US Geological Survey, 2008, Mineral commodity Summaries, US Geological Survey Publications, http://pubs.er.uses.gov./usgspubs/recentpubs.jsp.
21. Haridasan P.P., Maniyan C.G., Pillai P.M.B., Khan A.H., Dissolution Characteristics of 226Ra from Phosphogypsum. J. Environ. Radioac, 2002, Vol. 62, pp. 287–294.
22. Laiche Thomas P., Max Scott L., A Radiological Evaluation of Phosphogypsum. Health Phys, 1991, Vol. 60, pp. 691–693.
23. Rutherford, P.M., Dudas, M.J., Arocena, J.M.,1995, Radium in PhosphogypsumLeachates, Jr. of envirnmental Quality, 1995, Vol. 24, No. 2, pp. 307–314.
24. Paul A.C., Pillai K.C., (1991), Natural radionuclides in a tropical river subjected to pollution. Water, Air, Soil Pollut 1991, Vol. 55, pp. 305–319.
25. Haridasan, P.P., Pillai, PMB, Tripathi, RM and Puranik, VD, Radiation Protection Dosimetry, 2009, Vol. 135, pp. 211–215.
26. Bolivar, P., Garcia Tenorio, R. and Mas, J. L., 1998, Radioactivity of Phosphogypsum in the South West Spain, Radiat. Prot. Dosimetry, 1998, Vol. 76, pp. 186–189.
27. UNSCEAR, 2000, United Nations Scientific Committee on the Effects of Atomic Radiation, Sources and Effects of Ionizing Radiation (United Nations, New York).
28. Haridasan, P.P., Maniyan, C.G., Paul, Pillai, P.M.B. and khan, A.H., Journal of Environmental Radioactivity 2002, 62, pp. 287–294 www.elsevier.com/locate/jenvrad.
29. Haridasana, P.P., Paulb A.C., Desaib M.V.M., Natural Radionuclides in the Aquatic Environment of a Phosphogypsum Disposal Area, Journal of Environmental Radioactivity 2001, Vol. 53, pp. 155–165.

10

Market Potential Assessment of Gypsum Industry

10.1 Market Potential Assessment

The worldwide gypsum production has been reckoned as 159,000,000 metric tonnes in 2008 as compared to 167,000,000 metric tonnes in 2007 (Table 10.1). It can be seen that the production of gypsum has been reduced. The utilization of gypsum in different countries is variable. China produces maximum gypsum followed by USA, Iran and Spain. The production of gypsum in India has been estimated to only 25,50,000 metric tonnes as compared to other countries[1].

The global gypsum market was estimated at 187 million tonnes in 2009 and is projected to grow to 264 million tonnes by 2014, according to a new study by Intertech Pira published in association with PRo Publications[1]. According to the study, European (EU) demand for gypsum is expected to take until 2013 to recover to the peak level of 2006. It is expected that US demand will remain below peak demand but will recover to 2007 levels by 2014. Chinese markets have continued to grow and can be expected to grow more rapidly than the 5% growth in 2009, modest by Chinese standards. It is estimated that total gypsum consumption will grow by about 6% in 2010, with recoveries in Western economies and continued growth in China. The growth rate is likely to increase to about 7.8% in 2011 before leveling off at between 7.1% and 7.3% to 2014. Gypsum and anhydrite are very common minerals found in substantial quantities in evaporate rocks throughout the world. FGD gypsum is produced as a by-product of flue gas desulphurisation (FGD) at power stations fired by fossil fuel. Phosphogypsum is produced as a by-product at phosphate fertiliser plants. Total gypsum demand was estimated at 187 million tonnes in 2009 and is projected to grow to 264 million tonnes by 2014, largely driven by China. EU demand is expected to take until 2013 to recover to the peak level of 2006. It is expected that US demand will remain below peak demand but will recover to 2007 levels by 2014[2].

Table 10.1 Gypsum World Production

(Thousand metric tonnes)

Country	Year				
	2004	2005	2006	2007	2008
Afghanistan	3	2	2	2	2
Algeria	1,058	1,460	1,033	1,198	1,672
Argentina	836	1,073	1,203	1,227	1,200
Armenia	51	44	44	55	55
Australia	4,325	3,857	4,265	3,896	4,000
Austria	921	911	936	1,006	1,000
Azerbaijan	9	28	35	22	23
Bhutan	131	151	160	170	160
Bolivia	—	—	1	4	4
Bosnia and Herzegovina	140	153	132	154	154
Brazil	1,472	1,582	1,737	1,750	2,100
Bulgaria	176	188	216	234	234
Burma	71	68	69	69	69
Canada	9,339	9,400	9,036	7,562	5,740
Chile	630	661	845	773	774
China	29,000	32,000	42,000	48,000	46,000
Colombia	161	173	186	200	200
Croatia	148	196	170	170	170
Cyprus	255	260	250	250	250
Czech Republic	71	25	16	66	35
Dominican Republic	459	370	356	350	350
Ecuador	—	1	1	2	1
Egypt	2,000	2,000	2,000	2,000	2,000
El Salvador	6	6	6	6	6
Eritrea	1	1	1	1	1
Ethiopia	51	35	39	40	40
France	5,700	4,902	4,800	4,800	4,800
Germany	1,579	1,644	1,771	1,898	1,900

(Contd...)

Country	Year				
	2004	2005	2006	2007	2008
Greece	500	500	500	500	500
Guatemala	106	350	227	495	500
Honduras	6	6	6	6	6
Hungary	62	55	30	26	26
India	2,350	2,400	2,450	2,500	2,550
Indonesia	6	6	6	6	6
Iran	12,594	11,196	12,000	12,000	12,000
Ireland	450	450	450	450	450
Israel	125	107	111	83	83
Italy	2,488	2,905	2,860	5,459	5,400
Jamaica	283	302	364	228	225
Japan	5,865	5,913	5,796	5,850	5,800
Jordan	135	345	334	288	300
Kenya	9	9	9	10	10
Laos	201	774	775	775	775
Latvia	226	220	230	230	230
Lebanon	30	30	30	30	30
Libya	175	175	200	240	250
Macedonia	165	190	268	256	256
Mali	1	—	—	—	—
Mauritania	39	43	45	49	44
Mexico	9,221	6,252	6,076	6,080	5,135
Moldova	103	131	186	312	300
Mongolia	25	25	26	26	26
Morocco	600	600	600	600	600
Nicaragua	36	36	42	40	40
Niger	18	17	13	5	5
Nigeria	160	150	169	579	500
Oman	60	60	60	60	60

(Contd...)

Country	Year				
	2004	2005	2006	2007	2008
Pakistan	467	552	650	620	640
Paraguay	5	5	5	5	5
Peru	150	150	151	151	495
Poland	1,167	1,243	1,353	1,581	1,580
Portugal	461	389	400	400	300
Romania	490	502	615	707	705
Russia	2,077	2,200	2,200	2,300	2,300
Saudi Arabia	641	713	2,101	2,100	2,300
Serbia	45	45	45	45	45
Slovakia	127	107	110	110	110
South Africa	452	548	554	627	571
Spain	12,534	13,000	11,500	11,500	11,500
Su	11	9	7	8	8
Switzerland	250	250	250	250	250
Syria	432	467	444	448	573
Tajikistan	57	9	9	9	9
Tanzania	59	63	33	3	3
Thailand	7,169	7,113	8,355	8,569	8,000
Tunisia	108	113	151	157	165
Turkey	2,301	3,501	4,370	3,241	3,000
Turkmenistan	100	100	100	100	100
Ukraine	337	381	376	742	740
United Arab Emirates	110	120	130	150	200
United Kingdom	2,914	2,000	1,700	1,700	1,700
United States	17,200	18,800	18,500	17,900	14,400
Uruguay	1,150	920	—	—	—
Venezuela	4	6	7	7	7
Yemen	37	38	44	45	50
Total	145,000	148,000	159,000	167,000	159,000

Source: United States Geological Survey Mineral Resources Program

A very large fraction of gypsum production is used in the construction industry as a setting retarder in Portland cement, in the manufacture of plasterboard and other plaster products, or as wet plaster. Much of the plasterboard production is vertically integrated; major plasterboard producers are the largest producers of gypsum in the West, although there are many small and sometimes very small producers, especially in developing countries that produce gypsum mainly for the cement industry or local wet plaster production. In North America, Europe and Japan, the largest use for gypsum is to make plasterboard. Cement is the world's largest application for gypsum. The US is by far the largest producer of plasterboard; plasterboard is 90% of US gypsum consumption when housing markets are strong. China is a relatively small but rapidly growing producer and consumer of plasterboard; its biggest use for gypsum is cement.

According to the study, the largest declines in gypsum for plasterboard have been in the US; this is because the recession produced a big fall in new housing construction. It is expected that house building will not return to 2007 levels until 2014 and it will probably take even longer to return to peak levels. EU demand is smaller. It has declined by a lower fraction and is likely to return to normal levels by 2011, although there are variations from country to country.

China's plasterboard consumption continues to grow from a much lower base than in the US or the EU, but China has overtaken Japan as the third largest producer and consumer of plasterboard. China has a low consumption of plasterboard per head, so growth is expected to remain strong for many years. Japanese consumption of gypsum in plasterboard is quite mature and will tend to follow overall trends in the construction industry, which is expected to be growing again by 2012.

Gypsum use in cement is increasing with leaps and bound; outside North America, where plasterboard consumption is so high, gypsum use in cement is greater than gypsum use in plasterboard. China is the main growth driver in the cement sector. In the short term, demand for gypsum in cement is predicted to rise faster than demand for gypsum in plasterboard, due to the effect of fiscal stimulus packages on the consumption of cement in infrastructure or other government projects. Growth largely depends on how long the fiscal stimulus packages continue.

The US market favours wood-framed housing with plasterboard cladding and the US can be 50% of the plasterboard market in a good year. As the housing market is depressed - very low levels of housing starts, less commercial construction and less remodeling-consumption of plasterboard has been depressed. The main drivers of demand in the US are new residential construction plus commercial construction and remodeling. Cement has wider markets in road and other infrastructure projects plus residential markets; as much as 50% of cement use is for public works.

Markets for plasterboard are considerably down from their peaks in 2006 because of the financial crisis and particularly its effect on the housing market in the US and elsewhere. Recovery is expected to be slow and seems unlikely until well into 2010–12. Fiscal incentive programmes are expected to help but there are still concerns about how long they will remain. Gypsum use in cement is expected to benefit significantly from government stimulus spending on infrastructure programmes in many regions. There has been strong growth in China but lower than in previous years.

10.2 Markets Trend and Consumption

The worldwide market for wallboard is dominated by North America. Figure 10.1 shows the TOP 10 countries with worldwide capacities of approximate of 7900 million m^2 (Mm^2) or almost 45% of worldwide capacity, before second – placed Japan (680 Mm^2). The next places are taken by Germany (360 Mm^2), Canada (350 Mm^2), France (340 Mm^2) and Great Britain and China with 300 Mm^2 each. The TOP 10 countries, which also include South Korea, Scandinavia and Australia, are responsible for more than 82% of the world worldwide capacity.

The per-capita consumption of wallboard varies greatly around the world. At a production quantity of 7100 Mm^2, the average worldwide per capita consumption is 1.1 m^2. The USA takes the lead with 10.0 m^2 per capita before Canada with 9.5 m^2 and Australia with 8.0 m^2. South Korea's per capita consumption is 5.4 m^2, while Japan is 4.7 m^2. In Europe, Scandinavia has the highest consumption with 6.5 m^2, followed by France (5.0 m^2), Great Britain (4.5 m^2) and Germany (3.9 m^2). At present China is far below the world average with a per-capita consumption of 0.3 m^2. The greatest potential of wallboard exist in China and other Asian courtiers. In contrast, only small rates of increase can be expected in North America.

Figure 10.2 shows the development of wallboard capacity and production in the USA over the last 20 years. Figure 10.3 shows the relationship between capacity utilization and net factory price in the USA.

In Europe, the present growth of the gypsum industry is even higher than in North America. Even though the average plant capacity in Europe is only about

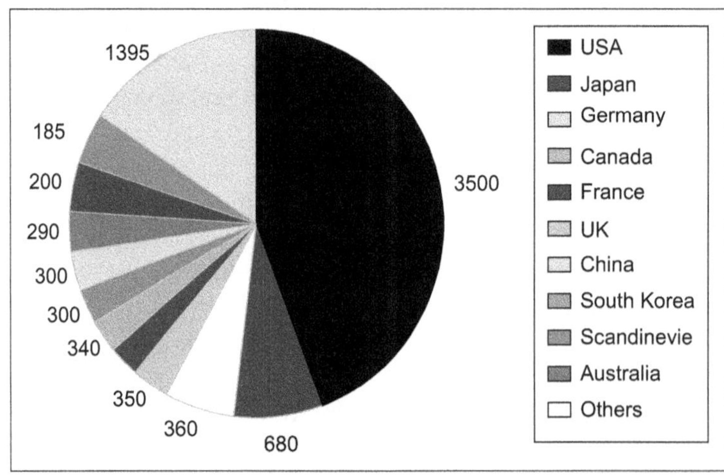

Fig. 10.1 Top10 Countries Sandwich-Type Gypsum Plaster Board (Mm^2 capacity)

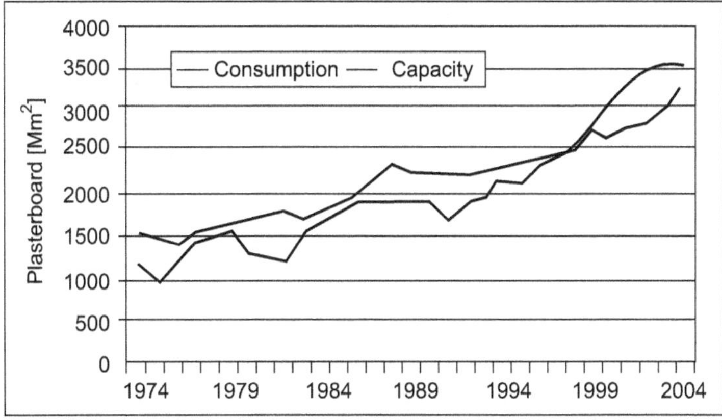

Fig. 10.2 Development of Capacity and Production in the USA

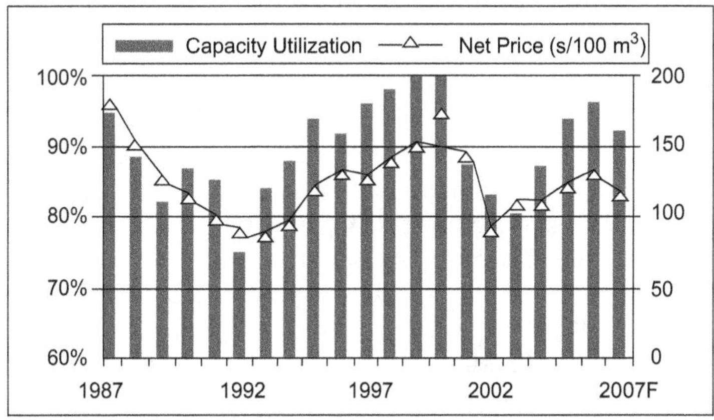

Fig. 10.3 Capacity Utilization and Net Factory Price in the USA

29 Mm^2/year, all the markets are experiencing high capacity utilization figures of above 90% due to stricter regulations governing thermal insulation and the execution of building work. In the far east, the already developed markets of Japan and South Korea are still dominant.

At a production quantity of approx. 600 Mm^2, Japan also achieves a capacity utilization of nearly 90% and an average plant capacity of 30% Mm^2/year. It uses more by-product gypsum than any other country in the world, consuming 6 Mta, compared to 2.1 Mta of natural raw gypsum, which is 100% imported from Australia, Thailand and Mexico, 5 Mta of gypsum go into Japan production of wallboard. However, the biggest Asian market growth is currently taking place in China and ASEAN countries such as Thailand and Malaysia. For China, an annual market growth of 20% is forecast for the coming years[3]. In 2005 alone, three new wallboard factories will be put into operation, and more are being planned. One plant for 30 Mm^2 is being built by Grenzebach Germany.

Intertech Pira expects growth to return in 2010 in North America after declines of about 13% in 2009 and close to 40% since the peak in 2005. Growth is expected to be fairly modest and, because of overbuilding in the US, it may be after 2014 that plasterboard consumption returns to its previous peak. Cement demand is expected to pick up faster in 2010 as the fiscal stimulus is spent on construction projects. European markets have not declined as far as those in the US and are expected to grow back to above previous peaks before 2014. Chinese markets have continued to grow and can be expected to grow more rapidly than the 5% growth in 2009, modest by Chinese standards. It is estimated that total gypsum consumption will grow by about 6% in 2010, with recoveries in Western economies and continued growth in China. The growth rate is likely to increase to about 7.8% in 2011 before leveling off at between 7.1% and 7.3% to 2014.

10.2.1 Top Producers

Worldwide, there are 250 gypsum fibre board and wallboard manufacturing plants[4]. The market leader is British plaster Board (BPB) with a capacity of 1.300 Mm^2 or 16.5% market share, ahead of US Gypsum (14.9%) and Knauf (13.3%).

In the last 3 years, BPB has almost doubled its sales.US Gypsum Company (USG) produces approx. one third of all wallboard in the USA. In 2004, the company achieved a record turnover of 4.5 billion US$. This corresponds to an increase of 0.843 billion US$ over the preceding year. USG current share price is 142 US$/100 m^2. In the USA the company owns 21 wallboard manufacturing plants. One of the largest is the Plaster City Plant in California, which has a capacity of approx. 105 Mm^2 and includes one of the biggest grinding and calcining system in the world.

Knauf currently operates 132 production facilities in 35 countries. In 2004, the company achieved a turnover of 3.5 billion US$. Its principal markets are West and East Europe, including Turkey. In recent years, the East Europe, the former CIS countries and China played leading roles in the company's business development.

For Lafarge Gypsum, the last two years have been a consolidation place after a period of rapid growth in which the company became the worldwide No. 3. The manufacturing of wallboard now makes up 85% of the firm's business of approx. 1.2 billion US$. 10 years ago, Lafarge Gypsum's activities were largely confined to Europe, where the firm now owns 52 plants. Presently Lafarge has captured market in India too.

In 2000, Lafarge entered into a Joint Venture with the Australian company Boral (Lafarge Boral Gypsum Asia), with both companies holding 50% each. In 2001, Siam Gypsum was incorporated into this Joint Venture. LBGA now owns eleven production plants in nine countries and has a wall board capacity of 280 Mm^2. Without taking account of Japan, the company has now became the market leader in Asia, with a market share of 31%. The capacity in China will be doubled to 70 Mm^2 with two new plants.

National Gypsum, No. 2 in the USA, owns 15 gypsum board manufacturing plants. In 2010–12, it will put a further 120 million US$ plant into service at Charlotte, North Carolina, with a capacity of 75 Mm^2.

Yoshino Gypsum Company owns 22 production plants for wallboard in Japan and has a production capacity of 520 Mm^2. The company has been in existence since 1901 and has grown substantially since the mid 90's. Its market share in Japan is 75%. The only producer in Japan is apart from Yoshino Gypsum is Chiyoda Ute., which was established in 1948 as a subsidiary of the Onida Cement. Co. Having started up as the roofing tile manufacturer It began producing gypsum boards in 1955. Meanwhile, the firm owns five production plants.

10.2.2 Technology Trends

The present high worldwide capacity utilization of 90% has brought plant makers a large number of new projects and orders. For wallboards plants, gypsum with short setting times are preferred. Kettles, direct flash calcining and integrated grinding and calcining processes and combination processes are generally used. For FGD –gypsum, calciners with specially designed hammer mills have been established as the chief process. Grinding and calcining systems equipped with vertical roller mills currently makes up approx. 40% of the market.

Wallboard plant may include parallel flow dryers and cross flow dryers. For the gypsum fibre board manufacturing process there is also the so called "screen" dryers. In order to get high gypsum board retention times of 32 to 40 minutes with the most compact dryers, the units are designed with 6 to 16 decks and 2 to 3 drying zones.

Importance is being made on the handling equipment in wallboard plants. The feeding, laying, stacking, de-stacking and transfer equipments received much attention and therefore new sophisticated state-of-art have been introduced time to time to handle different sizes of the boards.

10.2.3 Products Trends

Importance of fire and moisture resistant gypsum boards has increased in recent years, no doubt USA is the chief market. In Europe, a number of innovations have been introduced, such as fire protection and composite boards for improved sound insulation and thermal insulation. With their product line "Duraline", BPB Rigips introduced wallboard with strong hard surface. This board has hard surface and high impact resistance than the traditional boards and is more suited to tougher situations like public buildings such as schools, kindergartens and hospitals.

Knauf recently introduced two innovative products. The "La Vita" wallboard, 2 years old in the market, helps in reducing 95% of alternating electrical and high frequency radiation field of 200 MHz. The outer layer of the rear-side paper layer contains electrically conductive carbon fibres. The boards can be mounted on a metal and earthed. The wallboard "Cleaneo", which Knauf introduced in 2005, eliminates pollutants and smell from the air.

A family of by-product gypsum materials–such as flue gas desulfurization (FGD) gypsum, phosphogypsum (PG), fluorogypsum, titanogypsum, and desulfogypsum is produced as a result of stringent environmental regulations or are inherent to the industrial processes themselves. From a mineralogical viewpoint, all these materials are calcium sulfate.

10.3 Chemical Gypsum

10.3.1 Phosphogypsum

Out of the main 3 types of chemical gypsum available in India *i.e.,* phosphogypsum, fluorogypsum and marine gypsum, phosphogypsum is significant from the point of view of the quantities generated and the distribution of the recovery centres. There are over one dozen plants recovering phosphogypsum as a by-product during the manufacture of phosphoric acid by wet process in India. As already stated, the recovery of phosphogypsum stands at about 6.0 million tonnes per annum in India.

Bulk of phosphogypsum recovered is consumed in the manufacture of cement, ammonium sulphate fertilizer and a small quantity for reclamation of alkaline soil in the states of Punjab, Haryana and Uttar Pradesh. In fact, the applications of phosphogypsum is related to purity of the material as well as the impurity contents. Generally, the purity of Indian phosphogypsum varies from 80 to 95% $CaSO_4.2H_2O$ and the P_2O_5, F, lie in the range 0.12 to 1.8%. At present, four units *viz.,* E.I.D. (India) Ltd., Chennai, the Fertilizer Corporation of India Ltd., Sindri, Gujarat State Fertilizer Company Ltd., Vadodara and Fertilizer and Chemicals Trarancore Ltd., Kochi consume some of their waste gypsum for the manufacture of ammonium sulphate in their own plants.

Besides, major applications as listed under Chapter 3, the phospho-gypsum may be used in the manufacture of Sulphuric acid and cement clinker by the Muller-Kuhne Process and also as a filler in paint, rubber, paper etc. Calcined form of gypsum/gypsum plaster may be utilized in the preparation of a variety of building materials. Industry-wise uses of phospho-gypsum are reported a below:

Cement

About 4–5% phosphogypsum (beneficiated) may be used in Portland cement and the blended cements (PPC, PSC, SSC, LHSR, etc.) as a retarder to regulate the setting time of cement. Cement industry is the largest consumer of gypsum in India. It is mainly used for regulating the setting time of cement. Different types of cements produced in India are:

- Ordinary Portland Cement (OPC) (33 Gd, 43 Gd, 53 Gd)
- Portland Pozzolana Cement (PPC)
- Portland Slag Cement (PSC)
- High Strength Ordinary Portland Cement
- Low Heat Cement

- Hydrophobic Cement
- Sulphate Resistant Portland Cement
- Super-Sulphated Cement
- Oil-Well Cement
- White Cement
- Coloured Cement
- Masonry Cement

Gypsum is used as a set retarder in the manufacture of various types of cements listed above, except oil-well cement where rapid setting is necessary. Although, the production of oil well cement is not high. The production of main cements *i.e.*, OPC, PPC and PSC was 217 million tonnes in December, 2011.

Today, India is the second largest producer of cement in the world. The cement production rose from a mere 22.54 million tonnes per annum in 1982 to over 217 million tonnes per annum in 2011 registering tremendous growth. India has 171 largest and 365 mini cement plants.

Present Consumption of Gypsum

The total consumption of gypsum in the cement industry ranged from 6.98 million tonnes in 2009–10 and 7.14 million tonnes in 2010–11. Cement industry accounted for 90% of the total consumption of gypsum in the country. The respective share of mineral, marine and by-product phosphogypsum in the total consumption was 47%, 48% and 5%, respectively. The consumption of gypsum of different grades is shown in Chapter-I.

Future Demand for Gypsum

It is seen that the consumption of gypsum per tonne of cement varies from plant to plant. It ranges from 4 to 6% depending upon the purity of gypsum on an average it works out to be about 5%. By applying this norm, the future demand estimates of gypsum would be as shown in (Table 10.2).

At present, phosphogypsums having purity ($CaSO_4.2H_2O$) in between 85 to 95% with P_2O_5 : 0.8 to 1.5% and F : 0.7 to 0.8 are used as a retarder for regulating the setting time of cement by several cement industries in the country. Phosphogypsum after beneficiation has a great potential for replacing natural or marine gypsum in the manufacture of cement.

Table 10.2 Future Demand of Gypsum for Cement

Period	Demand for Cement (Million Tonnes)	Demand of Gypsum (Million Tonnes)
2009–2010	160–165	8.0–8.25
2012–2015	180–320	9.0–15

The consumption of phosphogypsum may be enhanced if the cements such as super-sulphated cement or low heat sulphate resistant cement can be produced. These cements require 15–25% of gypsum as dihydrate, hemihydrate or anhydrite. The use of phosphogypsum may be environment friendly.

Ammonium Sulphate

Phosphogypsum of purity, 80% minimum may be used as a raw material in the manufacture of ammonium sulphate fertilizer. After the cement, fertilizer industry is the largest consumer of gypsum. High purity gypsum is utilized for manufacturing ammonium sulphate fertilizer. The total consumption of gypsum (natural, marine, by-product, moulds etc.) in the manufacture of ammonium sulphate fertilizer was 1.20 million tonnes in 2010–11 which is likely to increase to 2.0 million tonnes in the near future.

Gypsum of properties - purity minimum 85–90%, $CaSO_4.2H_2O$, SiO_2 and other insoluble matter : max.6.0%, iron and aluminium oxide : max. 1.5%, MgO : max. 1.0%, NaCl : max. 0.03% is required. At present, FCI, Sindri, GSFC, Vadodara, Hindustan Lever Ltd., Haldia, FACT, Kochi and EID-Parry Ltd., Chengleput, Tamil Nadu are manufacturing ammonium sulphate fertilizer utilizing their phosphogypsum. GSFC, Vadodara is the major consumer of the phosphogypsum in the fertilizer industry which accounts for about 70% of the total consumption in this industry.

Sulphuric Acid

High purity phosphogypsum is a source of sulphur. The material has immense potential as a source of sulphur for making sulphuric acid. The P_2O_5 and F should be within the permissible limits. At present, no sulphuric acid is produced using gypsum in India. Some attempts were made in 1971 in Planning and Development (P&D) Sindri, to make cement clinker and sulphuric acid using phosphogypsum, but could not continue due to dilution of SO_2 from the rotary kiln to concentrator. However, two plants of capacity 250–350 TPD are functioning well at Phalobarwa, South Africa and Chemie Linz, Austria for the manufacture of cement and sulphuric acid. The major deterrents in their production are the impurities of P_2O_5 and F and the economics. The P_2O_5 and F should not exceed 0.5% and 0.15% respectively in the phosphogypsum.

The beneficiated phosphogypsum can be used to make cement and sulphuric acid by controlling the impurity level as well as other parameters to add to the economy of the country.

Land Plaster

Phosphogypsum can be used as a fertilizer for conserving moisture in soil and for aiding nitrogen absorption. Gypsum may improve soil permeability and may help in penetration of water and air. It may provide sulphur and calcium and may give catalytic support for maximum fertilizer utilization. Gypsum may reduce harmful effects of sodium salts present in the soil.

Fillers

Beneficiated Phosphogypsum, as a filler, may be utilized in paint, paper, textiles, insecticides, rubber, etc.

Paint

Ground phosphogypsum can be used in the paint industry as an extender or as a base in cold water paints and distempers. Sometimes, it can be used for diluting colours of pigments like chrome, yellow, iron oxides, etc.

Paper

Ground phosphogypsum (with and without calcination) may be a potential application for use as a filler in the manufacture of certain writing papers, boards, etc.

Textiles

Generally, gypsum is used for finishing of cotton and cloth to give weight to a product for the production of lustrous/glossy surface. Beneficiated phosphogypsum may be used for the same purpose.

Insecticides

Ground beneficiated phosphogypsum may be used as filler or distributor in many insecticides preferably white coloured material. Sometime China clay may be added to reduce the density.

Rubber

Beneficiated phosphogypsum may be used as a filler, molding agent in making rubber stamps.

Ceramics

Beneficiated phosphogypsum may be converted into high strength alpha plaster for use in making pottery moulds. The porous nature of plaster of Paris permits ready evaporation of the moisture from the clay.

Insulator

It can be used as a fire preventive material on account of 20% moisture present in it. Beneficiated phosphogypsum may be used as insulating material for the protection of column or beam of metal/wood from high temperature generated during the fire.

Medical

Beneficiated phosphogypsum with controlled level of P_2O_5, F and alkalis may be looked upon as a potential material for making surgical plasters. Care must be taken to reduce the impurities of P_2O_5 - Max 0.02%, F - Max 0.02% and alkalis - Max 0.01%. High purity phosphogypsum may be used in orthopedic surgery and pharmaceutical industry.

Building Industry

Gypsum is a well known building material. Gypsum as such is generally used as a retarder in the manufacture of cement. However, the industrial importance of gypsum is known as the calcined product *i.e.,* plaster of Paris.

Calcined gypsum/gypsum plaster particularly β-plaster can be produced from the beneficiated phosphogypsum. The following building materials may be produced from the calcined gypsum:

- Building plasters
- Gypsum plaster boards
- Partition Blocks
- Acoustic Tiles
- Gypsum Concrete
- Gypsum marble

Coal Mining

Phosphogypsum may be used for underground dusting in collieries to reduce explosion and siliceous hazards.

Black Board Chalk

Beneficiated phosphogypsum may be used in the manufacture of black board chalks.

Other Uses of Phosphogypsum

Phosphogypsum can be used as a flux in the smelting of nickel. It can be used for absorbing oil from factory flux. Calcined gypsum/plaster can be used in the polishing of rough plate glass. Beneficiated phosphogypsum may be used in the processing of sugar beet. It is a known practice in U.K., Germany, Netherlands and France. Beneficiated phosphogypsum can also be used in the manufacture of Ayurvedic medicines and toilet products. Calcined gypsum may be used as a fixing material for marble and granite blocks during cutting and polishing. It can be used as a substitute for elemental sulphur in the manufacture of certain Explosives.

10.4 Premixed Lightweight Gypsum Plasters

10.4.1 Vermiculite Plasters

Vermiculite is a natural mineral which is laminated in structure and very similar to mica in appearance (Fig. 10.4). Deposits of vermiculite are known to exist in various parts of the world but the best quality material produced in commercial quantities comes from the Palabora deposit situated in the North Eastern Transvaal, South Africa. Chemically the vermiculite is hydrated alumino silicate of Mg, Fe, Al metals $(Mg, Fe, Al)_3 [(Al, Si)_4O_{10}](OH)_2{\cdot}4H_2O)$.

Fig. 10.4 Crude Vermiculite

Crude vermiculite contains innumerable microscopic droplets of moisture between its lamiae, and, when particles of it are subjected to high temperatures, the moisture in these is converted to stream, and the particles caused to exfoliate (expand) up to fifteen times their original thickness into cellular accordion-like granules with myriad air cells. This characteristic gives exfoliated vermiculite its excellent insulating properties and is responsible for its very light weight.

1. Properties of Exfoliated Vermiculite

A brief review of its properties may be helpful and permit a better understanding to be obtained of the benefits which accrue from the use of the material as an aggregate for plaster. It is sterile and chemically inert; therefore, a most stable

Fig. 10.5 Exfoliated Vermiculite

and safe aggregate to use. Exfoliated vermiculite (Fig. 10.5) is completely incombustible, and has a fusing point in excess of 2,200°F.

According to the National Physical Laboratory, the 'K' factor of exfoliated vermiculite with a density of 5 lbs. to 7 lbs. per cubic foot expressed in B. Th. U/sq. ft./hour/1" thick/1:°F (difference of temperature), is 0.45 at normal atmospheric temperatures. It will efficiently insulate from sub-zero temperatures up to 2,200°F.

Due to its very low thermal capacity it is an excellent aggregate for all types of anti-condensation plasters. Vermiculite plaster aggregate for backing coats weighs only 6 lbs to 7 lbs. per cubic ft. and, for finishing coats, only 9 lbs. per cubic ft. Exfoliated vermiculite has a degree of resilience, and it is due to this property that vermiculite plasters have such great resistance to cracking and spelling. It will be appreciated from the description of vermiculite already given that vermiculite plasters have many excellent features[5–6].

The dry weight of vermiculite plaster *in situ* is less than half the weight of traditional sand plasters and is lighter to mix, to carry and to apply. Vermiculite plasters make a substantial reduction in the dead load of a structure. This is important when calculating steel specifications, and it is also an added safety factor.

Vermiculite plasters afford outstandingly high protection against fire. The Fire Research Station has carried out a number of tests which have amply demonstrated the excellent results that can be achieved. The four hour fire has been for rating 1½ thickness of vermiculite/gypsum plaster on a 3/4" gypsum plank used for protecting columns and beams.

2. Properties of Vermiculite Plaster

Other laboratory tests which have been conducted from time to time show that the vermiculite plasters can be relied on to give excellent fire protection for walls and floors. Typical results are given in Table 10.3.

Excellent results will also be obtained where vermiculite/Portland cement plasters are used.

The Underwriters' Laboratories in the U.S.A. and Canada list more than thirty approvals for vermiculite/gypsum plasters with fire resistance ratings for different standard types of construction of two, three and four hours.

Table 10.3 Fire Resistant Grading

Type of Structure	Fire Resistance Grading
½" Vermiculite/gypsum plaster applied both sides of 4½" brick wall.	4 hours
4" reinforced concrete floor with suspended metal lath ceiling plastered with minimum 1" vermiculite/gypsum plaster.	4 hours
Timber floor with metal lath ceiling fixed directly to joists and plastered with minimum ½" vermiculite/gypsum plaster.	1 hour

Table 10.4 Density vs Thermal Conductivity Values

Lbs./cu. Ft.	Kg/cu. M,	B-Th.U/sq.ft./1" thick/hour 1°F difference	Kg. cal/sq. m/ hour/1°C, difference
28	448	0.75	0.093
30	480	0.90	0.116
40	640	1.30	0.1612
48	768	1.40	0.1736
56	896	1.80	0.2232

3. Thermal Insulation

A wide variety of different vermiculite plasters have been tested from time to time. The following insulating values for vermiculite plasters are given in Table 10.4.

These compare with the following values published by the National Physical Laboratory (NPL), New Delhi (Table 10.5).

It will be seen that vermiculite/gypsum plasters have as much as two three times the insulation value of most traditional plasters and this feature, combined with the low thermal capacity of vermiculite plasters, reduces the rate at which heat is lost through walls and ceilings, and enables comfortable living and working temperatures to be reached quickly. This is particularly important in rooms, which are only in intermittent use.

Since the surface of vermiculite plaster closely follows the temperature of the atmosphere, condensation is also reduced. In this connection it should be mentioned that the surface should not be too heavily trowelled and the decorative finish should be permeable. Absorbent paints and distempers are more suitable than oil-bound paints.

A few examples of the improvements which can be obtained in the (B.Th. U/sq. ft /hour/1°F difference) value for walls and ceilings of buildings are given Table 10.6.

Table 10.5 Density and Thermal Conductivity Values

	Density expressed in		Thermal conductivity expressed in	
	Lb./ cu.ft.	Kg./ cu.m	B.Th. U/sq.ft./ 1" thick/hour 1°F difference	Kg. cal/sq.m./m/ hour/ 1°C difference
Sand, lime and cement plaster	90	1440	3.3	0.4092
Sand and Gypsum Plaster	88	1408	4.5	0.5580
Gypsum Plaster	80	1280	3.2	0.3968

Table 10.6 Thermal Conductivity of Plasters

	Uninsulated	Plastered internally with 5/8" vermiculite/ gypsum plaster
Ceilings		
4" reinforced concretete flat roof covered with three layers of bitumen felt which is in turn surfaced with ½" granite chippings	0.64	0.50
Walls		
11" cavity brick wall	0.31	0.26
9" brick wall	0.45	0.38

Resisting to Cracking

Due to the resilient nature of the vermiculite aggregate cracking normally associated with ordinary sand plasters is greatly reduced, vermiculite plasters will also accept nails and screws without cracking or spalling although they are naturally not so strong as traditional plasters.

Sound Absorption

A high degree of sound absorption can be obtained with vermiculite plasters, and will depend mainly on the mix proportions and surface treatment. In cases where particular acoustic properties are called for, a special vermiculite acoustic plaster designed for this use should be employed.

4. Specifications

Vermiculite should be looked upon as a normal internal plaster aggregate. The specification given below is designed for general plastering work and will take advantage of the special characteristics of the vermiculite aggregate. In general the Codes of Practice for plastering already published give excellent guides for the use of this aggregate, the vermiculite taking the place of sand. Vermiculite can be successfully used with gypsum plaster or Portland cement.

Vermiculite/gypsum plasters are also available as pre-mixes made by a number of different Companies and distributed throughout the Britain. Many plasterers, however, prefer to mix their own materials, and vermiculite plaster aggregates are now freely available from different Exfoliations in all parts of the country. In certain instances it is preferable to apply either, vermiculite/gypsum plasters or vermiculite/Portland cement plasters by means of a spray gun. There are a number of Companies who specialize in these applications, and several plaster spray guns are available.

The first specification below (Table 10.7) refers to retarded hemi-hydrate gypsum plasters (Class "B" in British Standard 1191 and IS : 9498–1980), as these plasters have given very satisfactory results in the experience of most uses. Many other formulae are in current use, and they will no doubt also give satisfactory results.

Table 10.7 Vermiculite Gypsum Plasters

Background	**Undercoat (nominal 1/2" thickness)Volumes**	**Setting Coat (1/2" thickness) Volumes**
Metal and wood Lathing (three coat application)	1½ coarse vermiculite plaster aggregate 1 special haired plaster	1 fine vermiculite plaster aggregate 2 hard-wall plaster
Plaster Board and Brick walls (two coat application)	1¼ - 1½ coarse vermiculite plaster 1 haired or fibred browning plaster	1 fine vermiculite plaster aggregate 2 hard-wall plaster

Table 10.8 Type of Plasters for Concrete

1¼–1½ coarse vermiculite plaster aggregate	1 fine vermiculite plaster aggregate
1 browning plaster	2 hard-wall plaster
For setting coats either Class "B" or Anhydrous Gypsum plaster Class "C" may be used. All plasters should be as per B. S. 1191.	

Table 10.9 Density of Plasters

1½ volume vermiculite plaster aggregate, 1 volume gypsum plaster	40 to 43 lbs. per cubic ft.
1½ volume vermiculite plaster aggregate, 1 volume gypsum plaster	46 to 49 lbs. per cubic ft.

Table 10.10 Vermiculite/Cement Plasters

Floating Coat	3 coarse vermiculite plaster aggregate 1 hydrated lime 1 Portland cement
Setting Coat	1 fine vermiculite plaster aggregate 2 Sirapite finishing plaster
Densities. 3 Volumes Vermiculite plaster aggregate 1 Volume hydrated lime 1 Volume Portland cement	52 to 55 lbs. per cubic ft.

Concrete

Dense smooth concrete surfaces require the application of a vermiculite bonding plaster or treatment with a suitable bonding primer as given below (Tables 10.8 to 10.10).

5. Densities

The normal densities for these mixes, set and dried *in a situ* are as per Table 10.9

Coarse Vermiculite plaster aggregate for undercoats may not always be supplied under this description. Some suppliers offer Grade 2 or mixture of Grades 2 and 3 Vermiculite.

Fine Vermiculite plaster aggregate for setting coats may equally be supplied as Grade I Vermiculite.

6. Coverage

Different backgrounds and the skill of individual operatives both influence the coverage obtained. The following, however, may be taken as a guide when ordering materials 1 ton coarse vermiculite plaster aggregate for backing coats when mixed with gypsum plaster in the proportion given in the specification (*i.e.,* 1¼ : 1 vermiculite/gypsum) will give an approximate coverage of 1,000 sq. yards at a nominal ½" thickness.

1 ton fine vermiculite plaster aggregate for setting coats when mixed with gypsum plaster in the proportions given in the specification (*i.e.,* 1:2 vermiculite/ gypsum) will give an approximate coverage of 3,000 square yards at a nominal 1/8" thickness.

7. Mixing Instructions

By Hand

Thoroughly mix the appropriate volumes of vermiculite and plaster dry, then slowly mix in the required quantity of clean water until the consistency is suitable for the type of background to be plastered. Only clean water should be used.

Mechanical Mixing

Place the Vermiculite aggregate into the mixer and add water to sufficiently moisten but not saturate it. Then gradually add the required amount of plaster to the slowly revolving mix. As above, the mixing time should not be longer than that required to obtain uniformity of colour and consistency. Mixing for too long a time tends unduly to compact the vermiculite and reduces volume. As a guide, mixing should not continue longer than three minutes.

Application

Vermiculite plaster is applied in much the same way as sand plasters, except that not quite so much pressure with the trowel is needed. Successive coats should be applied only when the previous coat has completely set. A vermiculite plaster undercoat should not be left to dry out before applying the next coat, since it develops quite a strong suction, independent of green suction, and accordingly requires a water floating or setting coat.

Perlite is an amorphous volcanic glass which has a relatively high water content and occurs naturally. It has one of the most unusual property, *i.e.,* it greatly expands when sufficiently heated.

10.5 Properties and Uses of Perlite

10.5.1 Perlite

Perlite becomes soft, as it is a glass, at a temperature of 850–900°C. The trapped water in the structure of perlite escapes and makes the material to expand 7 to 15 times of its original volume. The expanded material (Fig. 10.6) is brilliant white in colour because of the reflectivity of the trapped air bubbles.

Fig. 10.6 Expanded Perlite

The density of raw perlite or the unexpanded perlite is around 1100 kg/m^3 (1.1 g/cm^3). The density of typically expanded perlite is 30–150 kg/m^3

Since perlite has low density and is relatively economical, it is used commercially in following ways:

- Construction and manufacturing fields.
- Lightweight plasters and mortars, insulation, ceiling tiles and filter aids.
- In horticulture, it makes composts more open to air, while having still a good water-retention properties.
- It makes a good medium for hydroponics.
- It is also used in foundries and cryogenic insulations.

10.5.2 Analysis of Perlite

70–75% silicon dioxide: SiO_2, 12–15%, aluminum oxide: Al_2O_3, 3–4%, sodium oxide: Na_2O, 3–5%, potassium oxide: K_2O, 0.5–2% iron oxide: Fe_2O_3, 0.2–0.7%, magnesium oxide: MgO, 0.5–1.5%, calcium oxide: CaO, 3–5%, Rest Combined water.

Fig. 10.7 Ready-mix Gypsum Perlite Plaster (ISO 9001:2000)

10.5.3 Properties of Ready-Mix Gypsum - Perlite Plaster

- Density: 300–400 kg/m^3
- Heat capacity: 0.064–0.070 kcal/mh°C
- Packing: 50 lb kraft bags (or depends on the clients' request) (Fig. 10.7)

Perlite based plaster has endless unique properties accounting for its ready acceptance and widespread use by many leading architects, general contractors and plastering contractors.

10.5.4 Advantages

Lightweight: Perlite plaster is only one third the weight of sanded plaster. Lightness in weight also permits the plasterer to maintain his optimum production for the full working day with much less effort.

Fire Proofing: National bodies have been swift to recognize the advantages of superior fire protection and potential cost reductions afforded by perlite/plaster fireproofing and have modernized codes to take full use of this. Fire tests have resulted in ratings as high as 4 hours for plaster containing perlite.

Insulating: Perlite possesses millions of tiny air particles and in turn permanent insulation cavities that reduce heat transmission losses. Perlite adds a high thermal insulation value to plaster.

Better Workability: Its sharpness has similar workability to that of sand. Plasterers because of its ease of handling and simplicity appreciate it. Crack Resistant: Perlite/plaster has a high degree of resiliency ensuring less cracking.

Base Coat Plaster: Perlite is an ideal aggregate for all general (base coat) plastering. On many jobs it costs no more than ordinary sand plaster. It mixes with both gypsum plaster and cement for the scratch and brown coats.

Acoustics: Perlite acoustic plaster is an efficient, economical and attractive sound absorbent plaster that can be handled by any skilled laborer. It can be washed and redecorated, is of proper hardiness, and can withstand normal wear and tear without dusting and disintegration.

This natural white finish has an attractive no-glare surface texture remaining vermin-proof, with high insulation due to its low thermal conductivity. Combined with the use of INPRO perlite for loose-fill insulation in concrete blocks, true household energy efficiency is possible.

10.5.5 The Filling Compound for Dry Perlite Mixes

The perlite plaster mixtures are used to improve thermo technical, sound-proofing and acoustic properties in structures (walls, partitions) that enclose resident and industrial premises; cellars, as well as brick, concrete, reinforced concrete, expanded-clay concrete and other structures. Hardening of mixtures is occurring in natural conditions.

Table 10.11 Properties of Heat-insulating Perlite-based Mixes

Name of the mix	Content of Gd.200 Perlite in the mix, %	Boundary of Stability when Compressed, MPa	Boundary of Stability when Compressed, MPa	Bulk Weight in Dry Condition, Kg/m
Cement-Perlite	65 80 90	7.5 3.5 2,5	0.32 0.18 0.15	1000 700 550
Gypsum Perlite	70 80 90	2.5 1.5 1.0	0.23 0.20 0.16	800 700 500
Limestone	75 80	1.0 0.5	0.14 0.12	500 400

Perlite dry mixtures are made by mechanical interfusion cement or gypsum astringent with modifying polymeric additions and expanded perlite. It is possible to add mineral stuffs and pigments. Part of expanded perlite in dry mixtures is 1–25% on mass. The density of perlite solutions in the dry state is assumed from 400 to 900 kg/m^3, boundary of stability when compressed – 0,8–4,0 MPa; heat conductivity at (25,5) 0 – 0,08–0,20 Vt/(mk).

10.5.6 Advantages of Perlite Dry Mixtures

Heat Insulation → At construction of walls from blocks or bricks on an ordinary build solution generally created a 'Cold places'. Use of perlite dry mixture fully eliminates appearance of such defects (Table 10.11).

Sound Insulation → Due to natural form of perlite granule they closely adjoin to each other, that allows to decrease the transmission of sound waves through the walls.

Lightness → In dry climate density of 75–80 kg/m^3, that provides lightness of mixture and minimum loading on a construction is used.

Economy → 30 mm of plaster in thick substitute 150 mm of bricks.

10.5.7 Heat-Insulating Mixes, Method of Application

The plaster can be applied on bricks, concrete, slag concrete, metallic meshing, wood. Without any auxiliary works it can be painted or covered with wall paper; it can be used to insulate both heated and unheated premises.

Light perlite-conservation mixes are quite popular in construction business. Dry-mixed with gypsum or cement such mixtures are blended with water directly on the site and used immediately after that. Such light construction mixes are good at filling out spaces in walls, blocks and bricks. They can also be used to trowel seams and cracks.

Most frequently such mixtures are used to build structures of light bricks or foamed concrete, whose properties by their thermo technical parameters are close to those of the mixture. Besides the brickwork made on the basis if these mixtures does not have cold joints.

Perlite cement mixture is used for internal and outward works, perlite gypsum is used only for internal works. The layer of perlite plasters can be from 20 to 50 mm, laying solutions 10–15 mm, putties are 3–5 mm.

10.5.8 Applications of Expanded Perlite in Gypsum Products

Gypsum is extensively used in many countries. Pemixed lightweight plasters essentially consist of gypsum plaster and lightweight aggregate (*i.e.*, expanded perlite) and are characterized by low density, high thermal insulation and sound absorption properties. These plasters are batch mixed with water and applied in one coat to all types of concrete and masonry. With the exception of the necessary additives for good workability, only water needs to be added and the plasters can work without error independently of weather conditions.

In some countries, particularly Britain, the two or three coat method of plastering is still employed. Browning plaster is used as an under coat; it is hemihydrate gypsum plaster with factory-added retarder which is mixed with aggregate (*e.g.*, Sand or expanded perlite) either in the factory or at the building site. The setting time is several hours and is substantially longer than that of single coat plasters. The next day a smooth finishing coat of plaster of Paris and hydrated white lime is applied.

Applications of perlite – gypsum base coat plaster and finish coats should be in accordance with the American National Standard Institute (ANSI) Standard Specification for gypsum plastering. All metal lath surfaces and gypsum lath ceiling attached by resilient clips should be three coat work. Unit masonry and gypsum lath may be either three coat or two coat work. Table 10.12 provides proportions for perlite gypsum mixes.

Gypsum boards of good sound absorbing properties for use as ceiling panels were produced by mixing calcined gypsum ($CaSO_4$, 1/2 H_2O), perlite, glass fibre,

Table 10.12 Recommended Maximum Proportion Perlite per 45.35 kg of Gypsum Plaster

	Two coat work Double-up plastering	Three coat work	
		Scratch coat	**Brown coat**
Gypsum lath	57 litres	57 litres	**57 litres
Masonry*	—	57 litres	**57 litres
Metal lath	85 litres	85 litres	85 litres

*Except monolithic concrete

**Where plastering is 25 mm or more in total thickness the proportion for the second coat may be increased to 85 litres

10% polyvinyl alcohol solution, retarder and water for 10 minutes followed by vibration at 60 Hz for 50 seconds and hardening[7]. The hardened product was dried at 80°C for 4 hours followed by burnishing and coating with an alkyd resin paint and drying. The sound absorption was 13% and 58% before and after burnishing respectively.

Watanable and Mitsuhiko[8] produced lightweight (specific gravity 0.64), non-flammable gypsum plates by mixing B-hemihydrate plaster (65 parts), perlite (20), asbestos (8), glass fibre (4), pulp (3) and 2% solution of starch and water. The slurry was shaped into a plate and followed by pressing and curing The plate had a bending strength of 7.12 N/mm^2.

High-strength, water and weather-resistant lightweight gypsum composites have been manufactured by blending perlite, asbestors, pulp, non-ionic surfactant and water in a high speed mixer for 10 minutes, followed by mixing with 100 parts of calcined gypsum then moulding and curing at 50°C for 10 hours[9]. The composite has a specific gravity of 0.43 and bending strength of 3.0 N/mm^2.

High-strength, lightweight, load bearing insulating and internally coated blocks were produced by homogenizing moist expanding perlite with gypsum[10]. The blocks have compressive strength more than 30 N/mm^2 and density 1.5–8 kg/cm^3.

Glass fibre reinforced gypsum composites with density 570–830 kg/m^3 were prepared using a semi-dry method by mixing b-hemihydrate plaster, expanded perlite and glass fibre together and compacting at pressures of 345–1380 kN/m^2. Although these composites have good insulating and fire resistant properties they are weak (modulus of rupture 0.5–2.5 N/m^2). They are suitable for indoor applications *e.g.*, ceiling, tiles. This process of manufacturing composites requires less capital investment and labour than the traditional spray suction method[11].

Lightweight, high-strength fibre reinforced gypsum composites have been manufactured by mixing granulated slag (30%), gypsum (30%), cement (5%), fine SiO_2 (5%), fibres (20%) and perlite (10%) followed by moulding and curing in steam at 80°C for a period of 12 hours and then curing in humid air at 25°C for 14 days to obtain a building board[12]. The board has density of 1.1 g/cm^3 and bending strength of 17.0 N/mm^2 compared to 1.5 g/cm^3 and 11.5 N/mm^2 for that made without SiO_2 dust and perlite.

Kuper and Kalnajs[13] manufactured fibre reinforced lightweight stiff decorative ceiling tiles which had a density of about 3.25 g/cm^3 by mixing gypsum plaster, methyl cellulose, chopped glass fibre (length 1.2–2.5 cm, perlite and mica). Composites based on high strength gypsum plaster and vermiculite have also been manufactured for use as partitions[14].

1. Lighweight Gypsum-Perlite Plasters

The Effect of Expanded Perlite Aggregate on the Properties of Plain Plaster of Paris

Table 10.13 Properties of Lightweight Plaster Produced Using Litemix Perlite Aggregate

Litemix Content %	Bulk density (kg/m^3)	Compressive strength (N/mm^2)	Thermal conductivity (k Cal/m/h/°C)
5	1254	14.4	—
10	1238	9.3	—
15	1245	9.3	—
20	1120	9.0	—
25	1004	7.8	0.14
30	950	6.6	—
35	837	4.8	0.08
40	789	4.7	0.078

(*a*) Effect of Litemix Aggregate

The effect of litemix perlite aggregate on the properties of plain plaster of Paris are given in Table 10.13. The results indicate that with increase in perlite aggregate content, the bulk density and strength of plasters are reduced. Beyond 30% of perlilte, the fall in the level of strength is more or less the same, while the reduction in density is not appreciable. At 25% perlite addition, the dry set density of the lightweight plaster obtained is less than the 1040 kg/m^3 density specified in IS : 2547–1976 for the metal lathing plaster. Whereas at 35% and 40% perlite addition, the attainment of density is lower than the specified density for the browning plaster. However, the strength is higher than the minimum specified value of 0.93 – 1.0 N/mm^2 for all types of plaster.

Although IS : 2547 (Part II) – 1975 does not specify the thermal conductivity of the premixed lightweight plaster, the thermal conductivity of gypsum plaster containing 25%, 35% and 40% perlite aggregate was determined. The K-value is much lower than the K-value of conventional building materials like bricks, foam concrete, lightweight concrete.

Table 10.14 Properties of Lightweight Plaster Produced Using Fellite Perlite Aggregate

Fellite Content %	Bulk density (kg/m^3)	Compressive strength (N/mm^2)	Thermal conductivity (k Cal/m/h/°C)
5	1189	10.38	—
10	1031	7.85	0.12
15	874	5.12	—
20	777	4.90	0.09
25	685	3.30	—

(*b*) Effect of Fellite Aggregate

The effect of fellite perlite aggregate on the properties of plain gypsum plaster are given in Table 10.14. It can be seen that similar densities, strength results and K-value can be obtained using 10% and 20% fellite perlite aggregate contents as with 25%, 35% and 40% of litemix perlite aggregate for metal lathing and browning plasters. These lightweight premixed gypsum-perlite plasters are recommended for internal uses only.

Lightweight glass fibre reinforced gypsum binder composites

Investigations have been conducted using both litemix and fellite expanded perlight aggregates as filler in the water-resistant gypsum binder, to impart low density and thermal insulation to the composites for structural purposes the thermal conductivity of glass reinforced gypsum binder composites containing 10 to 15% of perlite has been found to be 0.086 to 0.09 k cal/m/h/°C[15].

10.6 Agricultural Uses of Phosphogypsum and Gypsum

10.6.1 Phosphogypsum

We know phosphogypsum is a by-product of the phosphate fertilizer industry and emanates from the production of phosphoric acid from rock phosphate. Production of phosphogypsum in India is estimated to be 6.0 million annually[16]. The composition of phosphogypsum varies depending upon the source of rock phosphate and the process for manufacturing phosphoric acid[17]. The approximate composition of phosphogypsum is shown in Table 10.15.

Phosphogypsum material normally has an aqueous pH between 4.5 and 5.0. One problem with using phosphogypsum in agriculture is that it contains radioactive radium and radon. In the late 1980's, the agricultural use of phosphogypsum

Table 10.15 Composition of Phosphogypsum

Major constituents (g kg^{-1})*	
Ca	200–240
p	1–5
S	150–190
F	5–3*
Minor constituents (mg kg^{-1})*	
K	100–800
Mg	8–400
Mo	65
Cd	0.23
Radioactive elements	
^{226}Ra	10–25 pCi g^{-1}

*Values are averages of those presented in Pavan *et al.* (1987), Alva and Sumner (1989), Alva *et al.* (1990), and Sumner (1990).

was suspended by the U.S. Environmental Protection Agency when the agency reduced the level of allowable radioactive radium and associated radon in phosphogypsum by a factor of five. This restriction put some of the phosphogypsum into the non-allowable category and therefore made phospho gypsum illegal for agricultural use[18]. Since then the U.S. Environmental Protection Agency (Federal Register 6/3/92) has permitted the controlled use of phosphogypsum in agriculture if radium-226 levels are <10 pCi g^{-I}. This restriction on the maximum radium radioactivity essentially eliminates the use of southern Florida phosphogypsum because its radium-226 levels are commonly in the range of 15 to 25 pCi g^{-1}. The restriction does not impact phosphogypsum from northern Florida or North Carolina, which generally have lower levels of radium-226.

The fate of radium-226 in Florida phosphogypsum was investigated by Mays and Mortvedt. They applied phosphogypsum containing 25 pCi g^{-1} ^{226}Ra at rates up to 112 Mg ha^{-I} to the surface of a silt loam soil and grew successive crops of cornZea mays L.), wheat (Triticum aestivum L.), and soybean (Glycine max L.). Application of phosphogypsum even at the 112 Mg ha^{-1} rate had no effect on the radioactivity levels in grain of corn, wheat, or soybeans. The 112 Mg ha^{-I} rate was more than 200 times the normal rate of gypsum used for peanut fertilization. Additionally, they noted no increase in grain Cd levels, but at the highest rate they found that corn growth slowed. They speculated that the slower growth was due to an imbalance of Ca and Mg.

Numerous studies have shown that phosphogypsum can alleviate some detrimental effects of subsoil acidity on plant growth when surface applied or subsoiled[19–21]. It was concluded that there was essentially no difference between mined gypsum and phosphogypsum regarding correction of subsoil acidity problems.

Huges[22] used a mesh bag technique to examine the effect of phosphogypsum versus lime on alleviating poor root growth in a Spodosol B_h horizon. Soil in mesh bags was amended with either lime or phosphogypsum, and the bags were implanted around mature orange trees and sampled for periods up to 139 days. The B_h horizon amended with lime had significantly higher root densities than control soils, but root densities in phosphogypsum-amended soil were not significantly different than those of controls.

Pavan *et al.*[23] compared the effect of applications of phosphogypsum, lime, calcium chloride, or magnesite (a magnesium-lime material) on apple trees (Malus domestica Borkh.) growing in Brazilian soils. Phosphogypsum and lime significantly increased rooting density in the surface of a high-aluminum soil, but this effect extended to a depth of 60 cm with the phosphogypsum application. Phosphogypsum or lime application significantly increased fruit size and yield compared to other treatments, reflecting the enhanced rooting and increased water supply to the trees.

2. Gypsum

Gypsum ($CaSO_4.2H_2O$) occurs geologically as an evaporate mineral associated with sedimentary deposits. The most important property of gypsum relating

Table 10.16 Benefits of Gypsum Application

Physical benefits	Chemical benefits
Increased infiltration	Increased subsoil Ca
Increased aggregation	Decreased subsoil acidity
Decreased Na adsorption	Reduced exchangeable Al
Reduced root impedance	Reduced restriction of hardpans

to agricultural applications is its solubility. Although gypsum is only slightly soluble in aqueous solution (solubility of 2.5 g L^{-1} in water), it is more soluble than calcite ($CaCO_3$, solubility of 0.15 mg L^{-1} in water). The benefits of gypsum on soil chemical and physical properties are as follows (Table 10.16).

Reviews on the use of gypsum in agriculture have been published. However, these reviews mostly discuss the effects on agronomic crops rather than the effects on soil properties. The ameliorative effect of increased surface infiltration from surface-applied gypsum on dispersive and sodic soils is well documented. Applied gypsum decreases the percentage of Na adsorbed on the soil and increases the free electrolyte concentration; these two effects lead to reduced dispersion and increased flocculation and aggregation of soils. In high-sodium soils with a pH between 8.5 and 10, applied gypsum raises the soluble Ca concentration to levels greater than that of calcite, thereby precipitating calcite. In turn, pH is reduced to 7.5 to 8.0, and calcite and gypsum coexist. The higher soluble Ca concentrations lead to enhanced flocculation of soil colloids.

The effect of surface-applied gypsum (104 kg ha^{-1}) on subsoil mechanical impedance was studied by measuring changes in cone penetrometer index for 2.5 year after application. A significant reduction in mechanical impedance and increase in root penetration was noted to a depth of 0.55 m within this relatively short time frame. The marked improvement in root penetration resulting from the gypsum appeared to be more directly related to increased Ca supplied by gypsum, which is known to be essential for rapid meristematic root growth. Greater root growth means that more organic matter is being produced in the soil, and this organic matter aids in aggregation and promotes the invasion of beneficial mesofauna such as earthworms. Earthworm burrows facilitate movements of water, oxygen, and carbon dioxide essential to crop growth.

Gypsum has received considerable attention because of its ability to ameliorate subsoil acidity and therefore improve plant rooting. The primary problem associated with subsoil acidity is the high level of phytotoxicity from soluble Al and, to some extent, from soluble Mn. In some cases these high levels of Al and Mn are related to deficiencies of Ca. Gypsum additions can lead to both negative and positive plant responses (Alva *et al.,* 1990 Ref. 18), indicating that the chemistry of gypsum in the soil system is not yet completely understood.

Gypsum provides both Ca and S for crop nutrition and has long been used as a Ca source for peanuts (Arachis hypogaea L.). Peanuts have a unique Ca requirement during pod development depending on peanut type soil Ca status and type and form of applied gypsum. Repeated annual applications of gypsum to peanuts, however, can cause a P deficiency since build up of excess Ca in soil may cause the P to be "tied up" in the form of calcium phosphate.

Gypsum increased Ca levels in cauliflower (Brassica oleracea botrytis L.) but had no effect on reducing tip burn, a physiological disorder commonly associated with Ca deficiency. The importance of gypsum is more soluble than calcite in brusselssprouts (Brassica oleracea gemmlfera) grown in a low-calcium soil. Gypsum raised tissue Ca levels and marketable yields significantly during the first growing season after application whereas the effect of calcite took longer.

Gypsum was added to blueberry (Vaccinium sp.) to study the effect of adding Ca on upland mineral soils without significantly affecting soil pH. Although the blueberry is considered to be acid loving, it showed at least a short-term tolerance for increased soil Ca from gypsum. The practicality of using gypsum to enhance root tolerance to high levels of Al in acid upland soils is under further study[24–25].

Continuous applications of gypsum or high rates of surface application can cause problems. One of the problems is excessive Ca buildup, which can induce P deficiency and cause excessive leaching of Mg and K from the surface, particularly in sandy soils. Koreak[26] applied a high-gypsum by-product between the rows in an orchard for 6 year and found that foliar Mg levels were becoming deficient. Deficiencies of P, Mg, or K may cause various plant symptoms, including reduced yield. Another problem with continuous or high application rates is increased soil salt content, which can also damage plants and stunt their growth. Sometimes the so-called disadvantages of a material, however, can be used to the grower's advantage, depending on the crop and the nutrient level of the soil.

The best way to avoid damaging a crop from gypsum applications is to develop standardized soil analyses that will allow for the determination of safe application rates. Soil test that is based on the soil's ability to absorb salt has been proposed. The test is based on the fact that soils showing a favourable response to gypsum are the ones capable of absorbing the most salt. The test, however, still needs to be calibrated standardized for a wide range of soil types.

Use of Gypsum in Wine

The use of gypsum was known to the ancients, and its addition to wine has the sanction of ages. The use of gypsum is worthy of consideration, more especially as the fault of the American wine seems to be the presence of too much acid.

The effect of gypsum in wine is well known. The solid residue of evaporation was burned and the ashes analyzed, as the most simple mode of determining the effects (Table10.17) produced by the gypsum. The analysis settled the most important point that the wine heated with gypsum contained no new ingredient, and that the gypsum added may be considered nil, because it is entirely changed into sulphate of potash.

Table 10.17 Salts Present in Wine

			Weight of Ashes (g)
Natural wine			9.048
Wine with pure gypsum			2.740
Salts present	Ashes of Wine	Natural Wine	Wine with Pure Gypsum
Soluble:	K_2SO_4	0.260	1.240
	K_2CO_3	1.092	0.40
	$K_2(PO_4)_2$	0.064	0.015
Insoluble:	$Ca_3(PO_4)_2$	0.376	0.980
	Al_2O_3	0.064	0.064
	CaO	0.064	0.064
	MgO	0.044	0.084
	$SiO_2 + Fe_2O_3$	0.080	0.080

If gypsum is used in moderate quantity, there could be no injury from it; but it is often used in the most censurable quantity. It is added in the proportion of two pints and a half to twenty-two gallons of wine. It is acknowledged that the effect of the plaster is to preserve the wine from acidity, and to increase the intensity of its colour.

According to Barry, *et al.*[27] the only beneficial effect of the gypsum is to preserve from acidity the-wines of the south, which are sweet and like saccharine and liable to degeneration. Wines, according to Barry, saturated with gypsum lose none of their good qualities, and may attain a great age, as is evidenced by the wines of Rossini and Spain.

It has been expressed that during fermentation process of the wine, the use of gypsum makes the red color stronger. It transforms the salts of potash in the wine into insoluble salts of lime and soluble sulphate of potash. This change may be of great importance, because many chemists attribute to the tartar or supersaturate of potash the property of holding the ferment in solution, while the sulphate or potash does not possess this power. Finally, the gypsum very often contains a certain portion of carbonate of lime, and this carbonate, in neutralizing the acid of the tartar (tartaric acid), assists, no doubt, to cause the deposition of the ferment which this salt held in solution.

Gypsum[28] is used as a water treatment in full grain brewing, where the calcium helps to lower the pH level (increase acidity) of the mash, coagulate proteins (reduce haze and promote clarity in the finished beer) and enhance the yeast action, whilst the sulphate contributes to the bitterness and can impart a sharp, dry flavour to the brew.

Many brewing recipes advocate the use of Gypsum when brewing Light Ale, Bitter, Pale Ale, Strong Ale or Barley Wine, usually at the rate of 1 teaspoon per 5 gallons. It can also be used as an aid to sherry fermentation where 1oz is added per gallon at the start of fermentation.

References

1. Source: United States Geological Survey Mineral Resources Programme (Google), 2011.
2. Web Site: http://www.StrategyR.com/ (Google). The Future of Gypsum - market forecasts to 2014.
3. Hayes D., Overview of China's Plasterboard Market, Global Gypsum Magazine, December 2004-January 2005, pp. 8–12.
4. Ma Caffrey, R, The Global Gypsum Industry-An Overview, Global Gypsum Magazine February 2004, pp. 9–10.
5. Low M.P. No. man, The Thermal Insulating Properties of Vermiculite, Jr. Building Physics, Oct. 1984, Vol. 8, Bi. 2, pp. 107–115.
6. Vermiculite, Sagepubs.com/Content/8/2/107 (Google).
7. Azuma Tomisa Buro, Ichimaru, Sept. 1975, Japan Kokai Tokyo Koho, Vol. 75, 1190, 18, 18, p. 4.
8. Watanabe Yatuka and Sito Mitsuhiko, Gypsum Based No. imflammable lightweight plate, 1976, Japan Kokai Tokyo Koho, 760, 42134, 16 Jan., p. 3.
9. Watanabe Yatuka and Sito Mitsuhiko, 1976, Japan Kokai 7669, 518, 16 Jan. 1976, p. 6.
10. Presztegi D *et al.* High Strength Building Blocks, Chemi. Abst., 1979, Vol. 91, 95776u.
11. Evans T.J., Mahumdar A.J., and Ryder J.F., A Semi Dry Method for the Production of Lightweight Glass Reinforced Gypsum, Intl. J. Chem. Compos. Lightweight Concrete, 1981, Vol. 3, No. 1, pp. 41–44.
12. Murase Yashumi, Oka Yoshinobu and Fukduda Yoshihasu, Lighweight Gypsum Panels, Chem. Abst., 1980, Vol. 92, 134173s.
13. Kupur N.M. and Kalnajs J.I., DecorativeTiles, Chem. Abst., 1987, Vol. 107, 160468q.
14. Magume Naomitsu, Building Materials from Vermiculite and Gypsum, Chem. Abst., 1980, Vol. 92, 98596 w.
15. Singh Manjit and Garg Mridul, Perlight based Building Materials- A Review of Current Applications, Constr. Build. Mater., June 1991, Vol. 5, No. 2, pp. 75–81.
16. Singh Manjit, Trends in Global Gypsum Industry, Indian Concrete Journal, Vol. 35, No. 10, pp. 49–60, 2011.
17. Kurt E Bhum, Straub Gay, Methods of Producing Gypsum Decorative Mouldings US Patents No. 5076978, 31 Dec. 1991 (US Patent Application - 07/318/965).
18. Alva, A.K., Gascho G.J., and Guang Y, 1989. Gypsum Material Effects on Peanut and Soil Calcium, Communications in Soild Science and Plant Analysis, 1989, Vol. 20, pp. 1727–1744.
19. Radeliffe, D.E., Clark R.L., and Sumner M.E., 1986. Effect of Gypsum and a Deep-Rooting Perennial on Subsoil Mechanical Impedance, Soil Science Society of America Journal, 1986, Vol. 50, pp. 1566–1570.
20. Alva, AK., and M.E. Sumner, Amelioration of Acit Soil Infertility by Phosphogypsum. Plant and Soil, 1990, Vol. 128, pp. 127–129.
21. Alva, AK., Gascho G.J., and Guang Y., 1991. Soil Solution and Extractable Calcium in Gypsum-amended Coastal Plain Soils Used for Peanut Culture, Communications in Soil Science and Plant Analysis, 1999, Vol. 22, pp. 99–116.

22. Hughes, J.C. 1988. The Disposal of Leather Tannery Wastes by Land Treatment: A Review. Soil. Use and Management, 1988, Vol. 4, pp. 107–111.
23. Pavan, M.A., F Bingham. T., and Peryea F.J., 1987. Influence of Calcium and Magnesium Salts on Acit Soil Chemistry and Calcium Nutrition of Apple. Soil Science Society of America Journal, 1987, Vol. 51. 4, pp. 1526–1530.
24. U.S. Department of Agriculture, 1954. Diagnosis and Improvement of Saline and Alkali Soils. U.S. Department of Agriculture, Agricultural Handbook, 1954, p. 0.
25. Lindsay, W.L., Chemical Equilibria in Soils. John Wiley Sons, New York. 1979.
26. Koreak, R.F. Short-term Response of Blueberry to Elevated Soil Calcium. Journal of Small Fruit Viticulture, 1992, Vol. 1, pp. 9–21.
27. Barry J.P., Downing A.J., Smith J. Jay, Peter B. Mead, Woodward F. W., Henry Williams. The Horticulturist and Journal of Rural Art and Rural Taste, Edited By P. Barry, Author of the "Fruit Garden". 1920.
28. www.colchesterhomebrew.co.uk/colchest

11

Standardisation of Gypsum and Gypsum Products

INTRODUCTION

Gypsum is one of the important mineral commodity in the modern world. It is a major rock forming sedimentary mineral that produces massive beds, usually from precipitation out of highly saline water. Gypsum contains several other minerals, trapped air and water.

In addition to mineral gypsum, sea water and phosphoric acid fertilizer plant form cardinal source of byproduct gypsum. Chemical plants produce hydrofluoric acid and refining borax are the other source of by product gypsum, but they are generated to little extent compared to by product phosphogypsum. Marine gypsum is recovered from salt pans during processing of common salt in coastal regions. The recovery of byproduct phosphogypsum, fluorogypsum and marine gypsum together is substantial and comparable with the production of mineral gypsum.

The production of mineral gypsum is approximately 4 million tonnes per annum in India. Rajasthan is the leading producer contributing above 90% to the Indian production of gypsum. The remaining gypsum is shared jointly by Jammu and Kashmir, Tamil Nadu and Gujarat.

Gypsum is useful as an industrial material as it readily looses its water of crystallization and forms calcined material called Plaster of Paris or hemihydrate. This gypsum plaster even totally dehydrated calcined gypsum as anhydrite, which on treatment with water, reverts back to the original dihydrate that is, set and hardned gypsum material. These two changes of dehydration and rehydration form the backbone of the gypsum technology.

Gypsum production is classified into four grades based on the calcium sulphate ($CaSO_4.2H_2O$) content.

1. Above 90%. 2. 85% to 90% 3. 80% to 85%. 4. Less than 80%.

High grade gypsum is mined in Bikaner and Jaisalmer district of Rajasthan. Rajasthan gypsum finds outlet in the cement plants in northern India, covering Rajasthan, Gujarat, Madhya Pradesh, West Bengal, Uttar Pradesh, Bihar, etc. Besides a substantial quantity of Rajasthan gypsum containing 60 to 70% $CaSO_4.2H_2O$ is supplied to Punjab, Uttar Pradesh, Haryana, Dehli etc. for reclaiming alkaline soil. A sizeable quantity of gypsum mined in Barmer, Bikaner, Shri Ganganagar, Jaisalmer, Utterlai and Nagaur district of Rajasthan and Tehri Gharwal district of Uttrakhand is supplied to the plaster of Paris unit in Rajasthan, Uttrakhand, Mumbai, Kolkata and New Delhi. Gypsum produced in Tamil Nadu and Gujarat is mainly of cement grade and hence dispatched to the cement plants in southern India.

11.1 Types of Gypsum

1. Mineral Gypsum
2. By product Gypsum
   - Chemical Gypsum
   - Marine Gypsum

11.1.1 Mineral Gypsum

Natural occurring gypsum is known as mineral gypsum. Natural gypsum is a common mineral in sedimentary environment. Gypsum has several variety named that are widely used in the mineral trade.

- Selenite
- Satin Spar
- Alabaster
- Gypsite
- Desert Rose

Selenite

is the colourless and transparent variety that shows a pearl like luster and has been described as having a moon like glow. The word selenite comes from the Greek for moon and means moon rock. Crystal of gypsum can be extremely large (up to the size of 11 meter long), among the largest on the entire planet in the form of selenite.

Satin Spar

Satin spar is a silky and fibrous aggregate. This variety has a satin like look that gives a play of light up and down the fibrous crystal. Finely it may also be quite compact or granular or transparent to opaque.

Alabaster

A very fine grained white or light tinted massive variety of gypsum is called alabaster. Which is prized for ornamental work of various sorts. It is used in fine carving for centuries.

Gypsite

It is porous gypsum of inferior quality generally mixed with clay and sand, dirt or some time calcite.

Desert Rose

In arid areas, gypsum can occur in a flower like form typically opaque with embedded sand grains called desert rose.

11.1.2 By-product Gypsum

It is the sulphate rich by-product of various industries. It includes chemical gypsum and marine gypsum. Chemical gypsum is by product of phosphoric acid industry, hydrofluoric acid industry, borax refining plants, dye industry, etc. Marine gypsum is obtained as a by-product during the production of common salt by solar evaporation.

Phosphogypsum

Phosphogypsum is obtained as a by-product during the manufacture of phosphoric acid by wet process from the phosphate rock. In wet process finely ground phosphate rock is dissolved in phosphoric acid to form monocalcium sulphate slurry. Sulphuric acid is added to the slurry to produce phosphoric acid (H_3PO_4) and a phosphogypsum (hydrated calcium sulphate) by-product).

$$Ca_5(PO_4)_3F + 5\ H_2SO_4 + 10\ H_2O \longrightarrow 3\ H_3PO_4 + 5\ CaSO_4.2\ H_2O + HF$$

As a general rule 4.5 to 5.5 tonnes of phosphogypsum are generated for every tonnes of phosphoric acid produced. Phosphate production generates a very large volume of phosphogypsum which is stockpiled near the plant or trucked away to rivers, ponds or sea.

Phosphogypsum is radioactive due to the presence of naturally occurring uranium and radium in the phosphate ore. The matter has been discussed in depth in Chapter 9.

The main impurities present in phosphogypsum (already discussed in detail in Chapter-III) are P_2O_5, fluorine (F), organic matter and alkalies. The P_2O_5 and F are impurities present in there different forms:

(*i*) On the surface of gypsum crystal as water soluble compound [H_3PO_4, $Ca(H_2PO_4)_2$, NaF).
(*ii*) Substituted in the lattice of gypsum crystal ($CaHPO_4.2H_2O$, Na_2SiF_6)
(*iii*) As insoluble compounds [$Ca_3(PO_4)_2$, CaF_2].

The purity of phosphogypsum ranges from 75.0 to 98% $CaSO_4.2H_2O$. It has about 0.5 to 1.5% total P_2O_5 content and 0.4 to 2.0% of F. The fluorine content of phosphogypsum cause land and water pollution. Phosphogypsum poses a serious problem of disposal. It is disposed of by dumping into river or sea or low lying area in the slurry form. Leach liquors from these deposit are acidic in nature and contain fluoride, P_2O_5, calcium sulphate and organic matter and find their excess into ground water and make it unsuitable for animal and human consumption and for agriculture purpose. Phosphogypsum is used in building industry, in reclamation of alkaline soil etc.

Fluorogypsum

Fluorogypsum is the by-product of hydrofluoric acid and aluminium fluoride base industries. It is generated during the production of aluminium fluoride and hydrofluoric acid from fluorospar and sulphuric acid. Fluorogypsum is obtained from 5 plants in India *viz.*; Navin Fluorine Industries Bhestan, Surat District, Gujarat; Tanfac Industries Ltd. (Formerly Tamil Nadu Fluorine and Allied Chemical Ltd.) Cuddalore, South Arot District, Tamil Nadu, Aegis Chemical Ltd., Dombivali, Thane, Maharashtra, M/s Everest Refrigerants, Mumbai, Gujarat Fluor alkali Ltd., Kalol, Gujarat recover fluorogypsum in their chemical plants. Fluorogypsum on analysis shows $CaSO_4.2H_2O$ 95–98%, CaF_2 1.5–3.0%, H_2SO_4 0.5–1%, P_2O_5 0.1%, and SiO_2 0.2%. The pH of fluorogypsum is about 3 to 5. The main impurities present in fluorogypsum are free acid and CaF_2.

Borogypsum

By product borogypsum which is obtained at a plant which refines calcium borate (colemanite and ulexite) to produce borax and boric acid. Borogypsum contains $CaSO_4.2H_2O$: 78.3%, $MgSO_4$: 4.8%, H_3BO_3: 4.5% and SiO_2: 9.3%. Borax Morarji Ltd, Ambernath, Thane District, Maharastra and Southern Borax Ltd, Chennai engaged in refining of borates, were reported production of by product borogypsum in the past.

Marine gypsum

Marine gypsum is obtained as a by-product during the production of common salt by solar evaporation in coastal regions, particularly in Gujarat and Tamil Nadu. The production of marine gypsum as reported by salt commissioner Jaipur was 165,694 tonnes in 2004 and 118,600 tonnes in 2005. This production was reported from Tamil Nadu and Gujarat.

It is reported that the marine gypsum recovered from Gujarat, Tamil Nadu, Karnataka contains 85.5 to 93.0 % $CaSO_4.2H_2O$, 0.48 to 2.08% NaCl, 0.57% $MgCl_2$ and KCl, 3.42% $MgSO_4$ and $MgCl_2$ 3.48 to 7.65% insoluble. As per the IS specification (IS : 12679) marine gypsum should contain 85% $CaSO_4.2H_2O$ and 0.1 NaCl.

Specification of By-product Gypsum for Use in Plaster, Block and Boards

Table 11.1 gives the specification of by-product gypsum for use in plaster, blocks and boards, as per IS : 12679–1989.

Table 11.1 Requirements of Byproduct Gypsum for Use in Plaster, Block and Boards

Sl. No.	Characteristic	Requirement		
		Phosphogypsum	Fluorogypsum	Marine gypsum
1.	P_2O_5, % by mass, max	0.40	—	—
2.	F, % by mass, max	0.40	1.0	—
3.	Na_2O, % by mass, max	0.10	—	—
4.	K_2O,% by mass, max	0.20	—	—
5.	Organic matter, % by mass, max	0.15	—	—
6.	$CaSO_4.2H_2O$, % by mass, min	85.0	90.0*	85.0
7.	Cl as NaCl, % by mass, max	0.10	—	0.10
8.	pH of 10% aqueous suspension of gypsum, min	5.0	5.0	6.0

* Fluorogypsum shall be in anhydrous form as $CaSO_4$

11.2 Aims of Study

In the foregoing matter the gypsum has been discussed in brief to have material and product knowledge of this important building material. However, the gypsum products are being produced in many forms and these materials are standardized to some extent. Different countries have different standard specifications depending upon the type of product and their applications. No standard specification or code of practice claimed of any country is complete in all respect. Indian Standard Specifications formatted by Bureau of Indian Standards (BIS) are on gypsum, gypsum plasters, concrete and products, etc.

Table 11.2 depicts different Indian Standards now available with BIS (Bureau of Indian standards). The Table 11.3 gives detailed standards on ASTM Specifications on Gypsum, Gypsum Plaster, Concrete and Products.

11.2.1 Standards Specifications for Gypsum

In gypsum, calcium or magnesium carbonates, chlorides, other sulphate minerals, clay minerals or silica are considered as deleterious constituents. As a result most mine production of gypsum will have the purity ranging between 70 and 95%. Often it is used as mined, although in certain cases, one or more methods of mineral beneficiation are employed to upgrade the product. The various Indian Standards specifications and International Standards specifications for gypsum in different industries are described as follows:

11.2.2 Indian Standards Specifications for Gypsum for Various Uses

Indian Standards Specifications for Mineral Gypsum for Use in Surgical Plaster, Ammonium Sulphate, Pottery and Cement Industries (IS : 1290:1973).

Table 11.2 Indian Standards (IS) on Gypsum, Gypsum Plaster, Concrete and Products

IS. No.	Title
2469	Glossary of terms relating to gypsum (first revision)
2542	Specification for methods of test for gypsum plaster, concrete and products
(Part 1/section 1 to 12) 1978 (Reaffirmed)	Methods of test for gypsum plaster, concrete and products
(Part 2/section 1 to 8) 1978 (Reaffirmed)	Methods of test for gypsum plaster, concrete and products
2547	Specification for gypsum building plaster
(Part 1) 1978 (Reaffirmed)	Part 1 Excluding premixed light weight plaster (first revision)
(Part 2) 1976 (Reaffirmed)	Part 2 Premixed light weight plaster (first revision)
2095	Specifications for gypsum plaster boards
(Part 1) 1978	Plain gypsum plaster boards (Third revision)
(Part 2) 2001	Coated/laminated gypsum plaster boards (second revision)
(Part 2) 1996	Reinforced gypsum plaster boards (second revision)
1288 – 1988 (Reaffirmed)	Methods of test for mineral gypsum
1289 – 1960 (Reaffirmed)	Methods of sampling mineral gypsum
1290 (Reaffirmed)	Specifications for mineral gypsum
2333 – 1992	Specifications for Plaster of Paris for ceramic industry
2849 – 1983 (Reaffirmed)	Specification for non-load beaing gypsum partitions blocks-code of practice
3630–1992	Construction of non-load bearing gypsum partitions blocks - code of practice
4738–1993 (Reaffirmed)	Bandage, Plaster of Paris-specifications
6046–1982	Specification for gypsum for agriculture uses
6237–1971(Reaffirmed)	Specification for handloom cotton cloth for Plaster of Paris bandages and cut bandages
6555–1972 (Reaffirmed)	Specification for dental laboratory plaster
6556–1972 (Reaffirmed)	Specification for dental impression plaster
8272–1994 (Reaffirmed)	Specification for gypsum plaster for use in the manufacture of fibrous plaster boards
9498–1980 (Reaffirmed)	Specification for inorganic aggregates for use in gypsum plaster
10170–1982 (Reaffirmed)	Specification for by - product gypsum
12654–1989 (Reaffirmed)	Low grade gypsum - use in building industry - code of practice
12679- 1989 (Reaffirmed)	By-product gypsum for use in plaster, blocks and boards - specification
13001–1991 (Reaffirmed)	Guidelines for manufacture of gypsum plaster in mechanized pan system

Table 11.3 ASTM Specifications on Gypsum, Gypsum Plaster, Concrete and Products

ASTM No.	Title
C 960 /C 960M M 04	Predecorated gypsum board
C 1264 – 99(2004)	Sampling, inspection, rejection, certification, making, shipping, handling and storage of gypsum board
C 630 /C 630M – 03	Water-resistant gypsum backing board
Test methods for:	
C 47 M – 01	Chemical analysis of gypsum and gypsum products
C – 472 – 99(2004)	Physical testing of gypsum plasters and gypsum concrete
C 473 – 03	Physical testing of gypsum panel products
Specifications for the application of gypsum and other products in assemblies	
Specifications for:	
C 840 – 04	Application and finishing of gypsum board
C 844 – 99	Application of gypsum base to receive gypsum veneer plaster
C 1280 – 04	Application of gypsum sheathing
C 843 – 99	Application of gypsum veneer plaster
C 842 – 99	Application of interior gypsum plaster
C 926 – 98a	Application of Portland cement based gypsum plaster
C 926 – 98a	Installation of cast-in-place reinforced gypsum concrete
C 956 – 04	Installation of cast-in-place reinforced gypsum concrete
C 841 – 03	Installation of interior lathing and furring
C 1063 – 03	Installation of lathing and furring to receive interior and exterior Portland cement based plaster
C 1007 – 04	Installation of load bearing (Transverse and axial) steel studs and related accessories
C 1467 / C 1467N – 00	Installation of molded glass fibre reinforced gypsum parts
C 754 – 00	Installation of steel framing members to receive screw – attached gypsum panel products
Guides for:	
C 1546 – 02	Installation of gypsum products in concealed radiant ceiling heating system
Terminology and editorial	
Technology for:	
C 11 – 03d	Gypsum and related building materials and system
Gypsum and related building materials and system	
Application of exterior insulating and finish systems and related products	

ASTM No.	Title
Practices for:	
C 1397 – 03	Application of class PB exterior insulation and finish systems
C 1516 – 02	Application of direct applied exterior finish systems
C 1535 -04	Application of exterior insulation and finish systems class PI
Specifications and test methods for accessories and related products	
Specifications for:	
C 1047 – 99(2004)	Accessories for gypsum wallboard and gypsum veneer base
C 897 – 00	Aggregate for job - mixed Portland cement based plasters
C 631 – 95a (2000)	Bonding compounds for use in gypsum plaster
C 35 – 01	Inorganic aggregates for use in gypsum plaster
C 475 C / 475M – 02	Joint compound and joint tape for finishing gypsum board
C 1397 – 03	Load bearing (transverse and axial) steel studs, runners (tracks), bracing or bridging for screw applications of gypsum panel products and metal plaster basis
C 847 – 95(2000)	Metal lath
C 514 – 01	Nails for the application of gypsum boards
C 645 – 04	Non-structural steel framing members
C 954 – 00	Steel drill screws for the application of gypsum panel products or metal plaster bases to steel studs from 0.033 in.(0.84 mm) to 0.112 in. (2.84 mm) in thickness
C 1002 –01	Steel self piercing tapping screws for the application of gypsum panel products or metal plaster bases to wood studs or steel studs
C 1513 – 01	Steel tapping screws for cold formed steel framing connections
C 932 –03	Surface applied bonding compounds for exterior plastering
C 933 – 04	Welded wire lath
C 1032 –04	Woven wire plaster base
Test methods for:	
C 474 - 02	Joint treatment materials for gypsum board construction
Specifications and test methods for gypsum products	
C 931 / C 93M –04	Exterior gypsum soffit board
C 1278 / C 1278M – 03	Fibre reinforced gypsum panels

ASTM No.	Title
C 1355 / C 1355M – 96(2001)	Glass fibre reinforced gypsum composites
C 1177 / C 1177 – 04	Glass mat gypsum substrate for use as sheathing
C 1178 / C 1178M – 04	Glass mat water resistant gypsum backing panel
C 442 / C 442M – 04	Gypsum backing board, gypsum core board, and gypsum shaft liner board Gypsum base for veneer plasters
588 / C 588M–03	Gypsum base for Veneer plasters
C 1396 / C 1396M – 04	Gypsum board
C 59 / C 59M – 00	Gypsum casting plaster and gypsum molding plaster
C 1395 / C 1395M – 04	Gypsum ceiling board
C 317 / C 317M – 00	Gypsum concrete
C 318 / C 318M – 00	Gypsum form board
C 61 / C 61M – 00	Gypsum Keene's cement
C 37 / C 37M – 01	Gypsum lath
C 28 / C 28M – 00	Gypsum plasters
C 79 / C 79 – 04a	Gypsum sheathing board
C 587 – 02	Gypsum Veneer plaster
C 1597M – 04	Gypsum wallboard (hard metric sizes)
C 36 / C 36M – 03	Gypsum wallboard
C 22 / C 22M – 00	Gypsum
C 1381 – 97 (2002)	Molded glass fibre reinforced gypsum parts

High purity mineral gypsum is used in the manufacture of ammonium sulphate fertilizer. Gypsum of less purity in crushed condition is utilized in Portland cement manufacture. In pottery, gypsum is used for moulding purposes. Calcined gypsum finds use in the manufacture of Plaster of Paris.

Mineral gypsum for use in the above industries has been divided into five grades, *viz*.

Grade 1 : for surgical plaster industry
Grade 2 : for ammonium sulphate
Grade 3 : for pottery industry
Grade 4 : for cement industry
Grade 5 : for soil reclamation

The material should be the natural mineral consisting essentially of hydrated calcium sulphate and free from added impurities.

The IS specifications for mineral gypsum are given in Table 11.4.

Table 11.4 Specification of Mineral Gypsum in Different Industries

Constituents	Surgical plaster	Ammonium sulphate fertilizer	Pottery	Cement	Reclamation of soil
$CaSO_4.2H_2O$	96% (min)	85–90% (min)	85% (min)	70–75% (min)	70% (min)
Free water	1.0%	—	1.0%	—	—
CO_2	1.0% (max)	—	3.0% (max)	—	—
SiO_2 & other insoluble	0.7% (max)	—	6.0% (max)	—	—
$Fe_2O_3+Al_2O_3$	0.1% (max)	1.5% (max)	1.0% (max)	—	—
MgO	0.5% (max)	1.0% (max)	1.5% (max)	3.0% (max)	—
NaCl	0.01% (max)	0.003% (max)	0.1% (max)	0.5% (max)	—
Na_2O	—	—	—	—	0.7% (max)
Fineness	—	—	—	—	Residue on 2 mm sieve-nil and on 0.2 mm sieve-50% (max)

1. Specifications for Gypsum Building Plaster (IS : 1247–1976)

Part I - Excluding Premixed Lightweight Plasters

This Indian Standard (Part I) (First Revision) was adopted by the Indian Standards Institution on 20th February, 1976 after the finalization of draft by Civil Engineering Division Council of Gypsum Building Materials Sectional Committee.

This standard was first published in 1963. It has now been revised in two parts: Part I deals with gypsum plaster excluding premixed lightweight plaster and Part II deals with premixed lightweight plasters. Gypsum plaster has been reclassified according to the latest method of classification and anhydrous plaster which was previously recommended as undercoat plaster and finishing plaster has now been recommended for only finishing purposes. Based on the changes in classification, changes in the requirements of plaster have also been made.

Gypsum building plasters are used extensively in many countries of the world including Australia, Canada, United Kingdom, Korea, United States of America, Middle East, Latvia, Germany, U.K. and Russia, for general building operations and for the manufacture of preformed gypsum building products which have the specific advantages of lightness and high fire resistance.

The various resources for gypsum in this country, of purity 70 per cent or less have great prospects of economic use mainly as building materials, namely, in the form of gypsum plaster, gypsum plaster boards, and gypsum blocks and tiles. This standard on gypsum plaster, which is one in the series, covers the various categories of gypsum plaster used in normal building construction.

Gypsum building plasters may vary widely in their properties partly because manufacturing processes differ and partly because adjustments are able to suit users requirements. Thus, the properties required of plasters for undercoat work differ to some extent from those required for finishing coats, a further variation is sometimes necessary in the latter class in order to control the hardness of finish or surfaces intended for specific purposes.

Keeping these points in view it has been attempted in this standard to classify gypsum plasters on the basis of partially dehydrated gypsum and anhydrous gypsum. This standard (Part I) covers the classification and chemical and physical requirements for gypsum building plasters which possess a definite set due to hydration of calcium sulphate, anhydrous or hemihydrates, to form gypsum and are used in the manufacture of gypsum building products.

Classification

Gypsum plaster shall be classified as follows:

- Plaster of Paris,
- Retarded hemihydrate gypsum plaster

Type I - Under coat

- Browning plaster,
- Metal lathing plaster,

Type II - Final coat plaster:

- Finish plaster,
- Board finish plaster,
- Anhydrous gypsum plasters are for finishing only, and
- Keene's plaster is for finishing only.

Keene's plaster is of the anhydrous type. It is characterized by being more easily brought to a smooth and clean finish associated with gradual set. In this Standard Keene's plaster is differentiated from a anhydrous gypsum plaster by a higher standard of purity ($CaSO_4$ not less than 80 per cent) and hardness. The special qualities traditionally associated with this type of plaster cannot be dealt with at present by any convenient direct test.

Part II - Premixed Lightweight Plasters

This Indian Standard (Part II) (First Revision) was adopted by the Indian Standards Institution on 22nd December, 1976, after the draft finalized by the Gypsum Building Materials Sectional Committee had been approved by the Civil Engineering Division Council.

Gypsum is a well-known building material. It has been extensively used in various countries. Premixed lightweight plasters essentially consists of gypsum plaster and lightweight aggregate which are characterized by low density, high thermal Insulation and sound absorption properties and can be easily used for building purposes.

In the formulation of this standard, due weightage has been given to International Co-ordination among the standards and practices prevailing in different countries in addition to relating it to the practices in the field on this country. This has been met by basing the standard on BS 1191: Part 2: 1973 'Specification for gypsum building plasters. Part 2 Premixed lightweight plasters', published by the British Standards Institution.

This standard (Part II) specifies requirements for premixed lightweight plaster consisting essentially of gypsum plaster and lightweight aggregate used in general building operations. For the purpose of this standard, the following definition shall apply.

Lightweight Plaster – A plaster consisting of suitable lightweight aggregates and retarded hemihydrates gypsum plasters complying with IS:2547 (Part I)–1976. Other additives may be incorporated to impart desired properties.

Classification

Premixed lightweight plaster may be divided into the following types:

Type A: Undercoat plasters:

(*a*) Browning plaster,
(*b*) Metal lathing plaster,
(*c*) Building plaster

Type B: Final coat plaster

2. Methods of Test for Gypsum Plaster, Concrete and Products

IS : 2542 (Part I/See 1 to 12) –1978

A number of Indian Standards on gypsum building materials Specifications, code of practices, etc., have been prepared with a view to assisting the gypsum industry in its development, The standard on Method of test on gypsum building materials has been prepared in two parts as follows:

The (Part I) was first published in 1964 and has now been in light of the experience gained in the use of the standard years and consequent to the revision of IS : 2547 (Part 1)-1976 of IS : 2547 (Part II)1976 a number of changes have been incorporated in this revision and new methods of test, such as determination of setting time by potentiometer, determination of bulk density, and determination of dry set density, have been introduced.

Methods of Test for Gypsum Plaster, Concrete and Products

Part II - Gypsum Products

This Indian Standard (First Revision) was adopted by the Indian Standards Institution on 30th June, 1981, after the draft finalized by the Gypsum Building Materials Sectional Committee, had been approved by the Civil Engineering Division Council.

A number of Indian Standards on building gypsum covering specifications, methods of test and codes of practice are being prepared with a view to assisting the industry.This standard, which is one in the series, covers the methods of test for evaluating the different physical properties of gypsum building materials. For the convenience of the users, this standard covering test methods on gypsum building materials has been prepared in two parts as follows.

This part (Part II) was first published in 1964. The revision was taken up with a view updating the methods of test in line with the current knowledge on the subject. A part from revising the methods of test already covered in the first version of this standard, the revision incorporates methods of test for fibrous gypsum plaster boards covered by IS : 8273–1976. Further, the method of test for determining the non-combustibility of gypsum partition blocks has been deleted since the Sectional Committee decided that the corresponding test method given in IS : 3808–1979 (Methods of test for non-combustibility of building materials) was suitable for determining the non-combustibility of gypsum partition blocks.

3. Indian Standards Specifications for Gypsum for Agricultural Use

Mineral Gypsum (IS : 6046–1982) : The major soil amendment for alkali soils available in the country is done by the mineral gypsum. According to this standard, all the material should pass through a 2 mm sieve but at least 50% of it should pass through a 0.25 mm (60 mesh) sieve. The material should contain not less than 70% calcium sulphate dihydrate ($CaSO_4.2H_2O$) and the sodium (Na) content of the mineral should not be more than 0.75%.

By-product Gypsum (IS : 10170–1982) : By-product gypsum is produced in the country in phosphoric acid plants following wet process technology. This by-product gypsum like mineral gypsum is also a major soil amendment for reclamation of alkali soils. According to this standard the material should pass through 0.25 mm sieve but 50% of it should pass through 0.25 mm (60 mesh) sieve. The material should conform to the following requirements (Table 11.5).

4. Indian Standards Specifications for Gypsum for Paints (IS : 69–1950)

Gypsum is used as an extender in the manufacture of paints. The material should be natural or artificial product consisting essentially of hydrated calcium sulphate ($CaSO_4.2H_2O$). The material should be supplied in the form of dry powder and

Table 11.5 Requirements for By-product Gypsum for Soil Amendment (IS : 10170–1982)

S. No.	Characteristic	Requirement (%)
1.	Calcium sulphate dihydrate ($CaSO_4.2H_2O$) (min.)	70.00
2.	Sodium (Na) (max.)	0.75
3.	Fluorine (F) (max.)	1.00
4.	Free moisture (max.)	15.00

consist almost entirely of the characteristic crystals of gypsum when examined under microscope. It shall contain not more than 0.05% of lead or lead compounds. The residue on sieve shall not be more than 0.5%. It shall contain not more than 0.5% of free water when heated for 2 hours at 45°C. After drying the material at 45°C for 2 hours, the $CaSO_4$ content of the material shall be not less than 75%.

5. Indian Standards Specifications for Gypsum Plaster for Use in the Manufacture of Fibrous plaster Boards (IS : 8272–1984)

This specification applies to calcined gypsum plaster for use in the manufacture of fibrous plaster boards which are used as covering for walls, ceilings and partitions in normally dry environments in buildings. The plaster shall consist essentially of calcium sulphate hemihydrate ($CaSO_4.1/2\ H_2O$) and should contain not less than 42% sulphur trioxide (SO_3). The residue retained on 600 microns IS sieve should not be more than 1% by mass. The compressive strength of the plaster should not be less than 7.6 N/mm^2. The time of initial set of the plaster should be 20–35 minutes.

11.3 Indian Standards Specifications for Plaster of Paris for Ceramic Industries (IS : 2333–1992)

This standard prescribes requirements and test for Plaster of Paris for use in ceramic and optics industries. There shall be four types of material as follows:

Type 1 - Suitable for moulds for slip casting,

Type 2 - Suitable for moulds for jiggering and case and block making,

Type 3 - Suitable for mounting optical glass items.

Type 4 - Suitable for automatic machine jiggering and for roller head.

The material should be in the form of a fine white powder of smooth texture, free from foreign matter and lumps. It should be calcined gypsum and shall correspond essentially to the formula $CaSO_4.1/2H_2O$.

The ceramic plaster is generally, made from the mineral gypsum of different purities. As the phosphogypsum retains higher purity, this material has great potential for use in ceramic plaster which may increase the use of phosphogup-sum mani-fold in near future.

IS : 2333–1992 covers the requirements of ceramic grade plaster. This standard was earlier a combined standard of dental laboratory standard, dental impression plaster and the gypsum building plasters up to 1972. Later, the standard was revised and since 1981 new standard *i.e.*, IS : 2333 was made as the standard for Plaster of Paris for Ceramic Industry.

The physical and chemical requirements for Plaster of Paris are given in Table 11.6.

Table 11.6 Specifications for Plaster of Paris for Use in Ceramics (IS : 2333–1992)

Physical Requirements for Plaster of Paris

Sl. No.	Characteristic	Requirement			
		Type-1	Type-2	Type-3	Type-4
(*i*)	Fineness				
	(*a*) Material retained on 150 micron IS sieve, % by mass (max.)	Nil	Nil	Nil	Nil
	(*b*) Material retained on 75 micron IS sieve, % by mass (max.)	7	Nil	Nil	Nil
(*ii*)	Normal consistency	60 to 80	45 to 60	55 to 65	40 to 55
(*iii*)	Setting time, Min.				
	(a) Initial (b) Final	8 to 15 15 to 30	8 to 15 10 to 30	8 to 15 10 to 15	8 to 15 10 to 15
(*iv*)	Temperature rise during setting, °C (min.)	12	12	12	12
(*v*)	Modulus of rupture, MPa (min.)	4.0	5.0	5.0	7.0
(*vi*)	Dry compressive strength, MPa (min.)	9	15	17	20
(*vii*)	Expansion after setting, %	0.2 to 0.4	0.2 to 0.4	—	—
(*viii*)	Water absorption, % by Mass	25 to 35	20 to 25	15 to 20	12 to 18

Chemical Requirements for Plaster of Paris

Sl. No.	Characteristic	Requirement			
		Type-1	Type-2	Type-3	Type-4
(*i*)	Free moisture (max.)	2.0	2.0	0.5	0.5
(*ii*)	Carbonates ($CaCO_3$), % by mass	3.0	3.0	1.0	1.0
(*iii*)	Matter insoluble in Hydrochloric acid, % by mass (max.)	7.0	7.0	2.0	2.0
(*vi*)	Calcium sulphate as $CaSO_4$, % by mass Min	85.0	85.0	90.0	90.0
(*v*)	Compound water, %	5.8 to 6.4	5.8 to 6.4	5.8 to 6.4	5.8 to 6.4

11.4. Requirements of Gypsum Building Plaster (IS : 2547–1976)

As per the specification the gypsum plaster for building purposes should be one of the following classes.

- Plaster of Paris
- Retarded hemihydrate gypsum plaster
- Anhydrous gypsum plaster
- Keene's or Parian gypsum plaster

Plaster of Paris

Plaster of Paris shall be such that its sulphur trioxide content should not be less than 35%. The calcium oxide content should not be less than two-thirds of the sulphur trioxide content. The soluble sodium and magnesium salts content (Na_2O) and magnesium oxide (MgO) should each be not greater than 0.3%. The loss on ignition should not be greater than 9% or less than 4%. The residue on IS sieve No. 1.18 (90 um) mm should not be greater than 5%. The transverse strength (modulus of rupture) of the set plaster should be not less than 0.5N/mm^2. IS : 2547 (1976) covers the specifications for Gypsum Building Plasters, Part-1: Excluding Premixed lightweight plasters. Set plaster pats should show no signs of disintegration, popping or pitting.

Retarded Hemihydrate Gypsum Plaster

It may be one of the following types (*a*) Undercoat plaster, (*b*) finishing plaster, (*c*) dual purpose plaster (undercoat and finishing). The chemical composition shall be the same as that of Plaster of Paris. The residue on IS sieve No. 1.18 mm shall not be greater than 1%. The transverse strength of the set sanded plaster shall not be less than 0.4 N/mm^2. Set plaster pats should show no signs of disintegration, popping or pitting. The expansion of plaster should be 0.20 at 24 hours.

Anhydrous Gypsum Plaster

It shall be one of the following types: (*a*) Undercoat plaster, (*b*) finishing plaster, and (*c*) dual plaster (undercoat and finishing). In this plaster the sulphur trioxide content should not be less than 40%; the calcium oxide content should not be less than two-thirds of the sulphur trioxide content; the Na_2O and MgO should not be greater than 0.3%; and the loss on ignition should not exceed 3%. The residue on IS sieve No. 1.18 mm should not exceed 2%. The set plaster pats should show no signs of disintegration, popping or pitting.

Keene's or Parian Gypsum Plaster

It is a plaster of anhydrous type but has a higher standard of purity and hardness. It should contain not less than 47% of SO_3 and the percentage of CaO shall be not less than two-thirds of that of SO_3. The loss on ignition is limited to 2%. The material retained on IS sieve No.1.18 mm must not exceed 2%. The setting time of plaster should be 20–360 minutes.

11.5 American Standards Specifications

For Gypsum - ASTM Desgn. C22/C22M-2012 (Reapproved 2010)

Specification for Gypsum

Gypsum shall contain not less than 70.0 weight percent $CaSO_4.2H_2O$ (when tested as per ASTM C471M - Test methods for chemical analysis of gypsum and gypsum products and ASTM C472 - Test methods of physical testing of gypsum plasters and gypsum concrete).

For Gypsum Plasters (ASTM-C28–92)

This specification covers 5 types, namely:

- Gypsum Ready-mixed plaster
- Gypsum neat plaster
- Gypsum wood-fibred plaster
- Gypsum bond plaster
- Gypsum gauging plaster

Gypsum Ready-mixed Plaster

It should contain not more than 4 cu. ft of mineral aggregate (vermiculite, perlite and sand) per 100 lb, of calcined gypsum plaster. Its setting time should not be less than 1.5 hours or more than 8 hours. The compressive strength of plaster should be 3.1–8.3 MPa.

Gypsum Neat Plaster

It is mixed at the mill with other ingredients to control working quality and setting time, may be fibred or unfibred, the necessary aggregate being added at the site. It shall set in not less than two or more than thirty two hours and it shall have a compressive strength of not less than 750 P.S. 1 (pounds per square inch) or 5.2 MPa.

Gypsum Wood-fibred Plaster

It should contain not less than 66% $CaSO_4$ 1/2 H_2O and not less than 0.75% wood fibre made from non-staining wood. It should have a setting time of not less than 1.5 hours or more than 16 hours and a compressive strength of not less than 1200 P.S.i. (8.3 MPa).

Gypsum Bond Plaster

It should consist of not less than 93% calcined gypsum and not less than 2% or more than 5% hydrated lime. It should not set in less than 2 hours or more than 10 hours. The calcined gypsum plaster used in this specification shall have a purity of not less than 60% $CaSO_4.1/2\ H_2O$.

Gypsum Gauging Plaster

It is used as finishing coat and it is prepared by mixing with lime putty. The plaster should contain not less than 66% $CaSO_4.1/2\ H_2O$, and should have a setting time when not retarded, of not less than 20 or more than 40 minutes and when retarded shall set in not less than 40 minutes. It should have a compressive strength of not less than 1200 P.S.i. (8.3MPa). Lime is frequently added to gypsum building plasters, hydraulic limes are unsuitable for use in gypsum building plaster.

For Gypsum Moulding Plaster (ASTM-C59–91)

These specifications cover gypsum moulding plaster, a material consisting essentially of calcined gypsum, for use in making interior embellishments and cornices as gauging plaster, etc. Gypsum moulding plaster shall contain not less than 85% of $CaSO_4.1/2\ H_2O$. All the material shall pass a 30 mesh sieve and 90% to pass a 100 mesh sieve. Its setting time must be not less than 10 minutes or more than 50 minutes, and its compressive strength not less than 1800 P.S.i. (12.4 MPa). The moulding plaster should be ground to completely pass 600 micron sieve and not less than 90% shall pass 150 micron sieve.

Others

The specifications for gypsum for use such as textiles, dental plaster, chemicals and cosmetics are as follows:

Textiles

The gypsum should not contain more than 4% of calcium carbonate and for some purpose the limit may be 1%. The amount of iron present should be low. A sample of gypsum suitable for use in cloth finishing was stated to have the following chemical composition (Table 11.7). It is stated that anhydrite is not suitable for use in cloth finishing owing to its hardness and darker colour.

Table 11.7 Composition of Plaster for Textile Finish

Constituents	Percent
$CaSO_4$	77.42%
$CaSO_3$	1.45%
Fe_2O_3+Al_2O_3	0.12%
SiO_2	0.34%
MgO	trace
NaCl	0.26%
Combined water	20.46%

Dental plaster

Plaster for dental purposes should as a rule, carry not less than 93% of calcined gypsum and it should all pass a 30 mesh sieve and 95% pass 100 mesh. Only the whitest and purest of gypsum can be employed for dental plaster. Indian Standard Specifications - IS : 6555–1998 and IS : 6556–1998 cover Dental Plaster and Dental Impression Plaster respectively.

Chemicals

There is no IS specification. Gypsum containing at least 94% $CaSO_4.2\ H_2O$ and a maximum of 3.6% NaCl is generally used.

Cosmetics

There is no IS specification. Selenite variety of gypsum containing 28.75% CaO, 46.5% SO_3, up to 3% SiO_2, up to 0.25% Fe_2O_3 and up to 18% moisture is generally preferred.

Bandage Plaster

Plaster for bandage purpose should have setting time 8 minutes and breaking strength should not be less than 350 N.

11.6 Specifications of Various Types of Gypsum Consumed in Cement Industry

The Cement industry consumes all the types of gypsum *viz.* mineral, phospho and marine, and generally uses the gypsum with the following specifications (Table 11.8).

As per the Indian Standards specifications the $CaSO_4.2H_2O$ in gypsum for use in cement manufacture should be not less than 70%. The limits prescribed, by the BIS for deleterious constituents like MgO (3% max.) and NaCl (0.5% max.) are fairly comparable with the user industry specifications. Fluorine

Table 11.8 Composition of Gypsum for Cement

Sl. No.	Type of gypsum	Size	$CaSO_4.2\ H_2O$	MgO	NaCl
1.	Mineral	Lumps upto 15 cm (max.)	60 to 90% but by and large Above 70%	3% (max.)	0.5% (max.)
2.	Phospho	Fine powder	75.5 to 97% but by and large above 70%	0.5% (max.)	0.5%(max.)
3.	Marine	Crystalline	60 to 95% but by and large above 85%	3% (max.)	1% (max.)
Moisture content in phosphogypsum varies from 12 to 32%.					

and P_2O_5 contents in by-product gypsum are considered deleterious. More than 0.3% P_2O_5 affects the setting properties of cement while fluorine with more than 0.15% renders ring formation in the kiln. In general the cement industry does not appear to be facing any quality problem with regard to gypsum, since it has been able to consume gypsum containing as low as 60–65% $CaSO_4.2H_2O$

11.7 Global Trends of Gypsum

World over different standards have been adopted by different countries depending upon the type of gypsum, plasters, building products, test methods and practices. All countries have adopted different standards and code of practices according to their requirements and market potential. India has many standard specifications but not for all the products. Indian standards are mostly based on British standards. The information collected and collated on different standard specifications have been put into comparative tables to have first hand knowledge on standards available in various countries. Tables 11.9 and 11.10 show comparative statement of different World Standards on gypsum, products, technologies and practices.

Table 11.9 Comparative Statement of Different World Standards on Gypsum, Products, Technologies and Practices

British Standards	Title	Relevant Indian Standard	Relevant ASTM Standard
Gypsum			
BS 1191–1:1973	Specification for gypsum building plasters. Excluding premixed lightweight plasters.	IS 2547 Part 1 1976	—
BS 1191–2:1973	Specification for gypsum building plasters. Premixed lightweight plasters.	IS 2547 Part 2 1976	—
BS 1230–1:1985	Gypsum plasterboard. Specification for plasterboard excluding materials submitted to secondary operations.	IS 2095: 1996	—
BS 4022:1970	Specification for prefabricated gypsum wallboard panels.	—	C 960
BS 8212:1995	Code of practice for dry lining and partitioning using gypsum plasterboard.	—	—
BS EN 520:2004	Gypsum plasterboards. Definitions, requirements and test methods.	IS 2095:1996	C 1396, C 1395, C 318, C79, C 1597, C 36, C 630

(Cond...)

British Standards	Title	Relevant Indian Standard	Relevant ASTM Standard
BS EN 12859:2001	Gypsum blocks. Definitions, requirements and test methods.	IS 2849:1983 IS 2542 Part1 SEC 1–12 1978	C 471, C 472, C 473
BS EN 12860:2001	Gypsum based adhesives for gypsum blocks. Definitions, requirements and test methods.	—	—
BS EN 13279–1:2005	Gypsum binders and gypsum plasters. Definitions and requirements.	IS 8272:1984	—
BS EN 132779–2:2004	Gypsum binders and gypsum plasters. Test methods.	IS 2542 Part 1 SEC 1–12 1978	C 28
BS EN 13815:2006	Fibrous gypsum plaster casts. Definitions, requirements and test methods.	—	—
BS EN 13915:2007	Prefabricated gypsum plasterboard panels with a cellular paperboard core. Definitions, requirements and test methods.	—	—
BS EN 13950:2005	Gypsum plasterboard thermal/ acoustic insulation composite panels. Definitions, requirements and test methods.	—	—
BS EN 13963:2005	Jointing materials for gypsum plasterboards. Definitions, requirements and test methods.	—	C475
BS EN 14190:2005	Gypsum plasterboard products from reprocessing. Definition, requirements and test methods.	IS 12679 - 1989	
BS EN 14195:2005	Metal framing components for gypsum plasterboard systems. Definitions, requirements and test methods.	—	C 645
BSEN 14209:2005	Preformed plasterboard cornices. Definitions, requirements and test methods.	—	—
BS EN 1426:2006	Gypsum elements for suspended ceilings. Definitions, requirements and test methods.	—	—
BS EN 14496:2005	Gypsum based adhesives for thermal/acoustic insulation composition panels and plasterboards. Definitions, requirements and test methods.	—	—

(Cond...)

British Standards	Title	Relevant Indian Standard	Relevant ASTM Standard
BS EN 14566:2008	Mechanical fasteners for gypsum plasterboard systems. Definitions, requirements and test methods.	—	C 954, C 1002, C 1513
BS EN 15283–1:2008	Gypsum boards with fibrous reinforcement. Definitions, requirements and test methods. Gypsum boards with mat reinforcement.	—	C 1278, C 1355, C 1177
BS EN 15283–2:2008	Gypsum boards with fibrous reinforcement. Definitions, requirements and test methods. Gypsum fibre boards.	IS 2095 Part 3: 1996	—
BS EN 15318:2007	Design and application of gypsum blocks.	—	—
BSEN 15319:2007	General principal of design of fibrous (gypsum) plaster works.	—	—

Relevant DIN (German Standard)	CNS Catalogue PR, China	AS Standards Australia	ISO
18550–4	—	—	—
—	—	—	—
18184	—	—	—
—	—	—	—
18163	—	—	—
18180, 18181	4460/A 2062 4643/A 2063 4965/A 20701, 9958/A 2145 4459/A 3073, 4644/3075	2588, 2589, 2590	6308
—	—	—	—
—	—	—	—
18555–1, 18550–2 18550–3, 18501, 18550–4	6533/A 2082 11760/A2 199	2591, 2592	1587, 3048, 3049, 3051, 3052

Table 11.10 Published European (EN) Standards on Gypsum and Gypsum Based Products

Standard reference	Title	Directive (Citation in OJEU*)
EN 12859:2011	Gypsum blocks - Definitions, requirements and test methods	89/106/EEC (Expected)
EN 12860:2001	Gypsum based adhesives for gypsum blocks - Definitions, requirements and test methods	89/106/EEC (C 319, 2005–12–14)
EN 12860:2001/AC:2002	Gypsum based adhesives for gypsum blocks - Definitions, requirements and test methods	89/106/EEC (C 167, 2010–06–25)
EN 13279–1:2008	Gypsum binders and gypsum plasters - Part 1: Definitions and requirements	89/106/EEC (C 321, 2008–12–16)
EN 13279–2:2004	Gypsum binders and gypsum plasters - Part 2: Test methods	89/106/EEC (No)
EN 13454–1:2004	Binders, composite binders and factory made mixtures for floor screeds based on calcium sulfate - Part 1: Definitions and requirements	89/106/EEC (C 319, 2005–12–14)
EN 13454–2:2003+A1:2007	Binders, composite binders and factory made mixtures for floor screeds based on calcium sulfate - Part 2: Test methods	—
EN 13658–1:2005	Metal lath and beads - Definitions, requirements and test methods - Part 1: Internal plastering	89/106/EEC (C 319, 2005–12–14)
EN 13658–2:2005	Metal lath and beads - Definitions, requirements and test methods - Part 2: External rendering	89/106/EEC (C 319, 2005–12–14)
EN 13815:2006	Fibrous gypsum plaster casts - Definitions, requirements and test methods	89/106/EEC (C 304, 2006–12–13)
EN 13915:2007	Prefabricated gypsum plasterboard panels with a cellular paperboard core - Definitions, requirements and test methods	89/106/EEC (C 290, 2007–12–04)
EN 13950:2005	Gypsum plasterboard thermal/ acoustic insulation composite panels - Definitions, requirements and test methods	89/106/EEC (C 304, 2006–12–13)

(Contd...)

Standard reference	Title	Directive (Citation in OJEU*)
EN 13963:2005	Jointing materials for gypsum plasterboards - Definitions, requirements and test methods	89/106/EEC (C 319, 2005–12–14)
EN 13963:2005/AC:2006	Jointing materials for gypsum plasterboards - Definitions, requirements and test methods	89/106/EEC (C 304, 2006–12–13)
EN 14190:2005	Gypsum plasterboard products from reprocessing - Definitions, requirements and test methods	89/106/EEC (C 319, 2005–12–14)
EN 14195:2005	Metal framing components for gypsum plasterboard systems - Definitions, requirements and test methods	89/106/EEC (C 319, 2005–12–14)
EN 14195:2005/AC:2006	Metal framing components for gypsum plasterboard systems - Definitions, requirements and test methods	89/106/EEC (C 304, 2006–12–13)
EN 14209:2005	Preformed plasterboard cornices - Definitions, requirements and test methods	89/106/EEC (C 304, 2006–12–13)
EN 14246:2006	Gypsum elements for suspended ceilings - Definitions, requirements and test methods	89/106/EEC (C 304, 2006–12–13)
EN 14246:2006/AC:2007	Gypsum elements for suspended ceilings - Definitions, requirements and test methods	89/106/EEC (C 290, 2007–12–04)
EN 14353:2007+A1:2010	Metal beads and feature profiles for use with gypsum plasterboards - Definitions, requirements and test methods	89/106/EEC (C 167, 2010–06–25)
EN 14496:2005	Gypsum based adhesives for thermal/acoustic insulation composite panels and plasterboards - Definitions, requirements and test methods	89/106/EEC (C 304, 2006–12–13)
EN 14566:2008+A1:2009	Mechanical fasteners for gypsum plasterboard systems - Definitions, requirements and test methods	89/106/EEC (C 309, 2009–12–18)
EN 15283–1:2008+A1:2009	Gypsum boards with fibrous reinforcement - Definitions, requirements and test methods - Part 1: Gypsum boards with mat reinforcement	89/106/EEC (C 309, 2009–12–18)

(Contd...)

Standard reference	Title	Directive (Citation in OJEU*)
EN 15283–2:2008+A1:2009	Gypsum boards with fibrous reinforcement - Definitions, requirements and test methods - Part 2: Gypsum fibre boards	89/106/EEC (C 309, 2009–12–18)
EN 15318:2007	Design and application of gypsum blocks	89/106/EEC (No)
EN 15319:2007	General principles of design of fibrous (gypsum) plaster works	89/106/EEC (No)
EN 520:2004+A1:2009	Gypsum plasterboards - Definitions, requirements and test methods	89/106/EEC (C 309, 2009–12–18)

Source: OJEU - Official Journal of the European Union (Google)

References

CBRI Library, Roorkee, India
Bureau of Indian Standards Library, Manak Bhavan, New Delhi, India
IIT Bombay Library, Powai, Mumbai, India
Google Net
BIS (India), British, Chinese, German, Australian, European and ASTM (USA) Standards

Index

For Product Safety Concerns and Information please contact our EU representative GPSR@taylorandfrancis.com
Taylor & Francis Verlag GmbH, Kaufingerstraße 24, 80331 München, Germany

www.ingramcontent.com/pod-product-compliance
Lightning Source LLC
Chambersburg PA
CBHW081045220326
41598CB00038B/6994

* 9 7 8 1 0 3 2 3 8 4 2 7 6 *